Speech and Audio Processing

With this comprehensive and accessible introduction to the field, you will gain all the skills and knowledge needed to work with current and future audio, speech, and hearing processing technologies.

Topics covered include mobile telephony, human–computer interfacing through speech, medical applications of speech and hearing technology, electronic music, audio compression and reproduction, big data audio systems and the analysis of sounds in the environment. All of this is supported by numerous practical illustrations, exercises, and hands-on MATLAB examples on topics as diverse as psychoacoustics (including some auditory illusions), voice changers, speech compression, signal analysis and visualisation, stereo processing, low-frequency ultrasonic scanning, and machine learning techniques for big data.

With its pragmatic and application driven focus, and concise explanations, this is an essential resource for anyone who wants to rapidly gain a practical understanding of speech and audio processing and technology.

Ian Vince McLoughlin has worked with speech and audio for almost three decades in both industry and academia, creating signal processing systems for speech compression, enhancement and analysis, authoring over 200 publications in this domain. Professor McLoughlin pioneered Bionic Voice research, invented super-audible silent speech technology and was the first to apply the power of deep neural networks to machine hearing, endowing computers with the ability to comprehend a diverse range of sounds.

Speech and Audio Processing

A MATLAB®-based Approach

IAN VINCE MCLOUGHLIN
University of Kent

CAMBRIDGE
UNIVERSITY PRESS

University Printing House, Cambridge CB2 8BS, United Kingdom

Cambridge University Press is part of the University of Cambridge.

It furthers the University's mission by disseminating knowledge in the pursuit of education, learning and research at the highest international levels of excellence.

www.cambridge.org
Information on this title: www.cambridge.org/9781107085466

© Cambridge University Press 2016

This publication is in copyright. Subject to statutory exception
and to the provisions of relevant collective licensing agreements,
no reproduction of any part may take place without the written
permission of Cambridge University Press.

First published 2016

Printed in the United Kingdom by TJ International Ltd. Padstow Cornwall

A catalogue record for this publication is available from the British Library

Library of Congress Cataloguing in Publication data
Names: McLoughlin, Ian, author.
Title: Speech and audio processing : a Matlab-based approach / Ian Vince McLoughlin, University of Kent.
Description: New York, NY : Cambridge University Press, 2016. | © 2016 |
Includes bibliographical references and index.
Identifiers: LCCN 2015035032 | ISBN 9781107085466 (Hardcopy : alk. paper) |
ISBN 1107085462 (Hardcopy : alk. paper)
Subjects: LCSH: Speech processing systems. | Computer sound processing. | MATLAB.
Classification: LCC TK7882.S65 M396 2016 | DDC 006.4/5–dc23
LC record available at http://lccn.loc.gov/2015035032

ISBN 978-1-107-08546-6 Hardback

Additional resources for this publication at www.cambridge.org/mcloughlin and www.mcloughlin.eu

Cambridge University Press has no responsibility for the persistence or accuracy of URLs for external or third-party internet websites referred to in this publication, and does not guarantee that any content on such websites is, or will remain, accurate or appropriate.

Contents

Preface		page ix
Book features		xii
Acknowledgements		xv

1 Introduction — 1

1.1 Computers and audio — 1
1.2 Digital audio — 3
1.3 Capturing and converting sound — 4
1.4 Sampling — 5
1.5 Summary — 6
Bibliography — 7

2 Basic audio processing — 9

2.1 Sound in MATLAB — 10
2.2 Normalisation — 18
2.3 Continuous audio processing — 20
2.4 Segmentation — 24
2.5 Analysis window sizing — 32
2.6 Visualisation — 37
2.7 Sound generation — 44
2.8 Summary — 50
Bibliography — 50
Questions — 52

3 The human voice — 54

3.1 Speech production — 55
3.2 Characteristics of speech — 57
3.3 Types of speech — 67
3.4 Speech understanding — 71
3.5 Summary — 82
Bibliography — 83
Questions — 83

4 The human auditory system — 85

- 4.1 Physical processes — 85
- 4.2 Perception — 87
- 4.3 Amplitude and frequency models — 103
- 4.4 Summary — 107
- Bibliography — 107
- Questions — 108

5 Psychoacoustics — 109

- 5.1 Psychoacoustic processing — 109
- 5.2 Auditory scene analysis — 112
- 5.3 Psychoacoustic modelling — 121
- 5.4 Hermansky-style model — 132
- 5.5 MFCC model — 134
- 5.6 Masking effect of speech — 137
- 5.7 Summary — 138
- Bibliography — 138
- Questions — 139

6 Speech communications — 140

- 6.1 Quantisation — 140
- 6.2 Parameterisation — 148
- 6.3 Pitch models — 176
- 6.4 Analysis-by-synthesis — 182
- 6.5 Perceptual weighting — 191
- 6.6 Summary — 192
- Bibliography — 192
- Questions — 193

7 Audio analysis — 195

- 7.1 Analysis toolkit — 196
- 7.2 Speech analysis and classification — 208
- 7.3 Some examples of audio analysis — 211
- 7.4 Statistics and classification — 213
- 7.5 Analysing other signals — 216
- 7.6 Summary — 220
- Bibliography — 220
- Questions — 221

8 Big data — 223

- 8.1 The rationale behind big data — 225
- 8.2 Obtaining big data — 226

	8.3	Classification and modelling	227
	8.4	Summary of techniques	234
	8.5	Big data applications	263
	8.6	Summary	264
	Bibliography	264
	Questions	265
9	**Speech recognition**	267
	9.1	What is speech recognition?	267
	9.2	Voice activity detection and segmentation	275
	9.3	Current speech recognition research	282
	9.4	Hidden Markov models	288
	9.5	ASR in practice	298
	9.6	Speaker identification	302
	9.7	Language identification	305
	9.8	Diarization	308
	9.9	Related topics	309
	9.10	Summary	311
	Bibliography	311
	Questions	312
10	**Advanced topics**	314
	10.1	Speech synthesis	314
	10.2	Stereo encoding	324
	10.3	Formant strengthening and steering	334
	10.4	Voice and pitch changer	338
	10.5	Statistical voice conversion	346
	10.6	Whisper-to-speech conversion	347
	10.7	Whisperisation	354
	10.8	Super-audible speech	357
	10.9	Summary	363
	Bibliography	364
	Questions	365
11	**Conclusion**	366
	References	370
	Index	379

Preface

Humans are social creatures by nature – we are made to interact with family, neighbours and friends. Modern advances in social media notwithstanding, that interaction is best accomplished in person, using the senses of sound, sight and touch.

Despite the fact that many people would name sight as their primary sense, and the fact that it is undoubtedly important for human communications, it is our sense of hearing that we rely upon most for social interaction. Most of us need to talk to people face-to-face to really communicate, and most of us find it to be a much more efficient communications mechanism than writing, as well as being more personal. Readers who prefer email to telephone (as does the author) might also realise that their preference stems in part from being better able to regulate or control the flow of information. In fact this is a tacit agreement that verbal communications can allow a higher rate of information flow, so much so that they (we) prefer to restrict or at least manage that flow.

Human speech and hearing are also very well matched: the frequency and amplitude range of normal human speech lies well within the capabilities of our hearing system. While the hearing system has other uses apart from just listening to speech, the output of the human sound production system is very much designed to be heard by other humans. It is therefore a more specialised subsystem than is hearing. However, despite the frequency and amplitude range of speech being much smaller than our hearing system is capable of, and the precision of the speech system being lower, the symbolic nature of language and communications layers a tremendous amount of complexity on top of that limited and imperfect auditory output. To describe this another way, the human sound production mechanism is quite complex, but the speech communications system is massively more so. The difference is that the sound production mechanism is mainly handled as a motor (movement) task by the brain, whereas speech is handled at a higher conceptual level, which ties closely with our thoughts. Perhaps that also goes some way towards explaining why thoughts can sometimes be 'heard' as a voice or voices inside our heads?

For decades, researchers have been attempting to understand and model both the speech production system and the human auditory system (HAS), with partial success in both cases. Models of our physical hearing ability are good, as are models of the types of sounds that we can produce. However, once we consider either speech or the inter-relationship between perceived sounds in the HAS, the situation becomes far more complex. Speech carries with it the difficulties inherent in the natural language processing (NLP) field, as well as the obvious fact that we often do not clearly say what we mean or mean what we (literally) say.

Speech processing itself is usually concerned with the output of the human speech system, rather than the human interpretation of speech and sounds. In fact, whenever we talk of speech or sounds we probably should clarify whether we are concerned with the physical characteristics of the signal (such as frequency and amplitude), the perceived characteristics (such as rhythm, tone, timbre), or the underlying meaning (such as the message conveyed by words, or the emotions). Each of these aspects is a separate but overlapping research field in its own right.

NLP research considers natural language in all its beauty, linguistic, dialectal and speaker-dependent variation and maddening imperfect complexity. This is primarily a computation field that manipulates symbolic information like phonemes, rather than the actual sounds of speech. It overlaps with linguistics and grammar at one extreme, and speech processing at the other.

Psychoacoustics links the words *psycho* and *acoustics* together (from the Greek ψυχή and ἀκούω respectively) to mean the human interpretation of sounds – specifically as this might differ from a purely physical measurement of the same sounds. This encompasses the idea of auditory illusions, which are analogous to optical illusions for our ears, and form a fascinating and interesting area of research. A little more mundane, but far more impactful, is the computer processing of physical sounds to determine how a human would hear them. Such techniques form the mainstay of almost all recordings and reproductions of music on portable, personal and computational devices.

Automatic speech recognition, or ASR, is also quietly impacting the world to an increasing extent as we talk to our mobile devices and interact with automated systems by telephone. Whilst we cannot yet hold a meaningful conversation with such systems (although this does rather depend upon one's interpretation of the word 'meaningful'), at the time of writing they are on the cusp of actually becoming useful. Unfortunately I realise now that I had written almost the same words five years ago, and perhaps I will be able to repeat them five years from now. However, despite sometimes seemingly glacially slow performance improvements in ASR technology from a user's perspective, the adoption of so-called 'big data' techniques has enabled a recent quantum leap in capabilities.

Broadly speaking, 'big data' is the use of vast amounts of information to improve computational systems. It generally ties closely to the field of machine learning. Naturally, if researchers can enable computers to learn effectively, then they require material from which to learn. It also follows that the better and more extensive the learning material, the better the final result. In the speech field, the raw material for analysis is usually recordings of the spoken word.

Nowhere is the 'big data' approach being followed more enthusiastically than in China, which allies together the world's largest population with the ability to centralise research, data capture and analysis efforts. A research-friendly (and controversial) balance between questions of privacy and scientific research completes the picture. As an illustration, consider the world's second-biggest speech-related company, named iFlytek, which we will discuss in Chapter 8. Although largely unknown outside China, the flagship product of this impressive company is a smartphone application that understands both English and Chinese speech, acting as a kind of digital personal assistant. This cloud-based system is used today by more than 130 million people, who find it a useful, usable and perhaps invaluable tool. During operation, the company tracks both correct and incorrect responses, and incorporates the feedback into their machine learning model to continuously improve performance. So if, for example, many users input some speech that is not recognised by the system, this information will be used to automatically modify and update their recognition engine. After the system is updated – which happens periodically – it will probably have learned the ability to understand the speech that was not recognised previously. In this way the system can continuously improve as well as track trends and evolutions in speech patterns such as new words and phrases. Launched publicly a few years ago with around 75% recognition accuracy for unconstrained speech without excessive background noise, it now achieves over 95% accuracy, and still continues to improve.

It is approaches like that which are driving the speech business forward today, and which will ensure a solid place in the future for such technologies. Once computers can reliably understand us through speech, speech will quickly become the primary human–computer interfacing method – at least until thought-based (mind-reading) interfaces appear.

This book appears against the backdrop of all of these advances. In general, it builds significantly upon the foundation of the author's previous work, *Applied Speech and Audio Processing with MATLAB Examples*, which was written before big data and machine learning had achieved such significant impact in these fields. The previous book also predated wearable computers with speech interfaces (such as Google's Glass), and cloud-based speech assistants such as Apple's Siri or iFlytek's multilingual system. However, the hands-on nature and practical focus of the previous book are retained, as is the aim to present speech, hearing and audio research in a way that is inspiring, fun and fresh. This really is a good field to become involved in right now. Readers will find that they can type in the MATLAB examples in the book to rapidly explore the topics being presented, and quickly understand how things work.

Also in common with the previous work, this text does not labour over meaningless mathematics, does not present overly extensive equations and does not discuss dreary theory, all of which can readily be obtained elsewhere if required. In fact, any reader wishing to delve a little deeper may refer to the list of references to scientific papers and general per-chapter bibliographies that are provided. The references are related to specific items within the text, whereas the bibliography tends to present useful books and websites that are recommended for further study.

Book features

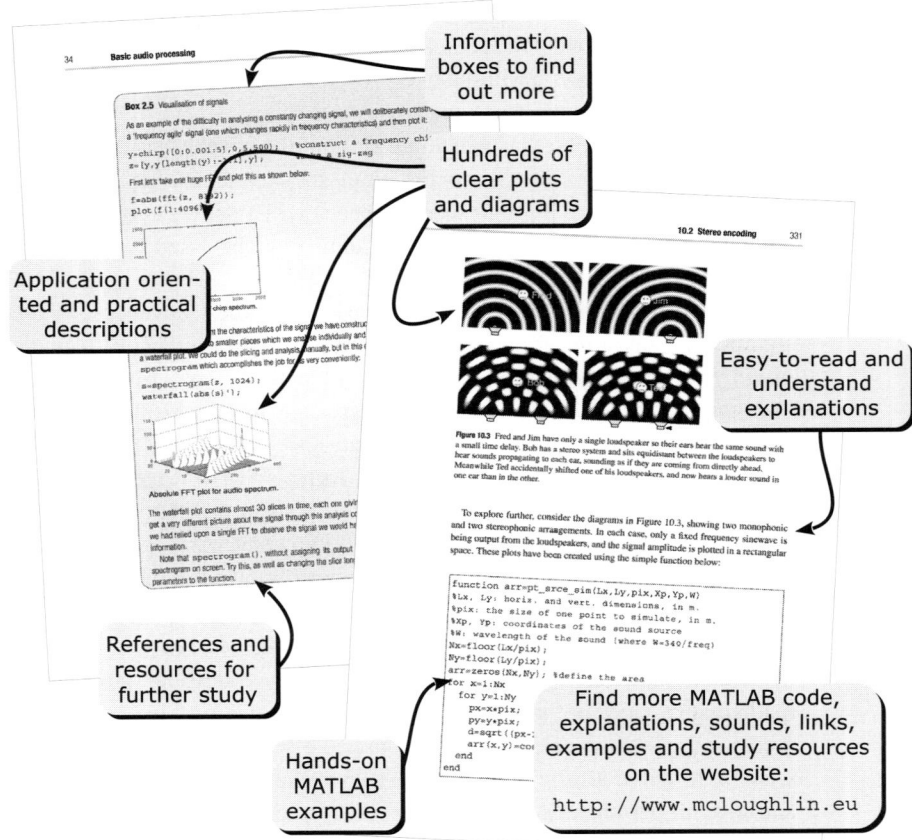

> **Box 0.1** What is an **information box**?
>
> Self-contained items of further interest, useful tips and reference items that are not within the flow of the main text are presented inside boxes similar to this one.

Each chapter begins with an **introduction** explaining the thrust and objectives being explored in the subsequent sections. MATLAB **examples** are presented and explained throughout to illustrate the points being discussed, and provide a core for further self-directed exploration. Numbered **citations** are provided to support the text (e.g. [1]) where appropriate. The source documents for these can be found listed sequentially from page 370. A **bibliography** is also provided at the end of each chapter, giving a few selected reference texts and resources that readers might find useful for further exploration of the main topics discussed in the text.

Note that commands for MATLAB or computer entry are written in a `computer font` and code listings are presented using a separate highlighted listing arrangement:

```
This is Matlab code.
You can type these commands into MATLAB.
```

All of the commands and listings are designed to be typed at the command prompt in the MATLAB command window. They can also be included and saved as part of an m-file program (this is a sequence of MATLAB commands in a text file having a name ending in .m). These can be loaded into MATLAB and executed – usually by double clicking them. This book does not use Simulink for any of the examples since it would obscure some of the technical details of the underlying processes, but all code can be used in Simulink if required. Note also that the examples only use the basic inbuilt MATLAB syntax and functions wherever possible. However, new releases of MATLAB tend to move some functions from the basic command set into specialised toolboxes (which are then available at additional cost). Hence a small number of examples may require the Signal Processing or other toolboxes in future releases of MATLAB, but if that happens a Google search will usually uncover free or open source functions that can be downloaded to perform an equivalent task.

Companion website

A companion website at http://mcloughlin.eu has been created to link closely with the text. The table at the top of the next page summarises a few ways of accessing the site using different URLs.

An integrated search function allows readers to quickly access topics by name. All code given in the book (and much more) is available for download.

Book features

URL	Content
mcloughlin.eu/speech	Main book portal
mcloughlin.eu?s=Xxx	Jump to information on topic Xxx
mcloughlin.eu/chapterN	Chapter N information
mcloughlin.eu/listings	Directory of code listings
mcloughlin.eu/secure	Secure area for lecturers and instructors
mcloughlin.eu/errata	Errata to published book

Book preparation

This book has been written and typeset with LaTeX using **TeXShop** and **TeXstudio** front-end packages on Linux Ubuntu and OS-X computers. All code examples have been developed on MATLAB, and most also tested using **Octave**, both running on Linux and OS-X. Line diagrams have all been drawn using the OpenOffice/LibreOffice drawing tool, and all graphics conversions have made use of the extensive graphics processing tools that are freely available on Linux. Audio samples have either been obtained from named research databases or recorded directly using Zoom H4n and H2 audio recorders, and processed using Audacity.

MATLAB® and Simulink are registered trademarks of MathWorks, Inc. All references to MATLAB throughout this work should be taken as referring to MATLAB®.

Acknowledgements

Anyone who has written a book of this length will know the amount of effort involved. Not just in the writing, but in shuffling various elements around to ensure the sequence is optimal, in double checking the details, proofreading, testing and planning. Should a section receive its own diagram or diagrams? How should they be drawn and what should they show? Can a succinct and self-contained MATLAB example be written – and will it be useful? Just how much detail should be presented in the text? What is the best balance between theory and practice, and how much background information is required? All of these questions need to be asked, and answered, numerous times during the writing process. Hopefully the answers to these questions, that have resulted in this book, are right more often than they are wrong.

The writing process certainly takes an inordinate amount of time. During the period spent writing this book, I have seen my children Wesley and Vanessa grow from being pre-teens to young adults, who now have a healthy knowledge of basic audio and speech technology, of course. Unfortunately, time spent writing this book meant spending less time with my family, although I did have the great privilege to be assisted by them: Wesley provided the cover image for the book,[1] and Vanessa created the book index for me.

Apart from my family, there are many people who deserve my thanks, including those who shaped my research career and my education. Obviously this acknowledgement begins chronologically with my parents, who did more than anything to nurture the idea of an academic engineering career, and to encourage my writing. Thanks are also due to my former colleagues at The University of Science and Technology of China, Nanyang Technological University, School of Computer Engineering (Singapore), Tait Electronics Ltd (Christchurch, New Zealand), The University of Birmingham, Simoco Telecommunications (Cambridge), Her Majesty's Government Communications Centre and GEC Hirst Research Centre. Also to my many speech-related students, some of whom are now established as academics in their own right. I do not want to single out names here, because once I start I may not be able to finish without listing everyone, but I do remember you all, and thank you sincerely.

However, it would not be fair to neglect to mention my publishers. Particularly the contributions of Publishing Director Philip Meyler at Cambridge University Press,

---

[1] The cover background is a photograph of the famous Huangshan (Yellow Mountains) in Anhui Province, China, taken by Wesley McLoughlin using a Sony Alpha-290 DSLR.

Editor Sarah Marsh, and others such as the ever-patient Assistant Editor Heather Brolly. They were responsive, encouraging, supportive and courteous throughout. It has been a great pleasure and privilege to have worked with such a professional and experienced publishing team again.

Finally, this book is dedicated to the original designer of the complex machinery that we call a human: the architect of the amazing and incomparable speech production and hearing apparatus that we often take for granted. All glory and honour be to God.

1 Introduction

Audio processing systems have been a part of many people's lives since the invention of the phonograph in the 1870s. The resulting string of innovations sparked by that disruptive technology have culminated eventually in today's portable audio devices such as Apple's iPod, and the ubiquitous MP3 (or similarly compressed) audio files that populate them. These may be listened to on portable devices, computers, as soundtracks accompanying Blu-ray films and DVDs, and in innumerable other places.

Coincidentally, the 1870s saw a related invention – that of the telephone – which has also grown to play a major role in daily life between then and now, and likewise has sparked a string of innovations down the years. Scottish born and educated Alexander Graham Bell was there at their birth to contribute to the success of both inventions. He probably would be proud to know, were he still alive today, that two entire industry sectors, named telecommunications and infotainment, were spawned by the two inventions of phonograph and telephone.

However, after 130 years, something even more unexpected has occurred: the descendents of the phonograph and the descendents of the telephone have converged into a single product called a 'smartphone'. Dr Bell probably would not recognise the third convergence that made all of this possible, that of the digital computer – which is precisely what today's smartphone really is. At heart it is simply a very small, portable and capable computer with microphone, loudspeaker, display and wireless connectivity.

1.1 Computers and audio

The flexibility of computers means that once sound has been sampled into a digital form, it can be used, processed and reproduced in an infinite variety of ways without further degradation. It is not only computers (big or small) that rely on digital audio, so do CD players, MP3 players (including iPods), digital audio broadcast (DAB) radios, most wireless portable speakers, television and film cameras, and even modern mixing desks for 'live' events (and co-incidentally all of these devices contain tiny embedded computers too). Digital music and sound effects are all around us and impact our leisure activities (e.g. games, television, videos), our education (e.g. recorded lectures, broadcasts, podcasts) and our work in innumerable ways to influence, motivate and educate us. Beyond music, we can find examples of digital audio in recorded announcements, the beep of electronic devices, ringing mobile telephones, many alarm sirens, modern hearing aids, and even the sound conveyed by the old-fashioned telephone network (named POTS for 'plain old telephone service'), which is now a cyborg of interconnected digital and analogue elements.

All of this digital audio impinging on our sensitive ear drums should not cause our mouths to feel neglected: digital speech processing has seen an equally impressive and sustained upward worldwide trend in performance and popularity. The well-charted cellular communications revolution begun by the European GSM (Global System for Mobile communications) standard has now led to mobile phones, cellphones, handphones and smartphones being virtually ubiquitous. Tremendous numbers of mobile phones are sold annually, even in the world's poorest regions, and few would disagree today that a mobile phone – a digital speech computer – is an essential consumer or business device. However, these devices have become more than just mobile telephones, as evidenced by the growing importance of smartphones with large colourful screens; a feature that would be unlikely to be much appreciated by one's ear, which gets the best view of the screen when the device is operating as a more traditional telephone.

It seems at first glance that, as these intelligent, desirable and useful devices worm their way further into our daily life, the relative importance of using the devices to convey 'speech' is reducing, compared with the importance of the visual display. In other words, they are moving away from being primarily telephone devices. In fact, a non-scientific survey of teenagers' attitudes to the essentials of human existence yields something like Figure 1.1, which might support the conjecture that telephony and speech are declining in importance. However, this impression would be untrue for two interesting reasons. The first is that the increasing adoption patterns of smartphones indicate that many of the newer uses (e.g. playing games, watching episodes of television or films, instant chatting) require and rely upon sound and speech just as much as a telephone does. The second is in the growing role of speech-based interfaces. These probably first became prominent with the general public through Apple's Siri, on the iPhone 4s, which could be spoken to using relatively natural language, and which would respond accordingly. Siri was able to order tickets, book flights, find directions, contact friends and relatives, advise on weather, traffic, restaurants and so on. In fact, Siri could be asked almost any question and would attempt to provide an appropriate answer. To many, this began to sound like a kind of electronic personal assistant, or butler. In fact

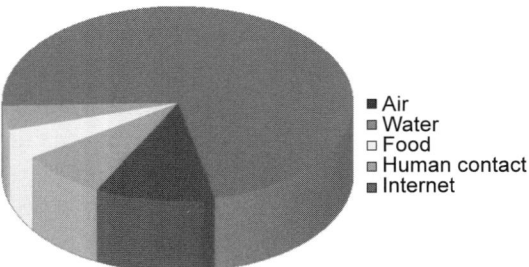

Figure 1.1 Essential human needs – according to teenagers.

the 2011 release of Siri was pre-dated by about 3 years by the Marvel Comics film *Iron Man* in which hero Tony Stark's suit of armour communicates with him using spoken natural language. The system, named J.A.R.V.I.S., is obviously a dramatisation played by an actor rather than a real system; however, it is a placeholder for the archetypal electronic butler of the future. It might not surprise readers to find out that there are currently a significant number of university research groups around the world who are working on making such systems a reality, as well as some large companies such as Google, IBM, iFlytek, Amazon, Samsung, and probably several telecommunications giants too. With this much brainpower and research expenditure, few would doubt that it is purely a matter of time until something (or somebody?) like J.A.R.V.I.S. becomes reality.

1.2 Digital audio

Digital processing is now the method of choice for handling audio and speech: new audio applications and systems are predominantly digital in nature. This revolution from analogue to digital has mostly occurred over the past two decades, as a quiet, almost unremarked upon, change.

It would seem that those wishing to become involved in speech, audio and hearing related research or development can perform much, if not all, of their work in the digital domain these days, apart from the interface which captures sound (microphone) and outputs it (loudspeaker). One of the great benefits of digital technology is that the techniques are relatively device independent: one can create and prototype using one digital processing platform, and then deploy using another platform, and the behaviour of both systems will be identical. Given that, the criteria for a development platform would then be for ease-of-use and testing, while the criteria for a deployment platform may be totally separate: low power, small size, high speed, low cost, etc.

In terms of development ease-of-use, MATLAB running on a computer is chosen by many of those working in the field. It is well designed to handle digital signals, especially the long strings of audio samples. Built-in functions allow most common manipulations to be performed very easily. Audio recording and playback are equally possible, and the visualisation and plotting tools are excellent. A reduced-price student version is available which is sufficient for much audio work. The author runs MATLAB on both Mac OS-X and Linux platforms for much of his own audio work.

Although there is currently no speech, audio or hearing toolbox provided by The MathWorks® for MATLAB, the Signal Processing Toolbox contains most of the required additional functions, and an open source toolbox called VOICEBOX is also available from the Department of Electrical and Electronic Engineering, Imperial College, London, which contains many additional useful functions.[1]

---

[1] VOICEBOX, released courtesy of Mike Brookes of Imperial College, can be downloaded from www.ee.ic.ac.uk/hp/staff/dmb/voicebox/voicebox.html

4　Introduction

All of the audio and speech processing in this book can also be executed using the open source Octave environment,[2] although some of the MATLAB examples may require a few small changes – usually some vectors will need to be transposed. Octave is less common than the industry standard MATLAB, and lacks one or two of the advanced plotting and debugging capabilities, but is otherwise very similar in capabilities. It is also highly efficient, easily handles parallel processing on a cluster computer, and can be integrated with other languages such as Python for script- or web-based automation.

1.3　Capturing and converting sound

This book is all about sound. Either sound created through the speech production mechanism, or sound as heard by a machine or human. In purely physical terms, sound is a longitudinal wave which travels through air (or a transverse wave in some other media) due to the vibration of molecules. In air, sound is transmitted as a pressure which varies between high and low levels, with the rate of pressure variation from low, to high, to low again determining the frequency. The degree of pressure variation (namely the difference between the high and the low pressures) determines the amplitude.

A microphone captures sound waves, often by sensing the deflection caused by the wave on a thin membrane, transforming it proportionally to either voltage or current. The resulting electrical signal is normally then converted to a sequence of coded digital data using an analogue-to-digital converter (ADC). The most common format, pulse coded modulation, will be described in Section 6.1.1.

If this sequence of coded data is fed through a compatible digital-to-analogue converter (DAC), through an amplifier to a loudspeaker, then a sound may be produced. In this case the voltage applied to the loudspeaker at every instant of time is proportional to the sample value from the computer being fed through the DAC. The voltage on the loudspeaker causes a cone to deflect in or out, and it is this cone which compresses (or rarefies) the air from instant to instant, thus initiating a sound pressure wave.

In fact the process, shown diagrammatically in Figure 1.2(a), identifies the major steps in any digital audio processing system. Audio, in this case speech in free air, is converted to an electrical signal by a microphone, amplified and probably filtered, before being converted into the digital domain by an ADC. Once in the digital domain, these signals can be processed, transmitted or stored in many ways, and indeed may be experimented upon using MATLAB. A reverse process will then convert the signals back into sound.

Connections to and from the processing/storage/transmission system of Figure 1.2 (which could be almost any digital system) may be either serial or parallel, with many possible connectivity options in practice.

Variations on this basic theme, such as those shown in Figures 1.2(b) and (c), use a subset of the components for analysis or synthesis of audio. Stereo systems would

---

[2] GNU Octave is an open source alternative to MATLAB, and is available for download from www.gnu.org/software/octave

1.4 Sampling

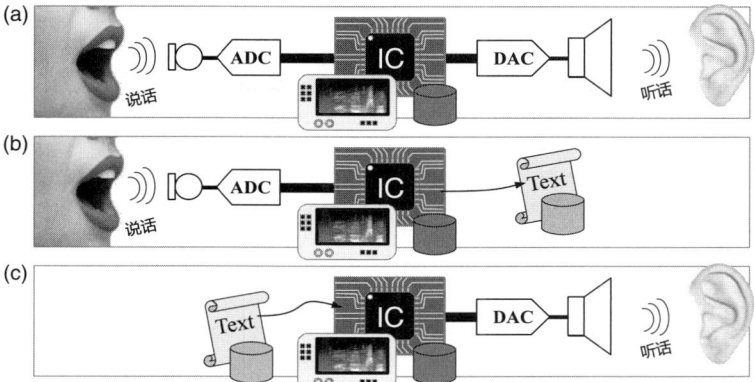

Figure 1.2 Three classes of digital audio system: (a) a complete digital audio processing path incuding (from left to right) an input microphone, amplifier, ADC, processing system, DAC, amplifier and loudspeaker. (b) A system recognising audio or speech. (c) A system that synthesises speech or audio.

have two microphones and loudspeakers, and some systems may have many more of both. The very simple amplifier, ADC and DAC blocks in the diagram also hide some of the complexities that would be present in many systems – such as analogue filtering, automatic gain control and so on, in addition to the type (class) of amplification provided.

Both ADC and DAC are also characterised in different ways: by their sampling rates, conversion technology, signal-to-noise ratio, linearity and dynamic range (which is related to the number of bits that they output in each sample).

1.4 Sampling

Considering a sequence of audio samples, first of all we note that the time spacing between successive samples is almost always designed to be uniform. The frequency of this timing is referred to as the sampling rate, and in Figure 1.2 would be set through a periodic clock signal fed to the ADC and DAC, although there is no reason why both need the same sample rate – digital processing can be used to change the sample rate. Using the well-known Nyquist criterion, the highest frequency that can be unambiguously represented by such a stream of samples is half of the sampling rate.

Samples themselves as delivered by an ADC are generally fixed point with a resolution of 16 bits, although 20 bits and even up to 24 bits are found in high-end audio systems. Handling these on computer could utilise either fixed or floating point representation (fixed point meaning each sample is a scaled integer, while floating point allows fractional representation), with a general rule of thumb for reasonable quality being that 20 bits fixed point resolution is desirable for performing processing operations in a system with 16-bit input and output.

> **Box 1.1** Audio fidelity
>
> Something to note is the inexactness of the entire conversion process: what you hear is a wave impinging on the eardrum, but what you obtain on the computer has travelled some way through air, possibly bounced past several obstructions, hit a microphone, vibrated a membrane, been converted to an electrical signal, amplified, and then sampled. Amplifiers add noise, create distortion, and are not entirely linear. Microphones are usually far worse on all counts. Analogue-to-digital converters also suffer linearity errors, add noise, create distortion, and introduce quantisation error due to the precision of their voltage sampling process. The result of all this is a computerised sequence of samples that may not be as closely related to the real-world sound as you might expect. Do not be surprised when high-precision analysis or measurements are unrepeatable due to noise, or if delicate changes made to a sampled audio signal are undetectable to the naked ear upon replay.

In the absence of other factors, an n-bit uniformly sampled digital audio signal will have a dynamic range (the ratio of the biggest amplitude that can be represented in the system to the smallest one) of, at best:

$$\text{DR (dB)} = 6.02 \times n. \qquad (1.1)$$

For telephone-quality speech, resolutions as low as 8–12 bits are possible depending on the application. For GSM-type mobile phones, 14 bits is common. Telephone-quality, often referred to as toll-quality, is perfectly reasonable for vocal communications, but is not perceived as being of particularly high quality. For this reason, more modern vocal communication systems have tended to move beyond 8 bits sample resolution in practice.

Sample rates vary widely from 7.2 kHz or 8 kHz for telephone-quality audio to 44.1 kHz for CD-quality audio. Long-play style digital audio systems occasionally opt for 32 kHz, and high-quality systems use 48 kHz. A recent trend is to double this to 96 kHz. It is debatable whether a sampling rate of 96 kHz is at all useful to human ears, which can typically not resolve signals beyond about 18 kHz even when young – apart from the rare listeners having *golden ears*.[3] However, such systems may be more pet-friendly: dogs are reportedly able to hear up to 44 kHz and cats up to almost 80 kHz.

Sample rates and sampling precisions for several common applications, for humans at least, are summarised in Table 1.1.

1.5 Summary

The technological fine detail related to the conversion and transmission process for audio is outside the scope of this book, which is more concerned with experimentation, analysis and digital speech/audio manipulation. Today's audio processing specialist is very fortunate to be able to work with digital audio without being too concerned with

---

[3] The die-hard audio enthusiasts, *audiophiles*, who prefer valve amplifiers, pay several years' salary for a pair of loudspeakers, and often claim they can hear above 20 kHz, are usually known as having *golden ears*.

Table 1.1 Sampling characteristics of common applications.

Application	Sample rate, resolution	How used
Telephony	8 kHz, 8–12 bits	64 kbps A-law or μ-law
Voice conferencing	16 kHz, 14–16 bits	64 kbps SB-ADPCB
Mobile phone	8 kHz, 14–16 bits	13 kbps GSM
Private mobile radio	8 kHz, 12–16 bits	<5 kbps, e.g. TETRA
Long-play audio	32 kHz, 14–16 bits	Minidisc, DAT, MP3
CD audio	44.1 kHz, 16–24 bits	Stored on CDs
Studio audio	48 kHz, 16–24 bits	CD mastering
Very high end	96 kHz, 20–24 bits	For *golden ears* listening

how it was captured, or how it will be replayed. Thus, we will confine our discussions throughout the remainder of this text primarily to the processing/storage/transmission, recognition/analysis and synthesis/generation blocks in Figure 1.2, ignoring the ADCs, DACs, transducers and other messy analogue detail.

However, it is important to remember that sound, as perceived by humans, is a real, continuously varying, noisy, analogue signal that has several attributes. These include time-domain attributes of duration, rhythm, attack and decay, but also frequency-domain attributes of tone and pitch. Other, less well-defined attributes include quality, timbre and tonality. Very often, a sound wave conveys meaning: for example, a fire alarm, the roar of a lion, the cry of a baby, a peal of thunder, a national anthem or a spoken word.

However, as we have seen, sound sampled by an ADC (at least the more common pulse coded modulation-based ADCs) is simply represented as a vector of samples, with each element in the vector representing the sound amplitude at that particular instant of time. The remainder of this book attempts to bridge the gap between such a vector of numbers representing audio and an understanding or interpretation of the meaning of that audio, as well as how that vector of numbers can be manipulated to apply meaningful processing (usually inside MATLAB).

Bibliography

- *Principles of Computer Speech*
  I. H. Witten (Academic Press, 1982)
  This book provides a gentle and readable introduction to speech on computer, written in an accessible and engaging style, along with some simple examples (in BASIC though). By now, this text is definitely a little dated in the choice of technology presented, but the underlying principles discussed remain unchanged.

- *The Art of Electronics*
  P. Horowitz and W. Hill (Cambridge University Press, 2nd edition 1989)
  For those interested in the electronics of audio processing, whether digital or analogue, this book is a wonderful introduction. It is clearly written, absolutely

packed full of excellent information (on almost any aspect of electronics), and is a hugely informative text. Be aware, though, that its scope is large: with over 1000 pages, only a fraction of the book is devoted to audio electronics issues.

- *Digital Signal Processing: A Practical Guide for Engineers and Scientists*
  S. W. Smith (Newnes, 2002)
  Also freely available from www.dspguide.com
  This excellent reference work is available in book form, or directly from the website above. The author has done a good job of covering most of the required elements of signal processing in a relatively easy-to-read way. In general the work lives up to the advertised role of being practically oriented. Overall, a very large amount of good information is presented to the reader, although this may not always be covered gradually enough for those without a signal processing background.

2 Basic audio processing

Most speech and audio researchers use MATLAB as a preferred tool for audio processing, although many of us will make use of other specialised tools from time to time, such as *sox* for command line audio processing[1] (particularly when there are a large number of files to convert or process, something it can do with a single command line option), and the sound capture and editing tool *audacity* which can record, edit, manipulate, convert and play back numerous types of audio file.[2] In fact both of these programs are extremely capable open source tools, having far more options than could be described here. However, while very useful, neither tool can replace the abilities of MATLAB to easily develop scripts that make use of hundreds of built-in functions and operators, and can plot or visualise speech and other sounds in a multitude of ways.

Recorded speech or other sounds are stored within MATLAB (as well as in many other computer-based tools) as a vector of samples, with each individual value being a double precision floating point number. A sampled sound can be completely specified by the vector of these numbers as long as one other item of information is known: the sample rate at which the data was recorded. To replay the sampled sound, it is only necessary to sequentially output a voltage proportional to the stored vector information, with a gap between samples equivalent to the inverse of the sample rate.

General audio programs and tools store audio information similarly, except that they tend to use fixed point numbers rather than floating point, which can reduce the storage requirement by a factor of four at the expense of very little degradation – assuming the system is correctly designed. In particular, a consideration of overflow and underflow effects is usually needed when designing a system that uses fixed point storage for audio, whereas in floating point-based tools such as MATLAB this is rarely a concern in practice.

Any operation that MATLAB can perform on a general vector can, in theory, be performed on stored audio. In fact, this is how we typically perform audio processing within MATLAB, and the audio vector can be loaded and saved in much the same way as any other MATLAB variable. Likewise it can be processed, added, plotted, inverted, transformed and so on.

---

[1] The sox tool can be freely downloaded from http://sox.sourceforge.net/sox.html
[2] Audacity is a free, cross-platform and open source sound audio editor and recorder. It is available for download from http://web.audacityteam.org

However, there are of course some special considerations when dealing with audio that should be discussed within this chapter, as a foundation for later chapters where we shall use MATLAB to implement some real speech and audio-based examples.

This chapter thus begins with an overview of audio input and output in MATLAB, including recording and playback directly within the program. We then consider scaling issues (including overflow and underflow), basic processing methods and the important issues of continuous analysis and processing. A section on visualisation covers the main time- and frequency-domain plotting techniques that we use to look at and understand the signals that we are working with. Finally, some methods of generating sounds and noise are presented, which will be useful in later chapters.

2.1 Sound in MATLAB

With a high enough sample rate, the double precision vector has sufficient resolution for almost any processing that may need to be performed. This means that one can usually safely ignore quantisation issues when processing in MATLAB. However, there are potential resolution and quantisation concerns when we need to input data into, or output data from, MATLAB, including the recording and replaying of sound. It is because input and output will normally be in a fixed point format, as will audio data stored in most files. We will thus look at an overview of audio input and output from MATLAB, starting with audio recording and playback, and then audio file handling.

2.1.1 Recording sound

Recording sound directly in MATLAB has changed in the latest versions, and now requires use of the `audiorecorder()` function. A much recommended alternative is to use a separate audio application to record sound (such as the excellent open source Audacity tool). Audacity is much easier to control and operate than the built-in MATLAB audio commands, and allows intuitive and fast editing of sounds which are displayed by waveform. Once the recorded data is ready to be used in MATLAB it can be exported by Audacity to a file in `.wav` format (or any other standard format), which can then easily be read into MATLAB as we will see in Section 2.1.3.

If Audacity is not used, the `audiorecorder()` function in MATLAB requires an audio recorder object to first be created, specifying sample rate, sample precision in bits, and number of channels, before recording can begin. This is done as follows:

```
aro=audiorecorder(16000,16,1);
record(aro);
```

At this point, after entering the `record()` command, although there may be no visible indication, the computer should be actively recording sound (always assuming a microphone is attached or built into the computer). When you wish to pause/resume or end the recording, simply use the commands shown below:

```
pause(10);     %waits 10 seconds before next command
pause(aro);    %this pauses the recording
resume(aro);   %this resumes recording
stop(aro);     %this ends the recording session
```

and then play back the resulting audio file as follows:

```
play(aro);
```

However, it should be noted that the `aro` object is not numerical data; in fact it is a structure that contains the numerical sample data along with various other items of information about the recording. To convert the stored recording within `arc` into the more usual vector of audio samples, it is necessary to use the `getaudiodata()` function as follows:

```
speech=getaudiodata(aro, 'double');
```

where we need to specify the resulting data type (double precision floating point in this example). The entire recording and playback sequences operate as background commands, making them a good choice when building interactive speech tools, or using MATLAB for conducting real-time experiments.

2.1.2 Storing and replaying audio

In the example given above, the recorded vector of sound (named 'speech') consisted of double precision samples, but was recorded with 16-bit precision. The maximum representable range of values in 16-bit format is between $-32\,768$ and $+32\,767$, but when converted to double precision is scaled to lie within a range of ± 1.0, and in fact this would be the most universal scaling within MATLAB so we will use this wherever possible. In this way, a recorded sample with integer value 32 767 (the maximum) would be stored with a floating point value of $+1.0$, and a recorded sample with integer value $-32\,768$ (the minimum) would be stored with a floating point value of -1.0.

Replaying a numerical vector of sound stored in floating point format is also easy in MATLAB:

```
sound(speech, 8000);
```

To play back the audio it is necessary to specify a sound vector by name and sample rate (e.g. 8 kHz in this case, or whatever was used during recording). If you have a microphone and speakers connected to your computer, you should take time to play with these commands a little to understand what they are capable of. For example, try recording a simple sentence and then increasing or reducing the sample rate by 50% to hear the changes that result on playback.

One handy command within MATLAB to find out what data is stored internally is whos. It gives a list of stored data, indicating the data type and number of items. So in this way, if you record some sound using any of the above techniques, you would be able to see the sound recording listed by name. Anything stored inside the memory, and shown in this list, will remain there until MATLAB is shut down, its memory is wiped (using a command like clearall), or it is overwritten by data having the same name. A quick way to remove some stored array data from memory is by overwriting the array with an empty one, indicated by square brackets. For example, to 'destroy' a recording called myspeech in the current memory space, you could do this:

```
myspeech = [];
```

Sometimes processing or other operations carried out on an audio vector will result in samples having a value greater than ±1.0, or in very small values that are close to zero. When replayed using sound(), this would result in clipping or inaudible playback respectively. In those cases we use an alternative playback command to automatically scale the output amplitude based upon the maximum amplitude element in the audio vector:

```
soundsc(speech, 8000);
```

This command scales in both directions so that a vector that is too quiet will be amplified, and one that is too large will be attenuated. Of course we could accomplish something similar by scaling the audio vector ourselves before playing back with the unscaled command:

```
sound(speech/max(abs(speech)), 8000);
```

It should also be noted that MATLAB is commonly used to develop audio algorithms that will be later ported to a fixed point computational architecture, such as a general purpose CPU like an ARM, an integer DSP (digital signal processor), or a microcontroller. In these cases it is important to ensure that the techniques developed using MATLAB are compatible with integer arithmetic instead of floating point arithmetic. It is therefore useful to know that changing the 'double' specified in the use of the wavrecord() and getaudio() functions above to an 'int16' will produce an audio recording vector of integer values scaled between -32768 and $+32767$. This happens to match the typical range used for audio digital-to-analogue converters and analogue-to-digital converters.

The audio input and output commands we have looked at here will be used during the first and last stages of much of the process of audio experimentation with MATLAB: graphs and spectrograms (a plot of frequency against time) can show only so much, and even many experienced audio researchers cannot repeatedly recognise words by looking at plots, whereas they can when listening to the sound. Perfectly audible sound, processed in some small way, might result in highly corrupt audio that plots alone will not reveal. Since the human ear is a marvel of engineering that has been designed for exactly the task of listening, there is no reason to assume that the eye can perform equally as well at judging sounds. Plots can sometimes serve as an excellent method of visualising or interpreting sound (and we will see many methods of plotting audio throughout this book), but often listening is quicker and better.

In fact, plotting a time-domain waveform representation of a vector of sound samples is easy in MATLAB:

```
plot(speech);
```

although we may wish the *x*-axis to display time in seconds (rather than counting the number of samples):

```
plot( [ 1: size(speech) ] / 8000, speech);
```

Again the sample rate (in this case 8 kHz) needs to be specified.

2.1.3 Audio file handling

In the audio research field, sound files are often stored in a raw PCM (pulse coded modulation) format, although most general audio files are stored in Wave file format (.wav) or in MP3 format (please refer to Infobox 2.1 on page 14 for details on these). We will consider both cases.

2.1.3.1 Loading a Wave file

Since the Wave file (i.e. a file with extension .wav) contains header information that specifies useful attributes like sample rate, number of channels and data format, it is very easy to read this kind of data into MATLAB. We simply use the wavread() or audioread() commands. We will make use of the latter (since wavread() has been marked to be removed in a future release of MATLAB). The syntax is simple:

```
[y,fs]=audioread('myfile.wav');
```

> **Box 2.1** Common audio file formats
>
> **Wave**: The Wave file format is usually identified by the file extension .wav, and actually can hold many different types of audio data identified by a header field at the beginning of the file. Most importantly, the sampling rate, number of channels and number of bits in each sample are also specified. This makes the format very easy to use compared with other formats that do not specify such information, and thankfully this format is recognised by MATLAB. Normally the wave file would contain pulse coded modulation (PCM) data, with a single channel (mono), and 16 bits per sample. The sample rate could vary from 8000 Hz up to 48 000 Hz. Some older PC recording hardware is limited in the sample rates supported, but 8000 Hz and 44 100 Hz are almost always supported. 16 000 Hz, 24 000 Hz, 32 000 Hz and 48 000 Hz are also reasonably common.
>
> **PCM** and **RAW** hold streams of pulse coded modulation data with no headers or gaps. They are assumed to be single channel (mono) but the sample rate and number of bits per sample are not specified in the file – the audio researcher must remember what these are for each .pcm or .raw file that he or she keeps. These can be read from and written to by MATLAB, but are not supported as a distinctive audio file. However, these have historically been the formats of choice for audio researchers, probably because research software written in C, C++ and other languages can most easily handle this format.
>
> **A-law** and μ**-law** are logarithmically compressed audio samples in byte format. Each byte represents something like 12 bits in equivalent linear PCM format. This is commonly used in telecommunications where the sample rate is 8 kHz. Again, however, the .au file extension (which is common on UNIX machines, and supported under Linux) does not contain any information on sample rate, so the audio researcher must remember this. MATLAB does support this format natively.
>
> **Other** formats include those for compressed music such as MP3 (see Infobox 2.2: *Music file formats* on page 15), MP4, specialised musical instrument formats such as MIDI (musical instrument digital interface) and several hundred different proprietary audio formats.

This will read in a file from the current directory named 'myfile.wav'. The vector `y` will then contain the audio data, while `fs` will specify the sample rate. If the file contained stereo data, then `y` will be a 2-by-*n* matrix (where *n* is the number of samples stored in the file). A single channel (monaural) file would return a one-dimensional column vector *y* with length equal to *n* samples.

There are several options to `wavread()` to allow it to read different file types (including Ogg Vorbis and MP3 files – see Infobox 2.2), as well as read in just a segment of a recording rather than the entire file; however, the simple example above is sufficient for the vast number of cases.

2.1.3.2 Loading a raw PCM file

Unlike a Wave file, a raw PCM (pulse coded modulation) file usually ends with an extension such as `.pcm`, `.dat`, or `.raw`. This file type consists of sample values only, with no information about sample rate, precision, number of channels and so on included in the file. In fact a raw PCM file stored on a hard disc without additional information is essentially useless unless the sample rate, number formant and number of channels can be either guessed or remembered.

> **Box 2.2** Common music file formats
>
> **MP3**, represented by the file extension .mp3, is a standard compressed file format invented by the Fraunhofer Institute in Germany. It has taken the world by storm: there is probably more audio in this format than in any other. The success of MP3, actually MPEG (Motion Pictures Expert Group) version 1 layer 3, has spawned numerous look-alikes and copies.
>
> One notable effort is the strangely named format **Ogg Vorbis**, which is comparable in functionality to MP3, but not compatible with it: it is solely designed to be an open replacement for MP3, presumably for anyone who does not wish to pay licence fees or royalties to the Fraunhofer Institute. As such it has enjoyed widespread adoption worldwide, but may require download of a separate audio codec for playback on computer.
>
> Fortunately for the audio researcher, compressed file formats tend to destroy audio features, and thus often may not be suitable for storage of speech and audio for research purposes, hence we will tend to confine ourselves to using raw PCM and Wave file formats.

There is also a potential endian problem for sample sizes greater than 8 bits if they have been handled or recorded by different types of computers (see Infobox 2.3: *The endian problem* on page 19).

Fortunately there are ways of guessing in most cases, and the number of choices is limited (so trial-and-error works too).

To read raw PCM sound into MATLAB, we simply use the general purpose `fread()` function, which has arguments that specify the data precision of the values to read in from a binary file, as well as the endianess. To use this command, we first open the file to be read by specifying its name (and path, if not in the current directory):

```
fid=fopen('recording.pcm', 'r');
```

Assuming that this was able to find and open the required file, we can now read the entire contents, all in one go, into a vector:

```
speech=fread(fid , inf , 'int16' , 0, 'ieee-le');
```

This command will now have read in the entire file (since we specified 'inf' or infinite values) of 16-bit integers. We specified that the format is IEEE little endian, which is the format typically used by a PC. Alternatively (but rarely these days) we could have done:

```
speech=fread(fid , inf , 'uint16' , 0, 'ieee-be');
```

which would read in an entire file of unsigned 16-bit integers, in big endian format (such as a large UNIX mainframe might use).

Finally it is good practice to close any file we had opened, after we have finished reading from it or writing to it:

```
fclose(fid);
```

Some other useful formats are 'int16' for signed 16-bit samples, and 'uint8' and 'int8' for the equivalent 8-bit samples. As mentioned, it is rare to find samples in big endian format these days, but 8-, 10-, 12-, 16-, 20- and 24-bit samples have been used for some common applications. Note that nothing so far has indicated the number of channels or the sample rate: we have just been reading the numbers into MATLAB at this point, but will need that further information once we start to actually *make use* of the data.

2.1.3.3 Saving data in MATLAB

It is possible to use `fwrite()` to save data in the same way as `fread()` above; however, it is much more sensible to save using Wave file format (i.e. with the `.wav` extension), since the resulting file is only slightly larger in terms of size, but includes the highly useful information regarding sample rate, data format and number of channels.

An array of audio samples named `y`, with sample frequency `fs`, can be saved using either the `wavwrite()` or `audiowrite()` commands (although the former will be removed in a future version of MATLAB). The syntax is simple:

```
audiowrite('myfile.wav',y,fs);
```

As with `audioread()` there are a number of options, including saving in Ogg Vorbis and MP4 formats.

2.1.3.4 Storing and reading data in MATLAB-friendly format

It is also useful to know how to save and load general arrays of numbers within MATLAB using its preferred data format. The built-in MATLAB binary storage format allows an array (of speech or anything else) to be saved to disc using the `save` command, and re-loaded later using the `load` command. The normal filename extension for the stored file is '.mat'.

For example, two vectors in the MATLAB memory space called `speech` and `speech2` could be saved to file 'myspeech.mat' in the current directory like this:

```
save   myspeech.mat   speech speech2
```

Later, the saved arrays can be reloaded into another session of MATLAB by issuing the command:

```
load   myspeech.mat
```

2.1 Sound in MATLAB

There will then be two new arrays imported to the MATLAB memory space called `speech` and `speech2`, having the original size and contents of the previously stored arrays. Unlike with the `fread()` command used previously, the name of the stored arrays is specified within the file. If this is a file containing audio, it would make sense to have saved the sample rate as one of the stored values:

```
save  myspeech.mat  speech speech2 fs
```

In general, we will use the abbreviation `fs` within our code snippets to refer to sample rate since it is easy to remember.

2.1.4 Problems with audio input and output

Given an audio file with unknown resolution, number of channels, sample rate and endianess, it is probably best to listen to it after it has been imported to MATLAB the first time, especially for beginners. The idea is to check it has been converted and imported correctly. But please learn at least one lesson from an experienced audio researcher – always turn the volume control right down to minimum the first time that you replay any sound you have processed or imported: sudden pops, squeaks and whistles, at painfully high volume levels, have surprised many of us at times, to the detriment of our ear drums.

It is also useful to plot a waveform as a fast check of its quality. Some common problems can be spotted from a visual examination. For example, Figure 2.1 shows an

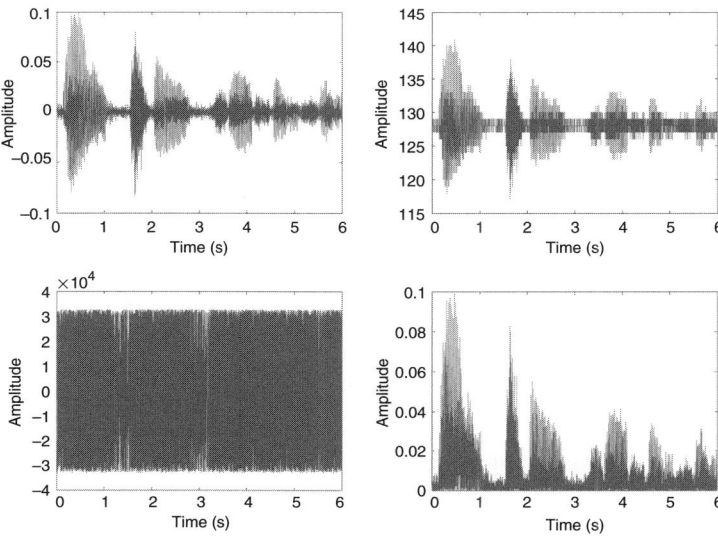

Figure 2.1 Four plots of an audio waveform shown unaltered on top left, correctly quantised to unsigned 8-bit number format on top right, with incorrect byte ordering on bottom left and converted as an absolute unsigned number on the bottom right.

audio recording plotted directly, and quantised to an unsigned 8-bit range at the top left and right of the figure respectively. On the bottom, the same sound is plotted on the left with incorrect byte ordering (in this case where each 16-bit sample has been treated as a big endian number rather than a little endian number), and as an absolute unsigned number instead of two's complement on the right. Note that all of these examples, when heard by ear, result in understandable speech – even the incorrectly byte ordered example that looks terrible (it is easy and quick to verify this; just try the MATLAB `swapbytes()` function in conjunction with `soundsc()`).

Other problem areas to look for when importing or processing sounds are recordings that are either twice as long or half as long as they should be. These are indicators of incorrect data type or an incorrect number of channels. A recording that seems twice as long in MATLAB as it should be might have been a 16-bit array that was read in as 8-bit samples, or may have been a stereo recording that was read in as mono. Of course, it could also be that the sample rate was incorrectly set to half of what it should be. Similarly, a recording that plays inside MATLAB for half as long as expected might be a 16-bit array imported as an array of double precision numbers, or maybe a mono file imported as stereo (or having an incorrectly doubled sample rate).

All of these issues are quite easy to spot for the experienced researcher, but could confuse newcomers to MATLAB or to speech/audio processing. Thankfully the problems discussed here are quite rare – mainly occurring when we use old data, or insufficiently documented audio data supplied by others.

As mentioned previously, the ear and human auditory system is often the best discriminator of sound problems: it may well be the best tool we have in our audio analysis toolkit! If you specify too high a sample rate when replaying sound, the audio will sound squeaky like a cartoon character. It will sound s-l-o-w if the sample rate is too low. Incorrect endianess will probably cause significant amounts of noise, and getting unsigned/signed mixed up will result in noise-corrupted speech (especially for loud sounds, while quiet sounds may be almost unaffected). Having specified an incorrect precision when loading a file (such as reading a logarithmic 8-bit file as a 16-bit linear) will often result in a sound playback that is noisy but recognisable. The intention of discussing this now is for readers to be aware of such problems if they occur in future (however, please rest assured that these issues will not occur when following the examples in this book).

2.2 Normalisation

There is one final step to basic audio handling, and that is normalising the sample vector. As you may have noticed when we discussed replaying sounds earlier, we sometimes had to normalise first to prevent clipping:

```
sound(speech/max(abs(speech)), 8000);
```

> **Box 2.3** The endian problem
>
> The two competing formats are big and little endian, supposedly named after two religious sects of Lilliputians in Jonathan Swift's 1726 book *Gulliver's Travels*, who ate their boiled eggs from different ends. Big endian means that the most significant byte is presented/stored first, and is used by computers such as Sun and HP workstations (which modern researchers are unlikely to come across, although their files are not uncommon). Little endian means that the least significant byte is presented/stored first, as used by Intel and AMD processors inside most desktop PCs and laptops, and most ARM processors (inside everything else).
>
> Unfortunately, endianess is complicated by the variable access-width of modern computers. When everything was byte-wide it was easier, but now there is an added dimension of difficulty. Given an unknown system, it is probably easier to check whether it is little endian, and if not classify it as big endian rather than working the other way around.
>
> **Example**
> We are given a 32-bit audio sample stored in a byte-wide file system (a very common scenario), with the stored word being made up of least significant byte (LSB), second most significant byte (B1), third most significant byte (B2) and most significant byte (MSB). Does the following diagram show a little or big endian representation?
>
> ```
>     4
>     3    MSB
>     2    B2
>     1    B1
>     0    LSB
>         7   0
> ```
>
> In the diagram, the storage address (in bytes) is given on the left, and the bit position shown below. In this case the bit positions are not really important.
>
> Checking for little endian first, we identify the lowest byte-wide address, and count upwards, looking at the order in which the stored bytes are arranged. In this case the lowest address is 0 and the lowest byte starts at bit 0. The next byte up holds B1, and so on. So counting the contents from lowest byte address upwards, we get {LSB, B1, B2, MSB}. Since this DOES follow least-to-most it must be little endian. By contrast, the following diagram shows a big endian representation of a 16-bit sample stored in 8-bit memory:
>
> ```
>     2
>     1    LSB
>     0    MSB
>         7   0
> ```
>
> These days, by using the Wave file format, the endianess is specified and is thus taken care of without any problem. It is also irrelevant with byte-wide formats such as 8-bit samples and A-law samples, and some older sound recordings.

Just to recap, the reason for this is that MATLAB expects each element in the sound vector to be scaled into a range of between -1.0 and $+1.0$. However, the audio that we import from a file is probably 16-bit signed linear fixed point format, which has a range several thousand times larger than this. Even if such an audio file was imported into MATLAB in the correct range of −1.0 to +1.0, subsequent audio processing often causes samples to exceed their original range. A simple example is the addition of two vectors each with range −1.0 to +1.0: the resulting audio data range is data dependent.

It is therefore good practice to comply with MATLAB's expectations and scale audio being processed so that it remains in the expected range after processing. The exception is when it is important to maintain bit exactness with some digital system, or when we are specifically trying to investigate quantisation and scaling issues. However, in general we can scale in two ways:

- **Absolute scaling** considers the format that the audio was captured in, and scales relative to that. In other words, we would divide each element in the input vector by the biggest possible value in that representation: for example, 32 768 for 16-bit signed integers. This scaling is independent of the *content* of the audio vector.
- **Relative scaling** scales relative to the largest data value found in the sample vector. This is the method we used when playing back audio earlier with the `sound()` function, and which is used by the `soundsc()` function.

In general the choice of scaling method depends on whether the absolute amplitude of the original sound is important. For example, when handling many music recordings for a performance, it is important to preserve the fact that some pieces of music should be quieter than others. Therefore we would use absolute scaling. On the other hand, if we wanted to detect the pitch signal within a recording of speech, we might use relative scaling since it probably doesn't matter how loud the original speech was, as long as we can find the pitch.

2.3 Continuous audio processing

One very important consideration with handling audio is the question of whether the sound vector can be processed in one go, or whether it needs to be split up and processed in pieces. In a real sound processing system, these questions would usually revolve around when output is required – a system able to split incoming audio and process it piece-by-piece or chunk-by-chunk would start to produce an output earlier than a system that needed to wait until input recording was finished before it started to process the sound.

Another issue that might influence the choice of chunk-by-chunk processing, as opposed to all-together processing, is the degree of processing power required. Chunk-by-chunk operation spreads the CPU processing into multiple smaller sequential operations, rather than leaving all processing to one massive block of data.

Within MATLAB, it is common for short sound recordings to be processed in their entirety, but processing longer recordings might easily consume all of the computer's memory, so they are split.

Where audio characteristics evolve over time, such as in a speech recording that contains two people speaking, it is often useful to split processing into either equal-sized regions or regions of similar characteristics (i.e. split into different speakers). This act of splitting in this way is usually called segmentation, and will be discussed further in Section 2.4.

Many time-domain operations, such as simple digital filtering, can be applied to a long audio vector in a single pass irrespective of its size (and irrespective of whether it has been segmented). For example, an FIR (finite impulse response) filter is achieved with:

```
y=filter(b, 1, x);
```

where b is a vector holding the filter coefficients. The x and y vectors are equal-length vectors corresponding to the arrays of input and output samples respectively, with each sample of y being calculated from the difference equation:

$$y(n) = b(1) \times x(n) + b(2) \times x(n-1) + b(3) \times x(n-2) + \cdots \\ + b(m+1) \times x(n-m). \quad (2.1)$$

An IIR or pole-zero filter is similarly achieved with:

```
y=filter(b, a, x);
```

where both a and b are vectors holding the filter coefficients. x and y are again equal-length vectors of input and output samples respectively, with each sample of y being calculated from the difference equation:

$$y(n) = b(1) \times x(n) + b(2) \times x(n-1) + b(3) \times x(n-2) + \cdots \\ + b(m+1) \times x(n-m) \\ - a(2) \times y(n-1) - a(3) \times y(n-2) - \cdots \\ - a(m+1) \times y(n-m). \quad (2.2)$$

Frequency-domain operations, by contrast, usually require the audio to be first converted to the frequency domain by using a Fourier transform (or similar, since there are other transforms and spectral estimation methods). An efficient method used by most researchers is the fast Fourier transform (FFT), which is used like this:

```
a_spec=fft(a_vector);
```

and implements a transformation that is equivalent to a discrete Fourier transform, $S(k)$, on an original time sequence $x(n)$ of length N points:

$$S(k) = \sum_{n=0}^{N-1} x(n) e^{-2\pi jkn/N} \quad \text{for} \quad k = 0, 1, \ldots, \{N-1\}. \quad (2.3)$$

Those who have studied digital signal processing will know that the original FFT algorithms were limited to using power-of-two sized input windows (vector lengths). In fact MATLAB uses a software library called FFTW [1][3] which allows arbitrary window sizes – although it is still generally quicker to compute an FFT when the window size is a power-of-two (such as 256, 512, 1024 and so on).

The fft() function can also zero-pad (or even truncate) an input window vector if the FFT size is specified. For example, the following illustrates a 256-element FFT, irrespective of the length of a_vector. If the vector is too short, it will be zero-padded to length 256. If it is too long, then only the first 256 elements will be used to compute the FFT:

```
a_spec=fft(a_vector, 256);
```

Another way of achieving the same thing when the input vector is longer than 256 would be:

```
a_spec=fft(a_vector(1:256));
```

One interesting question is 'how big is the output vector from that transform?' Since the input is a vector of time-domain samples, the output is a vector of frequency-domain information. We already know (from Chapter 1) that the highest frequency represented in our data is the Nyquist frequency (half the sample rate) and the lowest frequency is 0. So we already know the frequency *spread* of the output samples. However, it is the number of elements in the vector which determines the *frequency resolution*: the more elements in the output vector, the higher the frequency resolution.

For an FFT, the output vector elements are known as *frequency bins*, and there are half as many frequency bins as the number of time-domain samples.

In fact, the time and frequency resolution of the FFT will be explored further in Section 2.5.3, but, for the present, suffice it to say that a convenient power-of-two size is normally chosen for the frequency vector length and a longer vector means a higher output frequency resolution, but requires a longer duration of data to analyse.

In MATLAB, the resultant frequency-domain vector from fft() will be complex. Plotting the absolute value of the vector provides a double-sided frequency representation as shown in Figure 2.2. This is plotted using:

```
plot(abs(a_spec))
```

In this unusual plot, the frequency axis (if scaled correctly), would start at 0, progress to the Nyquist frequency at the centre point, and then decrease to 0 at the far right.

---

[3] FFTW is open source software that can be used directly in your own programs, if you need to calculate an FFT. It is available at www.fftw.org

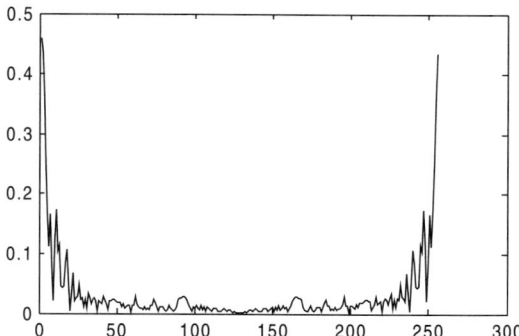

Figure 2.2 Absolute FFT plot for audio spectrum, with frequency index along the *x*-axis and amplitude along the *y*-axis.

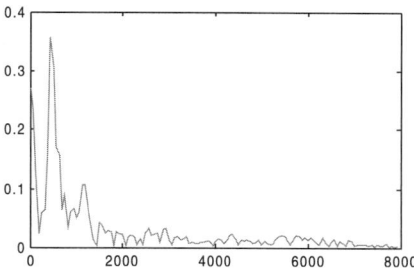

Figure 2.3 Single-sided absolute FFT plot for the same audio spectrum as shown in Figure 2.2.

Both positive and negative frequencies are shown – something which is not particularly useful. In fact MATLAB differs in this way from some of the many other FFT libraries in use for C and FORTRAN programmers. It is easy to produce a more standard plot with the low frequencies in the centre of the graph using:

```
plot(abs(fftshift(a_spec)))
```

However, in audio processing we tend to plot the single-sided spectrum – and give it more useful axes too. Plotting the same spectrum with variables Fs=8000 and Ns=256 describing the original sample rate and size of the FFT respectively, then a better plot can be achieved with:

```
plot([1:2*Fs/Ns:Fs], abs(a_spec(1:Ns/2)),'r')
```

This plots the spectrum as shown in Figure 2.3, which is clearly a more useful and physically relevant representation, with the 'r' argument to plot() meaning the plotted line is coloured red on a colour display. For more on plotting, visit http://mcloughlin.eu/plotting

When performing audio processing such as FFTs in the real world, it would be typical that some form of analysis would subsequently be performed on the frequency vector output from the FFT, and that would often require the single-sided real output samples.

Also, we normally need to perform some kind of conditioning on the input data before it is presented as a vector to the FFT function – the most common being to *window* the data, which we will discuss later in Section 2.4.2. We may also normalise, scale or filter the data.

2.3.1 Long vectors

Having seen how to perform an FFT to transform a series of sampled data into the frequency domain, and how to plot this, some questions might naturally arise.

For example, how should a very long vector of sound samples be analysed if the required frequency resolution is not high? The answer is that the longer vector will be split (or segmented) into several shorter vectors – we tend to call these frames – and each smaller vector analysed separately.

In fact, this kind of segmentation is not needed only because shorter vectors have a more convenient size, but when any of the following are true:

1. When the audio is continuous (i.e. you can't wait for a final sample to arrive before beginning processing), or maybe there *is* no final sample!
2. When the nature of the audio signal is continually changing, or short-term features are important and you want a snapshot of the current frequency information before it changes.
3. If the processing applied to each frame scales non-linearly in complexity (e.g. where a block twice as big would be four or even eight times more complex to process, instead of just twice as complex), then it would be more efficient to process smaller frames.
4. If memory space is limited in the computer that is doing the analysis, it would be necessary to keep the 'chunks' of processed data small.
5. When it is desirable to spread processing over a longer time period, rather than performing it all at the end of a recording.
6. When latency (the delay between the first sample in the block and the analysis output) is to be minimised – a common requirement for voice communication systems.

Almost all of these reasons are commonly found in real-world implementations. Most standalone audio or speech products in the world today will be designed around several of these restrictions. Therefore it is no surprise that almost all digital audio systems will split the sound into smaller chunks to process them.

2.4 Segmentation

Segmenting a long audio recording into smaller frames, which are then processed sequentially, is very common in practical audio systems. However, for a technique that is so important and used so frequently, there is surprisingly little discussion about it in the audio research literature. In fact most research papers fail to mention it at all – considering it an implementation technique which is perhaps not worthy of discussion in a research paper. This is even more surprising when we discover that the process of segmentation suffers from its fair share of problems, and is far from simple. To put it another way, when an engineer needs to implement an audio algorithm within a real-world product, this will usually involve the need to implement that algorithm in such a way that it works on segmented frames. At times, this may be quite a headache!

Consider an audio 'feature'. By that I mean some type of sound event, usually of short duration, or which is a one-off, and which is contained within a lengthy vector of samples.

When that vector is analysed it might be too long, and is therefore split into smaller parts or frames which are each analysed independently. During this splitting process, the feature might be divided into two: half appears in one audio frame, and the other half in another frame. The complete feature therefore does not appear in *any* analysis window, and may have effectively been hidden. In this way, features that are lucky enough to fall in the centre of a frame are emphasised at the expense of unlucky features spanning the edges of two frames, which are chopped in half. When windowing is considered (see later in Section 2.4.2), this problem is exacerbated further since audio at the extreme ends of an analysis frame will be de-emphasised even more. The solution to the lost-feature problem is to overlap frames.

2.4.1 Overlap

Overlap means that instead of straightforward segmentation of the audio vector into sequential pieces, each new frame contains a part of the previous frame and part of the next frame. Overlapping ensures that short audio features that happen to occur at a frame boundary are considered whole in the subsequent, overlapped, frame.

The degree of overlap (usually expressed as a percentage, or as a number of samples) describes the amount of the previous frame that is repeated in the following frame.

Figure 2.4 illustrates the process for 50% overlap, which is very common. The top graph shows a plot of a waveform over time. This is then split into a sequence of overlapping frames, numbered 0 to 6. Every feature in the original signal can now be found repeated in two frames.

Readers with experience of implementation might immediately notice one disadvantage with overlap: namely that we end up with more audio: a 50% overlap doubles the number of frames that need to be processed, which in turn uses more CPU power, and consumes more energy.

Both segmentation and overlap can also lead to problems with processing (i.e. when audio is analysed, modified and then reconstructed). Analysis means breaking down the audio into component parts, and overlap helps, as mentioned, to prevent features from being lost at frame boundaries. It also gives a benefit which can non-technically be described as being similar to taking two (or more) looks at every feature of the audio input, from different perspectives.

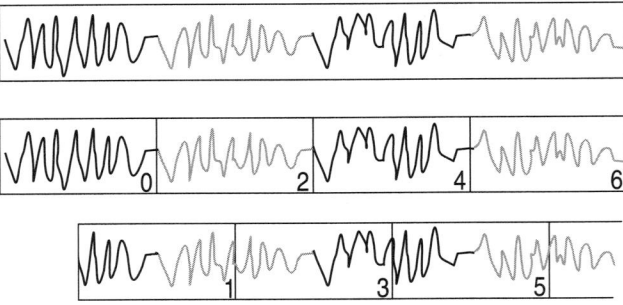

Figure 2.4 An audio recording (the upper waveform) is analysed by being divided into a sequence of seven frames (the two lower rows of waveforms) with 50% overlap. No short-term auditory feature is cut at the boundary between analysis frames. Instead it will appear unbroken in at least one analysis frame.

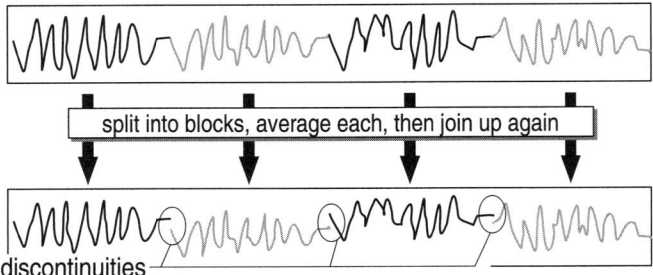

Figure 2.5 An audio recording (upper waveform) is split into equal-length analysis frames without overlap. Each frame is then normalised before being rejoined to create the lower waveform. The rejoined waveform exhibits obvious discontinuities at frame boundaries which would result in audible distortion.

Processing differs from analysis in that audio enters a processing chain which subsequently outputs audio again, and this reconstruction of output leads to complications. To visualise the issue, imagine a system that processes successive frames of audio. For example, maybe audio is divided into analysis frames, each frame is amplified with respect to the mean amplitude within that frame, then the amplified frames are later joined back together to produce averaged output audio.

The difficulty is how to cope with the overlap when re-joining the frames at the output. They cannot simply be concatenated because there would then be twice as many samples, and the time sequence would be incorrect. Simply adding them together won't work either. We will see the solution – windowing, overlap and add – in the next section. However, there is another problem with this arrangement that we need to be aware of first, as illustrated in Figure 2.5. This shows audio that has been split into frames, processed and then re-joined, leading to a sharp discontinuity between neighbouring frames which has been highlighted. This kind of discontinuity in an audio signal would often be very audible to a listener, perceived as a clicking sound. In fact it turns out

2.4 Segmentation

> **Box 2.4** Segmentation into overlapping frames
>
> There are a number of ways to segment and overlap audio in MATLAB. The most direct method is simply to step through an input array, collecting analysis frames until the end of the array is reached. Given an input audio vector Snd, a frame size of Ws samples and an overlap of Os samples:
>
> ```
> Ws=240;              %define frame/window size
> Os=120;              %step from one frame to the next
> Seg=[];
> L=length(Snd);       %determine length of audio array
> indx=1;              %index into the input array
> n=1;                 %frame number
> while indx+Ws-1 <= L
>    Seg=[Seg, Snd(indx:indx+Ws-1)]; %isolate the current frame
>    indx=indx+Os;                   %step to the next frame
>    n=n+1;
> end
> ```
>
> Some notes:
>
> - Array indices in MATLAB always start from 1 rather than zero, hence the several instances of '-1' in the code, which disappear if you do this in 'C'.
> - If the original audio array length is not an exact multiple of the frame size, then the final samples in the audio array will not appear in any analysis frame, so it is often wise to zero-pad the array prior to analysis.
> - The MATLAB Signal Processing Toolbox contains a function named buffer() which can conveniently accomplish much the same as the code above.

that almost any processing done separately frame-by-frame will result in discontinuities such as these in practice.

2.4.2 Windowing

The common solution for many of the re-joining problems in segmented audio is windowing and overlap: that is, to use overlapping frames, each of which is windowed so that its beginning and ending amplitudes are near zero. With a judicious choice of window, this method alleviates the discontinuity problem when joining together processed frames (see, for example, Chapter 18 of [2]). It turns out that windowing is also a prerequisite to prevent edge effects, such as Gibbs phenomena, from occurring during Fourier analysis [3].

To illustrate the overlap–window–add process, Figure 2.6 shows audio being segmented into 50% overlapping frames, windowed, and then reconstructed without creating discontinuities.

There are many predefined window types in frequent use, each of which has different characteristics in the frequency and time domains. If in doubt, beginners are probably well advised to use a Hamming window which, although it does not tail off all the way to zero at either end, has forgiving characteristics. It is often the default window used

28 Basic audio processing

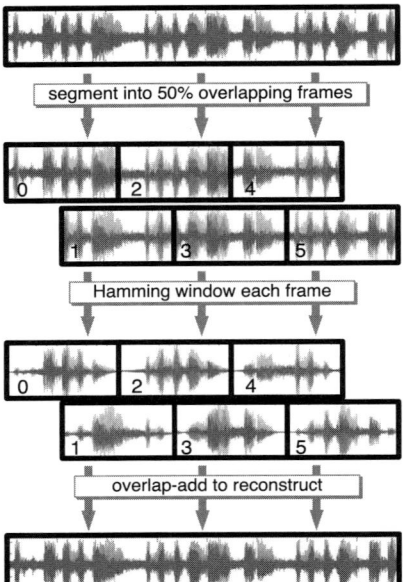

Figure 2.6 Illustration of an original audio recording (upper waveform) being split into 50% overlapped analysis frames, each of which is windowed before being summed together to reconstruct output audio which does not exhibit discontinuities (lower waveform).

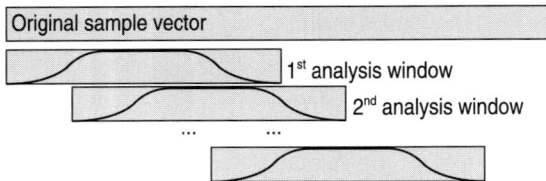

Figure 2.7 Illustration of 25% overlap analysis employing a custom window shape.

in MATLAB functions. Several examples of the more common window functions are given in Table 2.1. Altogether a vast number of other windows have been defined. Any reader looking for an activity to alleviate boredom might like to invent their own.

Finally, these existing windows can easily be skewed one way or another within an analysis frame, perhaps by multiplying with a ramp function. They can also be split into two and a flat-topped section can then be inserted between the two halves, which can be useful when performing 25% or 33% overlap. Figure 2.7 gives an example.

2.4.3 Continuous processing and filtering

Sections 2.4.1 and 2.4.2 discussed how to segment speech into overlapping windowed frames in order to 'smooth out' the effect of transitions between frames. As we had briefly mentioned previously, this is particularly important when those frames are to be processed and then reassembled.

2.4 Segmentation

Table 2.1 Common window functions illustrated for an $N = 200$ sized analysis frame. All functions are zero valued outside the plotted range.

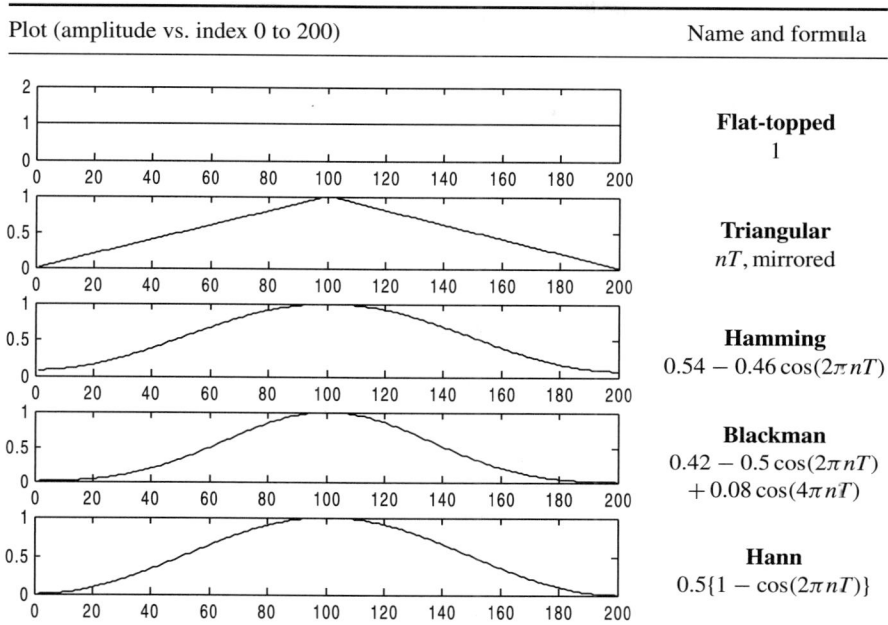

Plot (amplitude vs. index 0 to 200)	Name and formula
	Flat-topped 1
	Triangular nT, mirrored
	Hamming $0.54 - 0.46\cos(2\pi nT)$
	Blackman $0.42 - 0.5\cos(2\pi nT)$ $+ 0.08\cos(4\pi nT)$
	Hann $0.5\{1 - \cos(2\pi nT)\}$

We can easily illustrate the difference between continuous and discontinuous processing in MATLAB with some speech which we can filter and listen to. First, record a few seconds of speech at an 8 kHz sample rate as described in Section 2.1. We shall assume this is in an array named s.

As we already know, this could be replayed and listened to very simply:

```
soundsc(s);
```

Next we will define a small digital filter to act upon the speech. For the purpose of testing, we will just define a simple transfer function $h = (1 - 0.9375z^{-1})$ (which is actually a filter we will encounter later in Chapter 6, used for the pre-emphasis of speech). In the following MATLAB code, we will create the filter, apply it to the entire array of speech, then listen to the result, in vector y:

```
h=[1, -0.9375];
y=filter(h, 1, s);
soundsc(y);
```

This should produce a sound that is 'tinny' or 'nasal' but is otherwise clear speech. Next, we will reuse the same filter, but this time will split the speech recording into a succession of 240 sample frames, each of which will be filtered individually, and then reassembled:

```
w=240;
L=floor(length(s)/w);
for k=1:L
   seg=s(1+(k-1)*w:k*w);
   segf=filter(h, 1, seg);
   outsp(1+(k-1)*w:k*w)=segf;
end
soundsc(outsp);
```

The output should sound similar to the previous example, but will most likely be marred by a succession of audible clicks. Interestingly, if we were to plot the filtered output signals y and outsp, they would probably appear indistinguishable to the eye. It is only using methods like the spectrogram (see Section 2.6) that these clicks or discontinuities would become visible. In fact the reader is encouraged to experiment with plots and spectrograms of these signals – this is a very good example which illustrates the folly in trying to analyse or understand sounds using only frequency, or only time-domain information.

So now there are two questions which arise from this demonstration: firstly, what causes the discontinuities, and secondly, how can we prevent them from happening? The cause may be immediately obvious if one considers exactly what the filtering process is doing. In effect, an FIR filter is describing a process where 'the current output sample is a sum of the current input sample plus the previous P input samples, each scaled by some coefficients', where P is the filter order. Thus in the first-order ($P = 1$) filter shown, the current output sample will equal the current input sample minus 0.9375 times the previous input sample:

$$y[n] = s[n] - 0.9375 \times s[n-1], \qquad (2.4)$$

where the index n ranges from 0 to the length of the speech being filtered; which is either the entire speech vector in the continuous case, or the length of one frame in the segmented case.[4] The clicking problem comes at the join between frames. To understand this, consider the case of the *second* frame to be filtered. Since each frame has a length of 240 (starting from sample number 0), the second frame index 0 corresponds to input array sample number 240. When calculating $y[n]$ at the start of that frame, index $n = 0$ so $y[0] = s[0] - 0.9375 \times 0$ (since $s[-1]$ is outside the current frame

---

[4] Remember that, in MATLAB, the first index to an array is 1 rather than 0. Although we will use MATLAB to demonstrate the effects, we will use the common engineering approach of indexing from zero in our discussion and equations. Since this is an equation, n therefore starts at zero.

2.4 Segmentation

and therefore set to zero). However, when we calculate the same output value if the speech was *NOT* segmented, the index would be not zero but 240. Then we compute $y[240] = s[240] - 0.9375 \times s[239]$. Since $s[239]$ is known, the output y would differ from the result of the calculation in the segmented frame.

Now this is only a single tap filter. Imagine a filter having more taps: the number of samples affected at the start of each new frame would be greater, and thus the size of the discontinuity is likely to be greater. In a tenth-order filter the effect, if present, could very significantly degrade the quality of output speech.

2.4.4 Filter history

In fact, these past coefficients used within a digital filter are termed the filter *history*, or its internal state. When filtering an array continuously, the internal state is automatically updated sample-by-sample. But, when filtering the beginning of an array, the internal state needs to be specified, otherwise it will default to zero. In the segmented speech case, the filter history is actually being reset to zero at the boundary of each frame, rather than carried forward from the previous frame – no wonder the output is distorted.

MATLAB actually provides a convenient way to set and store internal filter history using the `filter()` command. We will illustrate this by very slightly adjusting the segmented example above:

```
w=240;
hst=[];      %define a history array
L=floor(length(s)/w);
for k=1:L
   seg=s(1+(k-1)*w:k*w);
   %use history array
   [segf, hst]=filter(h, 1, seg, hst);
   outsp2(1+(k-1)*w:k*w)=segf;
end
soundsc(outsp2);
```

In the updated code, a history array is now passed to the `filter()` command. Each time the filter is applied, the resulting changed internal history is output and stored. By specifying the filter history at the input to each subsequent frame, the internal coefficients are being reset to the state that they were in when the filter completed its operation on the previous frame. We can thus ensure a smooth filtering operation across frames. The resulting output speech should be free of the clicks and discontinuities evident with the previous segmented code. In fact, the output should be identical to the output obtained by filtering the entire speech array in one pass (in MATLAB you can check this by subtracting one array from the other – the results should be zero if the arrays were identical).

2.5 Analysis window sizing

Towards the end of Section 2.3, we discussed several motivations for splitting audio into segments or frames for processing, but we did not talk about what size the segments or frames should be. At one extreme, a single segment could contain an entire audio recording. Then we could divide this into smaller and smaller pieces, but would eventually end up with every segment containing just a single audio sample. This raises an obvious question: how big should the segments be?[5]

To answer that we may need to consider (i) the time-domain characteristics of the audio signal being handled, (ii) the type of algorithm being used to analyse or process the audio, (iii) the resolution or processing precision required and (iv) memory sizes and computational capabilities of the computer being used. The following four subsections will look, in turn, at each of these aspects, but please bear in mind that there is usually no 'perfect' method to define the analysis window size. Instead it is usually necessary to come to some kind of trade-off between opposing factors.

2.5.1 Signal characteristics and stationarity

Most signals requiring analysis are continually changing. While a single sustained note played on a musical instrument may appear to be quite stationary (i.e. unchanging in its characteristics) for the duration of that note, the next musical note will have changed in some way. Definitely in frequency if it is a different note, but possibly also in amplitude, tone, timbre, and so on. Furthermore, even sustained notes can droop in amplitude, and change frequency or quality noticeably from beginning to end.

For an application analysing recorded music to determine which note is currently being played, it would make sense to segment the recording into analysis windows that match the duration of the shortest single note. For each analysis window we could perform an FFT, and look for peaks in the spectrum which would correspond to the fundamental frequency of that note. However, if we analysed longer duration windows, we may end up performing an FFT that spans across two or more notes, and be unable to determine precisely which note is present. By having an analysis window which is too large, we would end up with a confused 'picture' of the sound being analysed. This is the case in Infobox 2.5 on page 34, where a complex frequency-time pattern is being examined, but a too-big analysis window is initially used and consequently the frequency pattern is obscured. A finer resolution waterfall plot, by contrast, allows much smaller analysis windows to effectively track the changing pattern.

Another important reason to get the analysis window size correct is that the theory underpinning the FFT assumes that the frequency components of the signal are unchanging across the analysis window of interest. So it is better to have a smaller

---

[5] There is also a seldom-asked secondary question – should all segments be the same size? In fact, different sized segments are sometimes used, for example when doing a word-by-word analysis of speech, given that words have different spoken durations. However, more often, the analysis frame size is fixed, and analysis of longer features is done by combining a few analysis frames – with more frames being combined together for a longer word, and fewer for a shorter word.

analysis window where the frequency is less likely to be changing. Any deviation from this assumption results in an inaccurate determination of the frequency components, although in practice small changes are not generally problematic.

These points together reveal the importance of ensuring that an FFT analysis window is sized so that the signal is stationary across the period of analysis. However, since many real-world audio signals do not remain stationary for long, smaller analysis windows are necessary to capture the rapidly changing details, as well as a big overlap (which can help to track fast-moving time-domain variations).

In speech analysis, as will be described in Chapter 3, many of the muscle movements which cause speech sounds are relatively slow (computationally speaking), resulting in slowly varying speech spectral characteristics. A useful rule of thumb is that the speech signal can generally be assumed to be stationary, in fact pseudo-stationary, over a period of about 20–30 ms. Thus speech analysis often involves segmentation into 20 ms long frames [4]. Some aspects of speech are even slower varying than this – such as pitch fundamental, typically processed using an analysis frame of around 80 ms.

The stationarity requirement also extends to linear prediction (Section 6.2.1) as well as to many other forms of signal analysis and processing. Therefore, each method of analysis needs to be carefully matched against the known characteristics of the audio signals which are to be handled.

2.5.2 Type of analysis algorithm

Evidently, some analysis might be long-term, some might be short-term or somewhere in between. For example, measurement of loudness – as in a sound level meter – might be computed over one second of audio, which would be a good reason to use an analysis window that is one second in duration.

When analysis involves a Fourier transform or similar algorithm, like an FFT or discrete cosine transform (DCT), the window length has traditionally been a power-of-two in size, e.g. 256, 512, 1024 samples. This was due to the computation of the FFT, which is most efficient at power-of-two sizes as described previously. However, it is also perfectly possible to zero pad any sized window to the next biggest power-of-two before computing the FFT (at the expense of some additional computation, e.g. take a vector containing 100 samples of audio, add 28 zeros to the end of the vector, and compute a 128-point FFT).

Sometimes, the analysis window size is limited by processing delay. For example, where the output of the analysis is subject to a timing constraint, but the analysis cannot begin until an entire block of audio has been collected. In such cases, keeping the size of the analysis block small would result in the minimum delay between input and output. In fact, many continuous speech or sound processing systems, especially ones that are accompanied by video signals, are quite delay sensitive – either because lip synchronisation is required, or due to the need for two-way conversation. In this way, it could be correctly surmised that the analysis window length used for the real-time processing of audio and speech tends to be shorter than that used for off-line processing (i.e. not in real time) or those for standalone analysis.

Box 2.5 Visualisation of signals

As an example of the difficulty in analysing a constantly changing signal, we will deliberately construct a 'frequency agile' signal (one which changes rapidly in frequency characteristics) and then plot it:

```
y=chirp([0:0.001:5],0,5,500);    %construct a frequency chirp
z=[y,y[length(y):-1:1],y];       %make a zig-zag
```

First let's take one huge FFT and plot this as shown below:

```
f=abs(fft(z, 8192));
plot(f(1:4096));
```

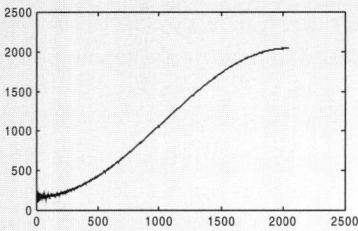

Absolute FFT plot of chirp spectrum.

Does this really represent the characteristics of the signal we have constructed? Lets explore further by slicing the signal up into smaller pieces which we analyse individually and then show the output using a waterfall plot. We could do the slicing and analysis manually, but in this case we can be lazy and use `spectrogram` which accomplishes the job for us very conveniently:

```
s=spectrogram(z, 1024);
waterfall(abs(s)');
```

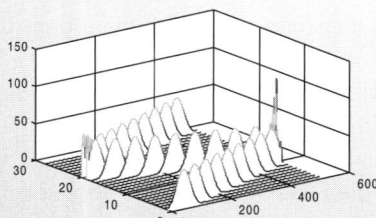

Absolute FFT plot for audio spectrum.

The waterfall plot contains almost 30 slices in time, each one giving a 512-bin FFT plot. Clearly we get a very different picture about the signal through this analysis compared to the single spectrum! If we had relied upon a single FFT to observe the signal we would have missed a significant amount of information.

Note that `spectrogram()`, without assigning its output to a variable, will directly plot the spectrogram on screen. Try this, as well as changing the slice length and exploring some of the other parameters to the function.

2.5.3 Time–frequency resolution

The length of an analysis window, typically expressed in milliseconds, is often chosen primarily for reasons of frequency resolution – with the general rule of thumb being that a longer time-domain window results in higher resolution in the frequency domain, whereas a shorter time-domain window results in lower frequency-domain resolution.

Consider the FFT, where the output frequency vector from an N-sample FFT of audio sampled at Fs Hz contains $N/2 + 1$ positive frequency 'bins'. Each bin collects the energy from a small range of frequencies in the original signal. The bin width is related both to the sampling rate and to the number of samples being analysed, Fs/N. Put another way, the bin width is equal to the reciprocal of the time span encompassed by the analysis window.

It therefore makes sense that a longer duration of samples is needed in the time domain to get more points into the frequency domain. However, for rapidly changing signals, a bigger time-domain window might result in some time-domain features being missed, as discussed in Section 2.5.1 and in Infobox 2.5: *Visualisation of signals* on page 34.

In fact, there is a basic uncertainty principle operating here: a single FFT can trade off between higher frequency resolution (more samples) or higher time resolution (fewer samples) but cannot do both simultaneously. Solutions vary with the requirements of the problem, but remember that there are many other frequency estimation alternatives to the FFT. It may even be possible to perform two FFTs, one using a long window and one using a short analysis window. Later in Section 7.2.2 we will describe more computationally intensive methods of attempting to satisfy the demands of both high frequency and high time resolution.

As a rule of thumb, sounds which are uniformly sampled at a frequency of Fs have a Nyquist frequency – the highest frequency that can be represented in that recording – of $Fs/2$ (as mentioned in Chapter 1). Meanwhile, a window of L samples represents a time span of L/Fs seconds.

When taking an FFT in MATLAB of this length-L vector, there will be a complex output vector also of L points, of which $L/2$ are the positive frequency components. So the frequency resolution is $L/2$ points across a frequency span from 0 to $Fs/2$ Hz. Thus the frequency resolution is simply Fs/L. Table 2.2 illustrates this by comparing the time and frequency resolution that can be expected from various combinations of common analysis window sizes and sample rates.

2.5.4 Processing memory and computational complexity

Generally, most audio algorithms (and this includes most MATLAB-based processing) operate more efficiently on larger blocks of data. In MATLAB this is because built-in functions often take advantage of inbuilt acceleration capabilities of the computer's CPU and minimise large data transfers. By contrast, if the same tasks are split into separate discrete commands written inside a .m file (or typed at the command line), the processing involves more data transfers, or copying of relatively large amounts of data.

Table 2.2 Frequency and time resolutions for various common sampling rates and analysis window lengths.

Sample rate (kHz)	Window size (samples)	Nyquist frequency (kHz)	Frequency resolution (Hz)	Time resolution (ms)
8	128	4	62.5	16
8	256	4	31.25	32
8	512	4	15.625	64
16	128	8	125	8
16	256	8	62.5	16
16	512	8	31.25	32
44.1	512	22.05	86.1328	11.61
44.1	1024	22.05	43.0664	23.22
48	512	24	93.75	10.6667
48	1024	24	46.875	21.3333

This is particularly true of complex iterative computations – and where a specialised MATLAB command exists which can replace several discrete steps, it is almost always faster and more efficient to use the single command.

The same could be said of other languages like C compared with using library functions: generally the library functions are better optimised, more efficient and faster than discrete code. So we could say that spending more time 'inside' such specialised functions and less time 'outside' them improves efficiency. Now imagine a program that loops around multiple times, each time analysing a chunk of audio. It would then follow that using larger analysis frames is better, because it means that fewer chunks are analysed, so that the program iterates through fewer loops.

A significant exception would be when issues such as latency (a critical consideration in telephony processing and similar applications) are important. Another major reason for reducing analysis window size is when either processing complexity or memory size are limited, such as when some very computationally intensive processing is performed, which uses a lot of CPU resources. In such cases, big analysis blocks may use too much memory, or be too complex to process in a given latency, and so smaller blocks need to be used instead.

As an example, consider the difference between an FFT and a discrete Fourier transform (DFT). The DFT, which simply implements the Fourier transform equation using N discrete points, has a computational complexity of $O(N^2)$. This 'Big O' notation means that the DFT complexity becomes four times bigger when the number of points doubles from 128 to 256. Compare that with some FFT algorithms, which can compute essentially the same thing with a complexity of $O(N \log N)$, meaning that complexity increases by a factor of only 2.3, compared with 4 for the DFT.

In general, any algorithm that has a complexity of much greater than $O(N)$ (which means the processing takes twice as long when the analysis window size doubles – in other words, the processing time scales linearly with the amount of data) would probably benefit from a smaller analysis window, whereas a smaller analysis window would make little difference for low-complexity algorithms (such as the FFT). The speech and audio processing domain contains a very wide variety of algorithms which span the range from better than $O(N)$ to worse than $O(N^3)$, so it can be very beneficial to understand the complexity of a processing algorithm.

2.6 Visualisation

Visualisation means transforming a signal under analysis to a visual representation. In essence, this allows our eyes to appreciate some of the complexity of sound. It can be a very good way of understanding the characteristics of a signal but – as any conjurer would attest – it can also be a very good way of obscuring something!

Obviously our eyes and our ears are different sense organs, but, because they interface to the brain in different ways, they have their own strengths and weaknesses. We will discuss the hearing system more fully in Chapter 4, but for now we can note that the hearing system is characterised by excellent relative precision over an exceptional dynamic range (i.e. it can sense small changes over a huge range of loudnesses and frequencies). It is good at picking up time-domain changes in a signal as well as other frequency discontinuities, but less good at discerning if a sound heard today is identical to one heard yesterday, or identifying absolute frequency and duration. Sound (assuming just one ear) is essentially a one-dimensional signal, so it is understood as a sequence in time. We won't consider moving images here, but even for simple static pictures, the eye allows us to pick out patterns and local relationships within the image, as well as overall structure, at both fine and gross resolution levels.

So we can 'view' visualisation as transforming a sound into a representation that takes advantage of different analytical capabilities of the human visual system. Different representations (through different transformations) can target different aspects of the visual system, and in turn allow us to examine different aspects of the sound. Let us now consider a few of the more common methods.

2.6.0.1 Time-domain waveform plot

This is the easiest and most basic method of visualisation, and can be very useful, especially for quickly scanning long recordings – for example, to detect periods of silence. But beware that much of the information is hidden, and what you expect to hear after viewing a waveform doesn't always tie up with what you actually hear.

A simple illustration of the problems inherent with a waveform view is given in Figure 2.8, where three different resolution views of the same signal (conversational speech) reveal very little visual similarity for what is really a fairly uniform audio signal.

2.6.0.2 Frequency spectrum

Again this was mentioned and illustrated previously, and is a basic and well-used tool to get a snapshot of the main frequency components occurring within a single analysis window. Unfortunately it is sometimes easy to miss an important feature by computing

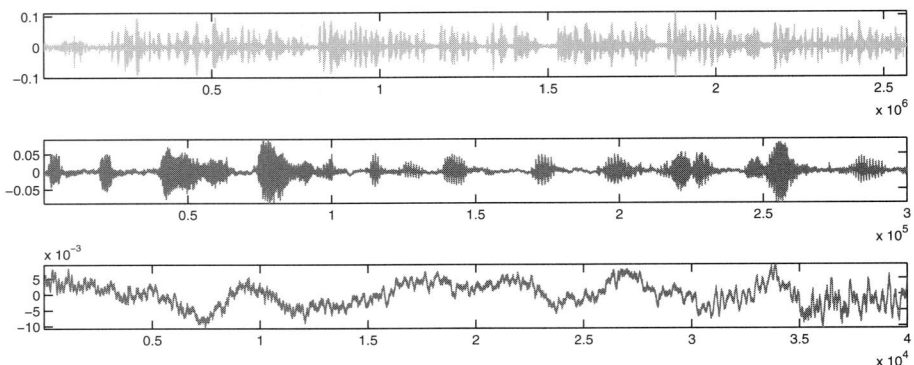

Figure 2.8 Amplitude against time plots of the same speech recording at three different time scales.

the spectrum from the wrong analysis window, or from a window selected at the wrong position in an audio file, or at the wrong resolution, wrong window size, and so on.

If an entire section of recording needs to be visualised in frequency terms, then a better tool is the spectrogram – an image made up from strips of spectra.

2.6.0.3 Short-time Fourier transform (STFT)

A short-time Fourier transform is a narrow Fourier transform over just part of a vector being analysed, then repeated sequentially along the vector of samples. This performs multiple time–frequency signal decomposition or analysis at discrete times stepping through the analysis vector. It results in a time sequence of individual spectra, which can be plotted against time, either in an x–y–z graph or as a spectrogram (which is actually the magnitude squared of the transformed data):

$$\text{spectrogram } x(x) = |X(\tau, \omega)|^2.$$

In MATLAB this used to be very easy to perform with the `specgram()` function. Unfortunately `specgram()` is shortly destined to be removed in a future version of MATLAB. A replacement called `spectrogram()`, which does much the same thing, is available (but only within the signal processing toolkit). It has many options regarding analysis window size, overlap and number of sample bins.

A spectrogram is essentially a set of STFTs plotted as frequency against time with the intensity (z-axis) denoted by brightness (when plotted in greyscale) or pixel colour (with various 'colour maps' being available to show intensity using different shades). For speech analysis, the spectrogram is an excellent method of visualising speech structure and how it changes over time.

An example of MATLAB spectrograms is shown in Figure 2.9, plotted in greyscale, and giving the same three periods of speech as shown in Figure 2.8. Some audio researchers prefer to plot their spectrograms in colour, but this is really just a matter of personal preference.

For reference, the bottom spectrogram in Figure 2.9 was plotted as follows, as a section from the long speech array called `sr`, sampled at 24 kHz:

2.6 Visualisation

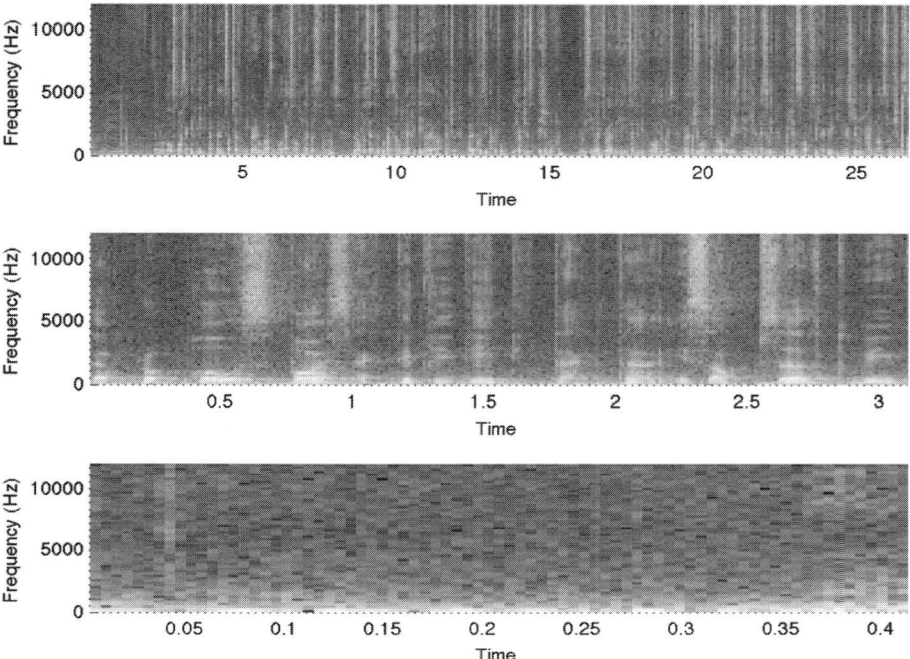

Figure 2.9 Spectrograms of the same single audio recording views as given in Figure 2.8 Louder sound components are shown with a darker grey shade and lower amplitudes with a lighter shade.

```
subplot(3,1,3)
spectrogram(sr(2500:12500,1),128,0,128,24000,'yaxis')
```

The section of audio vector `sr` began at sample 2500 and ended at sample 12 500. The analysis window size for each STFT was 128 samples in length. There was no overlap between windows, and the FFT size was also 128 samples. The 'yaxis' argument means that the frequency axis is plotted vertically and the time axis is horizontal – and this is definitely the most common arrangement for speech spectrograms.

2.6.1 A brief note on axes

The horizontal axis of an FFT plot, or single spectral plot, is traditionally used to represent frequency, whilst the vertical axis would display amplitude, as shown in Figure 2.3.

Due to the large dynamic range of audio signals, it is common to plot the logarithm of the absolute amplitude, rather than the amplitude directly; thus the spectrum will often be plotted as a power spectrum, using $20 \times \log_{10}(\text{spectrum})$, or with MATLAB:

```
semilogy(spectrum);
```

Note that this is also true of the `spectrogram()` function which plots the log of the spectral magnitude.

For a single spectral plot, the *x*-axis, showing frequency, is generally plotted linearly. The labelling of this axis defaults to the index number of the bins in the FFT output spectrum – not particularly useful in most cases. A far better approach is to specify this yourself either directly in Hz, scaled between 0 and 1 (DC to Nyquist frequency), or in radians between 0 and π:

```
res=pi/size(spectrum);
semilogy(res:res:pi,   spectrum);
```

This radian measure, normally denoted by the independent frequency variable being written as ω, represents 2π as the sampling frequency, 0 as DC and thus π to be the Nyquist frequency. It is referred to as *angular frequency* or occasionally *natural frequency*, and can be thought of as representing frequencies which are arranged around a circle. This notation is actually quite useful when dealing with systems that are over- or undersampled, but, apart from that, it is more consistent mathematically because it means equations can be derived to describe a sampled system that does not depend on having defined a fixed sample rate up-front.

2.6.2 Other visualisation methods

As you may expect, many other more involved visualisations exist. Some of these have evolved to be particularly suited for viewing certain features. One of the most useful for speech and audio work is the linear prediction coefficient spectral plot that will be described in Section 6.2.1. On the other hand, two other very general and useful methods now follow – namely the correlogram and the cepstrum.

2.6.2.1 Correlogram

A correlogram is a plot of the autocorrelation of a signal. Correlation is the process by which two signals are compared for similarities that may exist between them either at the present time or in the past (however much past data is available). Mathematically it is relatively simple. We will start with an equation, defining the cross-correlation between two vectors *x* and *y* performed at time *t*, and calculating for the past *k* time instants, shown in Equation (2.5):

$$c_{x,y}[t] = \sum_k x[k]y[t-k]. \tag{2.5}$$

In MATLAB such an analysis is performed using the `xcorr()` function over the length of the shortest of the two vectors being analysed:

```
[c,lags]=xcorr(x,y);
```

2.6 Visualisation

If just a single argument is provided to xcorr then the analysis results in an autocorrelation, which compares the single vector against each possible time-shifted version of itself. The comparison at time shift k is the sum of the product of each element of the input vector with the element k positions previous to it. Using this as a tool we can look for periodicities in the vector being analysed.

As an example, if the input vector contains two peaks, then the autocorrelation output will be large at time shifts where one peak ends up being multiplied by the other peak. In that case, the time shift k corresponding to the largest output value would be equal to the separation between the two peaks, in samples. Where the vectors being analysed are more complicated, autocorrelation analysis is a good way of looking for periodicities in the signal that might be hidden to the naked eye.

As an example, assume that we have a short recording of speech in MATLAB, imaginatively called speech. By plotting this, we notice that there is a section in there with a fairly regular structure, perhaps from a voiced segment of audio. For illustrative purposes, assume this lies between vector elements 9000 and 10 000 in our recording. We are going to 'cut out' that section for autocorrelation analysis:

```
segment=speech(9000:10000);
```

Next we perform an autocorrelation analysis and plot both the speech segment and the resulting correlogram. In this example, the plotting commands are also reproduced for reference:

```
[c,lags]=xcorr(segment);
subplot(2,1,1);
plot([0:1:1000]/8000,segment);
axis([0, 0.125. -0.05, 0.05]);
subplot(2,1,2);
plot(lags(1001:2001)/8000,c(1001:2001));
axis([0, 0.125. -0.2, 0.2]);
```

The argument to the axis commands to define the y-dimension (from -0.05 to 0.05 and from -0.2 to 0.2, respectively) will need to be changed depending upon the amplitude and composition of the speech segment that you are working with, but if chosen correctly will ensure that the plots fill the entire plot area. An alternative would be the command axis tight which ensures that the plot fills the axis area.

An example of this output is shown in Figure 2.10. It should be noted that the first major peak identified in the correlogram (which has been selected by the mouse pointer), at an x distance of 0.006 875, which is 6.875 ms, corresponds to the main period in the speech plot above. Try measuring the distance between speech peaks with a ruler, and then comparing this with the distance from the y-axis to the first identified peak

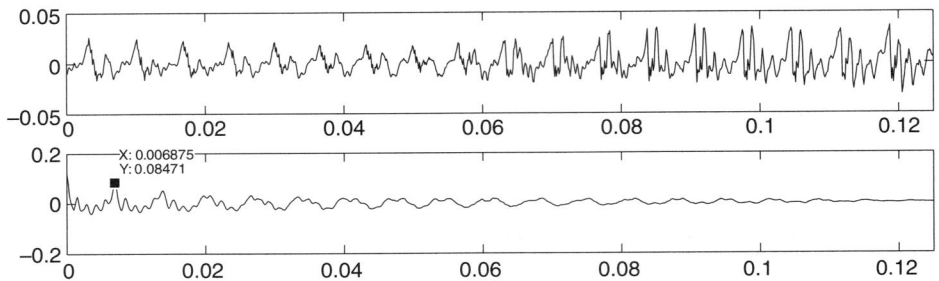

Figure 2.10 Plot of a segment of relatively periodic speech (above) and its autocorrelation analysis (below), both plotted as amplitude against time, with the first correlation peak highlighted.

in the correlogram. This illustrates the main use of the technique for audio analysis – detecting periodicities.

2.6.2.2 Cepstrum

The name 'cepstrum' comes about by reversing the first half of the word 'spectrum', and plots the amplitude of a signal against its 'quefrency' – actually the inverse frequency. Evidently neither word was chosen for ease of pronunciation. However, the technique is particularly good at separating the components of complex signals made up of several simultaneous but different elements combined together – such as speech (as we will see in Chapter 3).

The cepstrum is generated as the Fourier transform of the log of the Fourier transform of the original signal [5]. Yes, there really are two Fourier transform steps, although in practice the second one is often performed as an inverse Fourier transform instead [6].[6]

Using MATLAB again, a very simple example would be to plot the cepstrum of the speech analysed above with the correlogram. This is fairly simple to plot – not quite accurately as per the original meaning of cepstrum, but certainly useful enough, given a speech vector x:

```
ps=log(abs(fft(x.*hamming(length(x)))));
plot(abs(ifft( ps )));
```

In fact, the MATLAB signal processing toolbox contains the cepstral function cceps() and its inverse icceps(), which are much easier to use. For example, given a short section of speech in vector x we may do:

```
cx=cceps(x.*hamming(length(x)));
stem(abs(cx));
axis tight;
```

---

[6] Technically the method described here is the discrete time power cepstrum, arguably the most useful of the cepstral techniques.

2.6 Visualisation

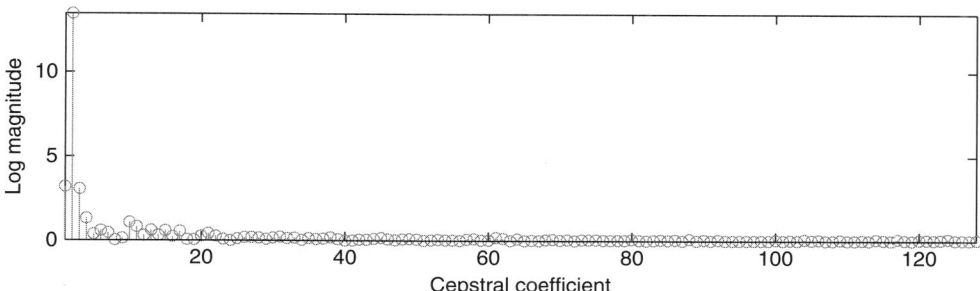

Figure 2.11 Example of cepstral analysis using `cceps()`.

```
xlabel('Cepstral coefficient')
ylabel('Log magnitude')
```

which would yield something like the plot in Figure 2.11.

Most likely, if the speech segment were as lengthy as that used for the previous correlogram example, the resulting cepstrum would have a huge DC peak and much of the detail in the plot would be obscured by that. Therefore the example we plot here is only 128 points in length.

If we *did* have a huge plot using the correlogram data we could use the MATLAB plot tools to zoom in to the region of interest on the plot. There should then be a peak visible at the same index position as the peak in the correlogram. For example, a peak in the 256-point cepstrum of 8 kHz speech at index 56 would relate to a frequency of $4000 \times 56/256 = 875$ Hz. In fact, this method of analysis will be illustrated later in Section 7.1.5.

Both the correlogram and the cepstrum can be used to determine the fundamental frequency. Both methods, while accomplishing similar tasks, have unique strengths: peak detection in the cepstrum may be easier to achieve automatically (assume a constrained search range), and certainly the cepstrum can highlight features missed in a correlogram. The correlogram, on the other hand, is significantly easier to compute and scale than the cepstrum. Both of these visualisation and analysis tools are available to speech researchers, and sometimes both can be consulted during a particular analysis to compare and contrast their different viewpoints.

Finally, note that the cepstrum has many uses in the speech community, including distortion and quality measurement. In practice, the cepstral coefficients are often obtained from linear prediction coefficients (described later), rather than computed directly as we have done here. Furthermore, they are very frequently calculated with a non-linear mel frequency band mapping which makes them better resemble the response of the human hearing system, as we will discover in Chapter 4. The resulting mel frequency cepstral coefficients (MFCC) are extremely important features for applications such as speaker identification, speech recognition, language identification and machine hearing, as well as many other speech analysis tasks [7].

2.7 Sound generation

This chapter has mainly been devoted so far to sound analysis in MATLAB, the handling of audio data and its visualisation. We have seen, in Section 2.1.2, how to output an array of sound so that (assuming our computer is equipped with loudspeakers or a pair of headphones) we can hear the audio. We have also seen how to save sounds in MATLAB into audio files in Section 2.1.3. Now we will turn our attention to the ability to generate or create new sounds. It turns out that this is quite useful – not least when experimenting with, or testing, audio and hearing responses. Some of the techniques and functions we will discuss below will come in handy during later chapters when we provide some hands-on demonstrations of various auditory phenomena.

2.7.1 Pure tone generation

The amplitude of a pure tone varies sinusoidally. So to generate such a tone, we just need to create a sinusoidal signal of the correct frequency. For example, the following MATLAB function which we will call tonegen() is able to generate a tone of frequency Ft at a sample rate of Fs, lasting for Td seconds:

```
function [s]=tonegen(Ft, Fs, Td)
    s=sin([1:Fs*Td]*2*pi*Ft/Fs);
```

The function code should be entered in the MATLAB editor and then saved somewhere (such as the default work directory) using a filename of tonegen.m, since we will be calling this function later, when we use it inside several other MATLAB functions and programs. From this point on we will assume that, when such a function is provided in the text, the reader will add the function to their slowly growing collection of MATLAB code.[7]

Once we have entered and saved the function, we can use it to generate tones. Here is an example to generate and then replay a 2 second long pure tone of frequency 440 Hz (corresponding to A_4 on the musical scale as described in Infobox 2.6 on page 45), at a sample rate of 16 kHz:

```
note=tonegen(440, 16000, 2);
soundsc(note, 16000);
```

These functions should work at any sample rate, so the 16000 Hz could be replaced with 8000 or 20000 – in fact anything greater than the Nyquist frequency (880 kHz) and below the maximum allowed sample rate for your computer and operating

---

[7] This function, as well as all others given in this book, can be downloaded from the associated website, www.mcloughlin.eu; in this case the function is at www.mcloughlin.eu/listings/tonegen.m

2.7 Sound generation

Box 2.6 Musical notes

The pitch of a musical note is determined by its perceived fundamental frequency of oscillation. Usually there will be a combination of different frequencies present in a musical note, which affect the timbre of the note through their presence.

Several musical scales exist, each defining the frequency of notes contained within them, although most people are familiar with the Western scale. This is tied to a frequency reference where A_4 (the A above middle-C) is set to 440 Hz. Defined by the International Standards Organisation (ISO), this is often called *concert pitch*. Any reader who has played an instrument in an orchestra would know the importance of all instruments playing in tune. Before the 1940s, however, lack of standardisation meant that A_4 could vary widely between orchestras and countries, and many composers insisted on setting their own (incompatible) standards. Both Germany and the United Kingdom created early standards, but it was the British Broadcasting Corporation (BBC) which chose the current 440 Hz standard, based on the very high precision oscillation of a piezoelectric crystal tied to some frequency division and multiplication circuitry. A popular earlier standard of 439 Hz, being a prime number, could not be easily generated in such a way, so it was superseded by the BBC standard in use today.

The Western scale, chosen (like most other scales) to match human perception of pitch differences, defines that the frequency relationship of one semitone is the twelfth root of two. Therefore since A_4 is 440 Hz, $A\sharp_4$, one semitone higher, should be $440 \times 2^{(1/12)} = 466.16$ Hz. Furthermore, since an octave consists of 12 semitones, A_5 would be:

```
440*(2^(1/12))^12=440*2=880
```

In this way, by knowing the semitone relationship between musical notes, we can determine the frequency, in Hz, of any note on the musical scale. Here are some examples:

C_1	32.70	C_2	65.41	C_3	130.81	C_4	261.63	C_5	523.25
$C\sharp_1$	34.65	$C\sharp_2$	69.30	$C\sharp_3$	138.59	$C\sharp_4$	277.18	$C\sharp_5$	554.37
D_1	36.71	D_2	73.42	D_3	146.83	D_4	293.66	D_5	587.33
$D\sharp_1$	38.89	$D\sharp_2$	77.78	$D\sharp_3$	155.56	$D\sharp_4$	311.13	$D\sharp_5$	622.25
E_1	41.20	E_2	82.41	E_3	164.81	E_4	329.63	E_5	659.26
F_1	43.65	F_2	87.31	F_3	174.61	F_4	349.23	F_5	698.46
$F\sharp_1$	46.25	$F\sharp_2$	92.50	$F\sharp_3$	185.00	$F\sharp_4$	369.99	$F\sharp_5$	739.99
G_1	49.00	G_2	98.00	G_3	196.00	G_4	392.00	G_5	783.99
$G\sharp_1$	51.91	$G\sharp_2$	103.83	$G\sharp_3$	207.65	$G\sharp_4$	415.30	$G\sharp_5$	830.61
A_1	55.00	A_2	110.00	A_3	220.00	A_4	440.00	A_5	880.00
$A\sharp_1$	58.27	$A\sharp_2$	116.54	$A\sharp_3$	233.08	$A\sharp_4$	466.16	$A\sharp_5$	932.33
B_1	61.74	B_2	123.47	B_3	246.94	B_4	493.88	B_5	987.77

You will find a simple MATLAB script named `musical_notes.m` to generate these notes and frequencies on the website.

system.[8] The sound frequency should not change when you alter the sample rate – although the *quality* might change quite noticeably.

---

[8] This should be at least 48 kHz, possibly even 96 kHz for modern operating systems like Linux and Mac OS. Otherwise, try the standard for CD audio, 44.1 kHz.

2.7.2 White noise generation

White noise is spread evenly in frequency from just above 0 Hz to the Nyquist frequency. It is easy to generate because it consists of uniformly distributed random numbers. In MATLAB, using the same notation as previously of `Td` seconds duration at sample rate `Fs`, we can generate a white noise vector as follows:

```
noise=rand(1,Fs*Td);
```

Later, we will show how white noise can be added to a recording of clean speech or audio to change the signal-to-noise ratio.

2.7.3 Chirp

We had actually used a chirp as an example in Infobox 2.5: *Visualisation of signals* on page 34, but did not explain it at the time. It is the name given to a signal which increases monotonically in frequency with time, or which decreases monotonically in frequency with time, and gets its name because it sounds like the noise a bird might produce. Sometimes a chirp is referred to as a 'frequency sweep' signal because it sweeps through a range of frequencies.

A **linear** chirp sweeps across frequency at a constant rate. To illustrate this, if a linear chirp starts at 100 Hz and then takes a second to sweep up to 200 Hz, we know that after half a second its frequency would be 150 Hz. After 9/10 of a second, it would have reached 190 Hz. The other main type of chirp is **exponential**, in which the rate of frequency change is itself increasing over time (or decreasing).

MATLAB can generate many types of chirp using the function `chirp()` from the signal processing toolbox, but we will only illustrate a linear one here:

```
bird=chirp([0:1/8000:1], 1000, 1, 4000);
soundsc(bird,8000)
```

The first argument to `chirp()` is a time array – specifying the vector of sound samples which we will create. The array has 8000 samples, and is 1 second long. Samples are spaced apart by $\frac{1}{8000}$ seconds, and so we can deduce that the sample rate is 8 kHz.

The chirp starts at a frequency of 1 kHz, and takes 1 second to sweep up to 4 kHz (the Nyquist frequency). When we listen to our artificial bird in the second line, we can hear how realistic it sounds. Readers who want a more realistic-sounding artificial bird might wish to experiment with non-linear chirps. Incidentally, we will be examining *real* bird chirps in Section 7.5.2.

2.7.4 Variable tone generation

When we experiment with sound, we often need a way to generate a signal which varies in frequency in an arbitrary way (i.e. not constant, linear or exponential). Many times, we want to create a sound that changes in frequency to follow some other signal.

It is actually quite easy to do this in a stepwise fashion, by generating some tones and then concatenating them together, but is not at all trivial to do it smoothly. We will illustrate the first (stepwise) method first, using our tonegen() function from page 44.

For the sake of illustration, let us try changing frequency sinusoidally: ±200 Hz around an average 1000 Hz frequency, with a period of 1 second. We will repeat this for 4 seconds. If we define a sample rate of 8 kHz, then our frequency should vary like this:

```
Fs=8000;        %define sample frequency
Tt=[0:1/Fs:4];  %create array of sample times
ModF=1000+200*sin(Tt*2*pi);
```

Just to be clear, this should have created an array called ModF which lasts for 4 seconds at a sample rate of 8 kHz, and defines what frequency we would like to hear, varying from 800 up to 1200 Hz (in MATLAB you can plot this array to be sure exactly what it contains, plot(ModF)).

Next, we could generate a sequence of tones, each one 0.1 s in duration, that have the correct frequency:

```
gensnd=[]; %initialise an empty array
for t=0:0.1:4
   fr=ModF(1+floor(t*Fs));
   % use floor() since index must be integer
   gensnd=[gensnd,tonegen(fr,8000,0.1)];
   % concatenates new tones to end of gensnd
end
soundsc(gensnd,Fs);   % listen then view it
spectrogram(gensnd,128,0,128,Fs,'yaxis');
```

If you look at the spectrogram, you will see something that varies sinusoidally, but is 'blocky' (shown in the top part of Figure 2.12). In addition, it definitely does not sound smooth to the ear, it is very much like a succession of short musical tones. Of course, it is possible to make the short tones even shorter – such as reducing the duration from 0.1 s down to 0.01 s or so, and the reader is invited to try that. But it doesn't work. The reason is hidden inside the waveform, and can be seen by looking at Figure 2.13. This plots part of the waveform of the 0.1 s step sound, and represents the audio sample

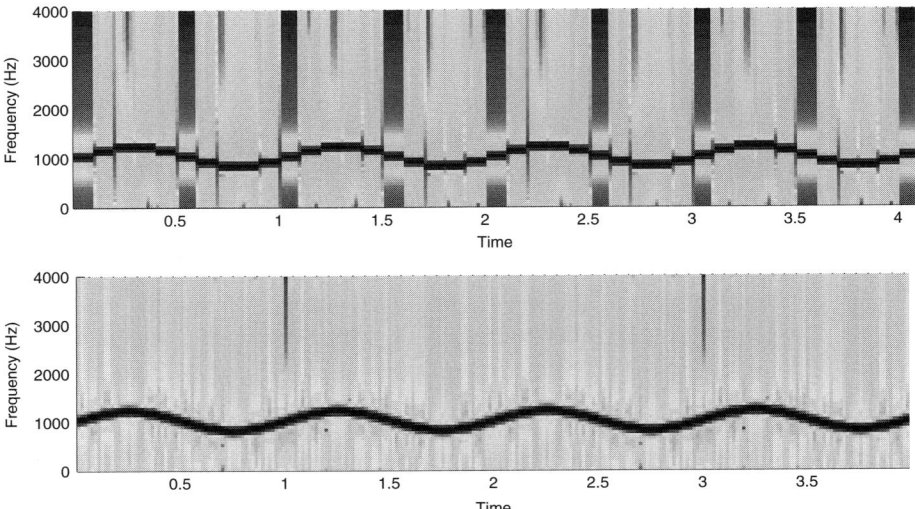

Figure 2.12 Two methods of generating a sound whose frequency varies sinusoidally with time; the upper plot shows a succession of 0.1 s tones, the lower plot shows a smoothly varying tone change.

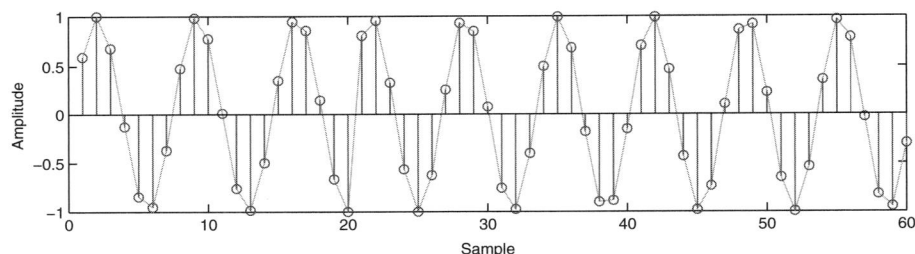

Figure 2.13 Part of the time-domain waveform of the signal plotted in the upper part of Figure 2.12, illustrating a discontinuity around sample 20.

points as small circles. If you look closely, you will see something strange at sample 20. It looks almost like there is a missing sample, which prevents the waveform from moving smoothly from low to high.

This is a discontinuity caused by the fact that sample 20 is a transition from one 0.1 s tone to the next one, of a different frequency. In fact, the discontinuity is due to a mismatch in *phase* between the two neighbouring tones, because the `tonegen()` function always starts at zero phase. Note that we could modify the function to take account of phase; however, a better way is to smoothly vary the sound frequency from sample to sample. The following MATLAB function does precisely that: it will smoothly transition frequencies when fed with an array specifying any desired frequency, sample-by-sample, in Hz:

2.7 Sound generation

```
function [snd]=freqgen(frc, Fs)
  th=0;
  fr=frc*2*pi/Fs;
  for si=1:length(fr)
    th=th+fr(si);
    snd(si)=sin(th);
    th=unwrap(th);
  end
```

We can test this function using the sound array we created earlier:

```
gensnd2=freqgen(ModF, Fs);

soundsc(gensnd2,Fs);   % listen, then view it
spectrogram(gensnd2,128,0,128,Fs,'yaxis');
```

The resulting spectrogram is shown in the lower half of Figure 2.12, and is obviously a lot smoother than the sound created by concatenating frequency steps. The time-domain waveform that is generated does not exhibit any of the discontinuity problems shown in Figure 2.13 – and it sounds smooth too!

While this method is excellent for smoothly varying frequencies, it can also be used to create discrete sequences, such as the following example which specifies a few musical note frequencies in a frequency array, then converts this into sound:

```
freq=[440*(1+zeros(1,1000)), 415.2*(1+zeros(1,1000)),
    392*(1+zeros(1,1000))];
music=freqgen(freq, 8000);
soundsc(music, 8000);
```

In fact, the only downsides with the freqgen() method are that it is extremely slow and can only generate one frequency for each instant of time. To get two frequencies at each time instant, we would probably need to mix two sound arrays together.

2.7.5 Mixing sounds

If we have two audio arrays of equal length we can combine or mix them together. We will illustrate this and listen to the result using soundsc() (we need to use soundsc() instead of sound() since the result of the addition may well cause individual sample amplitudes to exceed ±1.0, which would cause the sound() function to clip – to find out what that problem would sound like, you could easily try it and see).

Here are notes of a musical chord generated and replayed in MATLAB:

```
C=tonegen(261.63, 8000, 2);
E=tonegen(329.63, 8000, 2);
G=tonegen(783.99, 8000, 2);
B=tonegen(987.77, 8000, 2);
soundsc(C+E+G+B, 8000);
```

For those readers with a musical theory background, it should be relatively easy to create MATLAB functions able to create specified chords, minors or majors, and replay them on demand. The frequencies of different musical notes can be calculated as shown in Infobox 2.6 on page 45.

2.8 Summary

This chapter has covered many of the basic approaches related to obtaining and storing audio with MATLAB, some of the most important features to be aware of when working with audio, and methods of handling and visualising audio and finally of creating sounds. We could consider the methods in this chapter to be part of the basic skill-set for those working in fields related to speech and audio processing.

Subsequent chapters will gradually build upon these techniques, incorporating them into more involved and advanced methods of analysis and processing.

Bibliography

Readers wishing to learn more about using MATLAB itself, the underlying basics of signal processing or the theories and practicalities of audio applications may wish to consult some of the references given by category below:

MATLAB

- *An Engineers Guide to MATLAB*
  E. B. Magrab, S. Azarm, B. Balachandran, J. Duncan, K. Herold and G. Walsh (Prentice-Hall, 3rd edition 2007)
  This is an expensive and extensive textbook running to 750 pages of engineering applications in MATLAB. Although it does not mention audio and speech processing explicitly, it does very nicely cover techniques of data handling, visualisation and analysis. File input and output are covered, as are statistical techniques. This may well be a useful reference work, consulted in the library, rather than purchased for continuous use.

- *A Guide to MATLAB: For Beginners and Experienced Users*
  B. Hunt, R. Lipsman, J. Rosenberg, K. R. Coombes, J. E. Osborn and G. J. Stuck (Cambridge University Press, 2nd edition 2006)
  This far more concise reference book dispenses with the applications focus and spends more effort in covering the basic and underlying techniques of data handling, plotting and so on. It is pitched at a lower level audience, but does claim a relevance for experienced users who are upgrading to a newer MATLAB version. The focus of this work is primarily the applied mathematics foundation of handling signals in MATLAB.

Signal processing

- *Digital Signal Processing: A Practical Guide for Engineers and Scientists*
  S. W. Smith (Newnes, 2002)
  Also freely available from **www.dspguide.com**

- *Digital Signal Processing: A Practical Approach*
  E. C. Ifeachor and B. W. Jervis (Addison-Wesley, 1993)
  As the title implies, this work is reasonably practical in tone and covers most of the required areas of signal processing, including some simple coverage of speech and audio. There are very many practical examples included and stepwise methods given where appropriate. Good end-of-chapter problems are provided. This book would be useful for brushing up on DSP, but may not suit an absolute beginner. Unfortunately it is marred by a handful of scattered mistakes, so be sure to obtain the errata.

- *Signal Processing First*
  J. McClellan, R. W. Schafer and M. A. Yoder (Pearson Education, 2003)
  A text for beginners, this book starts with introductory descriptions related to sound, and follows through to show how this can be represented digitally, or by computer. The coverage of basic Fourier analysis, sampling theory, digital filtering and discrete-time systems is gentle yet extensive. It is also possible to obtain a set of MATLAB examples related to the material in this book.

Audio

- *A Digital Signal Processing Primer: With Applications to Digital Audio and Computer Music*
  K. Steiglitz (Prentice-Hall, 1996)
  There are few books that cover introductory audio and speech systems alone. This too covers audio as an application of digital signal processing rather than as a subject in its own right. The book thus spends time considering the DSP nature of handling

audio signals, which is no bad thing; however, the orientation is good, being practical and relatively focused on the applications.

- *Speech and Audio Signal Processing: Processing and Perception of Speech and Music*
B. Gold and N. Morgan (Wiley, 1999)
The epitome of speech and audio textbooks, this 560-page tome is divided into 36 chapters that cover literally every aspect of the processing and perception of speech and music. For readers wishing to purchase a single reference text, this would probably be the first choice. It is not a book for absolute beginners, and is not orientated at providing practical methods and details, but, for those already comfortable with the main techniques of computer processing of speech and audio, it would be useful in expanding their knowledge.

Questions

Q2.1 What is the dimension (size) of an array of speech in MATLAB that is stored in stereo, is 1.25 s long and has a sample rate of 8 kHz?

Q2.2 Which command would allow me to directly read sound from a file in the current directory, called `nowhere.wav`, into an array named `audio` within MATLAB?

Q2.3 How many bits do the A-law and μ-law formats use to represent each audio sample, and how do these differ from linear sampling?

Q2.4 When I replay my speech array in MATLAB using `sound(speech, 16000);` it sounds like Donald Duck and the speech seems to be twice as fast as it should be. What should I do to make the speech playback sound better?

Q2.5 One would normally obtain a spectral representation of a sound array of length 256 using a command such as `spec=fft(aary, 256);`. What would typically have been done immediately before this to prevent unwanted occurrences such as Gibbs' phenomena?

Q2.6 How can our `tonegen()` function be used to create a 1 second long tone corresponding to the musical note middle-C (known as C_4), at a sample rate of 8 kHz? How would we replay this in MATLAB?

Q2.7 Use the MATLAB `chirp()` function to create a 2 second long linearly increasing chirp from 100 Hz to 400 Hz at a sample rate of 16 kHz. Store the chirp in a variable called 'my_chirp'.

Questions

Q2.8 Using the MATLAB `fft()` function, write code to plot the spectrum of the chirp you created in Q2.7. Given that this was created at a sample rate of 16 kHz, ensue that the *x*-axis is linear in frequency from 0 to 8 kHz, and that the *y*-axis is plotted logarithmically. Note: use the `fftshift()` function, and begin your code with `n=length(my_chirp);`.

Q2.9 Use MATLAB code to repeat the chirp five times and then listen to it, then create a pre-emphasis filter with coefficients $[1, -0.98]$, filter the repeating chirp and listen to it.

Q2.10 Visualise the chirp using a spectrogram, as well as a waterfall plot of the spectrogram (see Infobox 2.5).

3 The human voice

In Chapter 2 we looked at the general handling, processing and visualisation of audio: vectors or sequences of samples captured at some particular sample rate, and which together represent sound.

In this chapter, we will build upon that foundation, and use it to begin to look at (or analyse) speech. There is nothing special about speech from an audio perspective – it is simply a continuous sequence of time varying amplitudes and tones just like any other sound – it's only when a human hears it and the brain becomes involved that the sound is interpreted as being speech.[1]

There is a famous experiment which demonstrates a sentence of something called sinewave speech. This presents a particular sound recording made from sinewaves. Initially, the brain of a listener does not consider this to be speech, and so the signal is unintelligible. However, after the corresponding sentence is heard spoken aloud in a normal way, the listener's brain suddenly 'realises' that the signal is in fact speech, and from then on it becomes intelligible. After that the listener does not seem to 'unlearn' this ability to understand sinewave speech: subsequent sentences which may be completely unintelligible to others will have become intelligible to this listener [8]. To listen to some sinewave speech, please go to the book website at http://mcloughlin.eu/sws.

There is a point to sinewave speech. It demonstrates that, while speech is just a structured set of modulated frequencies, the combination of these in a certain way has a special meaning to the brain. Music and some naturally occurring sounds also have some inherently speech-like characteristics, but we do not often mistake music for speech. It is likely that there is some kind of decision process in the human hearing system that sends speech-like sounds to one part of the brain for processing (the part that handles speech), and sends other sounds to different parts of the brain. However, there is a lot hidden inside the human brain that we do not understand, and how it handles speech is just one of those grey areas.

Fortunately speech itself is much easier to analyse and understand computationally: the speech signal is easy to capture with a microphone and record on computer. Over the years, speech characteristics have been very well researched, with many specialised analysis, handling and processing methods having been developed for this particular type of audio.

---

[1] These days, of course, computers are just as good at recognising speech as humans – maybe even better than humans in some situations.

3.1 Speech production

Initially turning our back on both the human brain as well as the computer processing of speech, this chapter begins by looking at the human speech production mechanism and its characteristics. This will be followed by an examination of the physical properties of speech itself and how these relate to its means of generation. We will then begin to study how these properties can be explored in speech research and exploited in new types of speech technology.

3.1 Speech production

The sound that we hear and recognise as speech begins with the lungs contracting to expel air. This air is forced up through the bronchial tract past a set of muscle folds at the top of the trachea called vocal chords, which are set vibrating by the passage of air. The air that encounters the vocal chords, and which has been squeezed out of the lungs, has a turbulent 'rushing' sound which is quite similar to Gaussian white noise [9].

After resonating the vocal chords, the air, now carrying the buzzing sound of that resonance, enters the rear of the mouth cavity where it follows one of two paths to the outside. The first path is over and around the tongue, past the teeth and out through the mouth. The second route is through the nasal cavity – and this is the only route possible when the velum is closed. Figure 3.1 shows a diagram of the speech production apparatus (more commonly known as the human head).

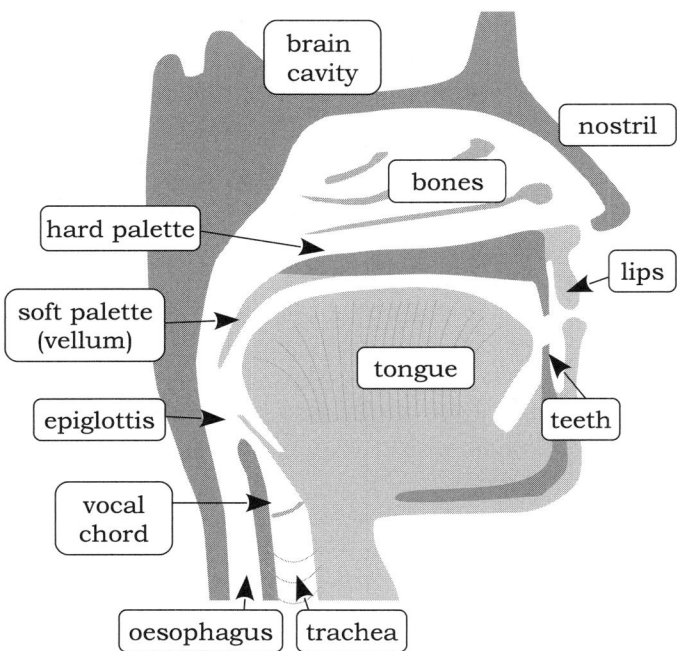

Figure 3.1 Sectional diagram of human vocal apparatus, showing the major articulators, resonators and features of the vocal and nasal tract.

The human voice

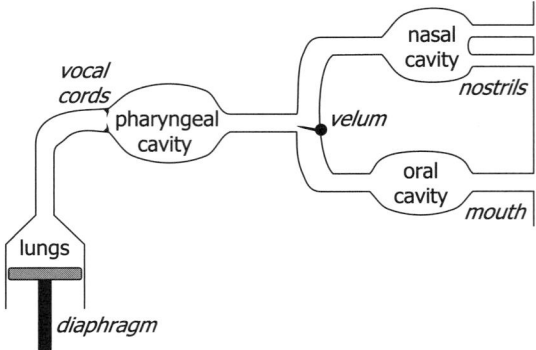

Figure 3.2 Functional diagram of the major parts of the human speech production mechanism.

The actual sound being produced at any particular time depends on many criteria, including the lung power and pressure modulation, the constriction at the glottis, the tension in the vocal chords, the shape of the mouth and the position of tongue and teeth. Each of these changes rapidly in complex patterns when we speak, largely automatically: if we desire to speak a word such as 'bob', we do not normally need to concentrate on opening and closing our lips, modulating our lung power, keeping the tongue low during the vowel, and tightening the glottis to form a voiced sound. In fact if we *do* try and consciously control each element of the speaking process we might produce some quite unnatural speech. A functional diagram of the interconnected articulators, as originators of sound pressure waves, is shown in Figure 3.2.

A brief overview of the articulators and how they relate to speech sounds follows, with more detail about components and classes of speech in Infobox 3.3: *Speech articulation*, on page 64.

(a) Lung power mostly affects the volume of the sound, with more rapid variation in lung pressure often being used to distinguish a boundary between syllables or words.

(b) If the glottis closes for a short time during speech, this temporarily halts the airflow, and is called a glottal stop. This does not occur in all languages, but a common dialect example is the /t/ in a Yorkshire-accented reading of 'I went t' shops'.

(c) Vocal chord muscle tension causes the chords to vibrate at different frequencies, causing a variation in the pitch fundamental frequencies. Voiceless sounds (e.g. /s/ in 'six') are produced when the vocal chords do not vibrate, so these contain little or no pitch structure.

(d) Some air can be diverted through the nose by the velum being lowered, otherwise it flows through the mouth. If the velum is lowered and the airway through the mouth is blocked (perhaps by the lips being closed or the tongue being raised), a nasal sound such as the /m/ in 'mad' is produced. Different timbre also results from the slight change in path length from lungs to nose compared with lungs to mouth (imagine two different length organ pipes which will produce different notes).

(e) If the air travels through the mouth, a humped tongue and opening then closing the lower jaw will cause a vowel sound (e.g. /a/ in 'card'). If the lower jaw does not begin to close, a glide (e.g. /w/ in 'won') is formed instead.
(f) Different sounds also result if the air is forced past the sides of a tongue touching the roof of the mouth or the teeth (e.g. /l/ in 'luck', and the /th/ sound).
(g) A plosive sound, like the /d/ in 'dog', is a short stop (i.e. the airflow is blocked for a short time) followed by an explosive release. The stop could occur at a number of places including the glottis, tongue or lips, to form different consonant sounds.

The above actions, and more, must be strung together by the speaker in order to construct coherent sentences, and this stringing together process depends on factors such as speaking rate, tiredness, volume and the need to breathe or swallow. In practice, sounds will often slur and merge into one another during speech, such as the latter part of a vowel sound changing depending on the following sound. This can be illustrated by considering how the /o/ sounds in 'or' and in 'of' differ.

In fact the merging of phonemes and words together is one major difficulty in speech processing – especially in the field of continuous speech recognition. For example, when speaking single syllable words, the obvious gaps visible in a waveform plot will correspond to demarcation points between words, but, as the complexity of an utterance increases, these demarcations become less and less obvious to spot. It would be extremely convenient for computer speech processing if there were gaps between words, but unfortunately the few noticeable gaps in practice are sometimes mid-word rather than between words. Such difficulties have led to speech segmentation becoming a flourishing research area (see also Section 9.1.3).

Finally, speech and pronunciation are very much context sensitive. When talking to another person in a location that is experiencing high levels of background noise, we might speak loudly or shout to be heard. Conversely, when speaking to someone in a very quiet location, we may choose to whisper to prevent the sound from carrying too far. These are natural responses to environmental conditions. However, such responses may not work well over man-made communications channels: imagine a situation where a man in a quiet office calls his wife in a noisy shopping mall. The husband will probably talk fairly quietly in order not to disturb (or be overheard by) his colleagues, but the wife will probably need to shout for her voice to be picked up by her mobile phone. There is clearly a mismatch in the environmental conditions at either end, and thus a mismatch in the way the two speakers respond to each other, with the wife shouting while the husband whispers. Evidently the wife will then ask her office-bound husband to speak up a little while the husband will request his wife to stop shouting – the author relates this example from personal experience.

3.2 Characteristics of speech

Despite many differences between individuals, and the existence of many languages, speech follows general patterns that have well-defined average characteristics in terms of loudness, frequency distribution, pitch rate and syllabic rate [10]. These will be explored in this section, and will be applied later when we begin to analyse and process speech.

The three main attributes of speech communications, namely generation, propagation and reception, are well matched with the physical characteristics of the speech itself. Generation refers to the speech production apparatus outlined previously. Propagation refers to the physical environment and the fact that speech frequencies travel through air with quite low levels of absorption and distortion.[2]

As we have discussed above, speech characteristics adapt with regard to environment, hearing and voice production limitations and context (situation). In the short term we react to distance, background noise, emotion, urgency, the content of the message and non-verbal communication cues. While most of our verbal communication uses normal speech, both shouting and whispering are relatively common, and will be discussed below in Section 3.3.

3.2.1 The physical components of speech

Physically, the sounds of speech can be described quite well in terms of a pitch contour, formant frequencies and airflow. We will learn in Chapter 6 how this fact is used by most speech compression algorithms to decompose speech into simpler components.

Formants are resonance frequencies of the vocal tract which appear as amplitude peaks in the speech spectrum. As an example, three distinct formant peaks at frequencies of about 900 Hz, 3.6 kHz and 4.6 kHz can be seen in the frequency-domain plot of a short speech recording of the vowel /a/, in Figure 3.3.

Formants have been described by the famous researcher Klatt and others as being highly important to speech communications [12]. Each utterance will contain several or many formants. The formant frequencies are one of the main ways in which we distinguish between phonemes. As an example, note the vowel frequencies plotted in Figure 3.4 as measured for a number of male speakers living in the City of Birmingham, UK. The vowels were extracted from recordings in framing words (for example, the vowel /i/ was found inside the word 'hid'), and, being vowels, each was spoken with the lips and vocal tract open. Speech researchers often plot data using strange axes, and this is no exception, with the first formant frequency being plotted against the difference between the first and second formants. It should be noticeable that different vowels, even when produced by different speakers, are generally clustered into distinct regions. Some of the regions overlap, and that is why, for example, the words 'had' and 'hod', spoken in isolation by a stranger, might be confused.

---

[2] For example, if our main speech frequencies were located around 30 kHz instead of 300 Hz, then our voices would carry a distance about 50 times shorter in dry air (which might appear to be a benefit when we are disturbed by over-loud talkers who are nearby). Speech at 300 kHz would arrive at our ears distorted by the dispersive effects of air in that frequency region. Moving down in frequency, speech at 30 Hz would be too slow to sustain our rate of talking, as well as being heavily attenuated by air and requiring more lung exhalation to sustain.

3.2 Characteristics of speech

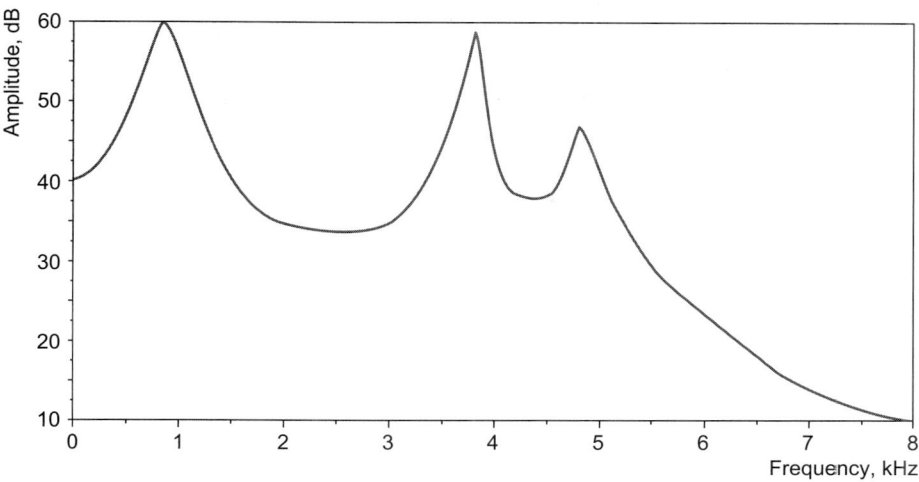

Figure 3.3 The spectral envelope of a 20 ms recording of voiced speech, showing three distinct formant peaks.

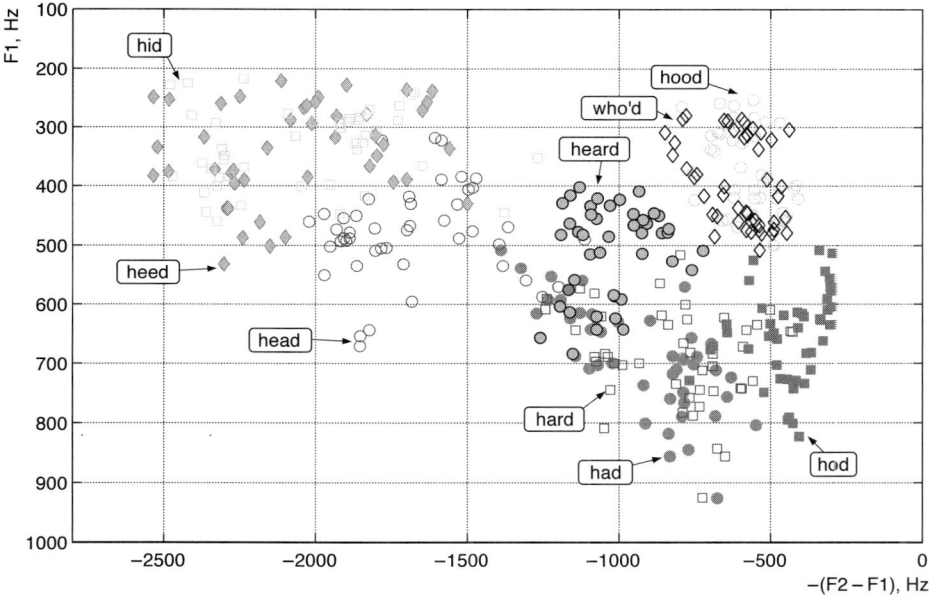

Figure 3.4 A plot of the first and second formant frequencies for various vowels as spoken by male speakers of Midlands English. Data courtesy of Sharifzadeh et al. [11].

The frequencies of formants vary strongly over time as the shape of the mouth changes to make different phonemes. Normally, formants are counted from the lowest frequency upwards, and usually only the first three (F1, F2 and F3) contribute significantly to the intelligibility of speech – although some fricative sounds such as /s/ or /ch/ can produce a lot of formants spreading quite high in frequency. Generally

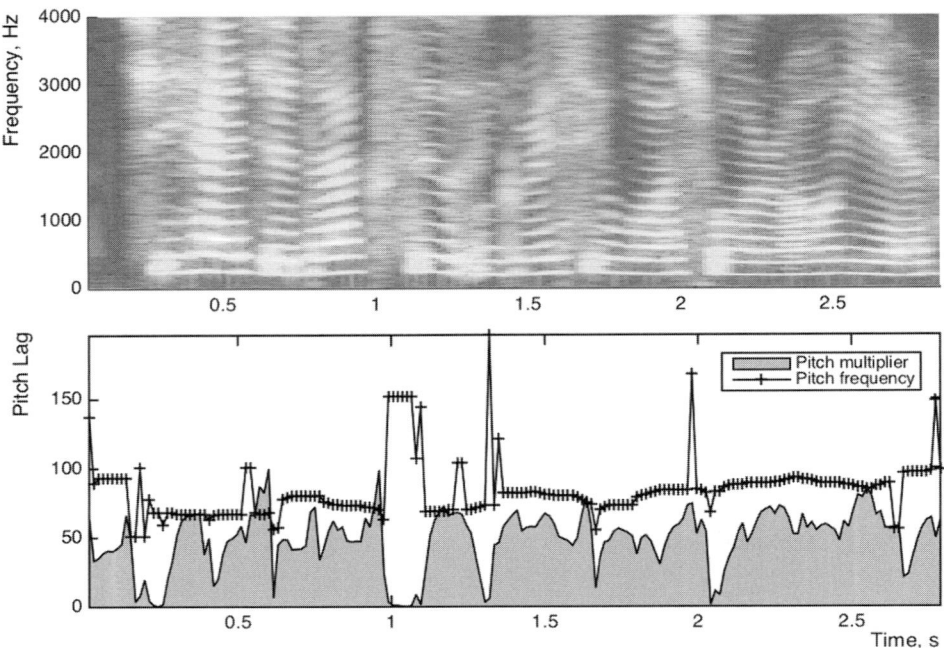

Figure 3.5 Pitch frequency and amplitude alongside a spectrogram of a short sentence of speech.

speaking, F1 contains most of the speech energy while F2 and F3 between them contribute more to speech intelligibility [13]. Many formants can be seen as parallel horizontal bands in the spectrogram at the top of Figure 3.5. This plots the spectrogram of a recording of a short spoken phrase. Sibilants are shown as diffuse clouds without any obvious sign of formants. Energy, represented by brighter shades, is very obviously concentrated around the formants, and they tend to change quite slowly over time during words, but have sharper transitions between words.

Meanwhile, the bottom part of Figure 3.5 shows the pitch lag of the same recording as extracted by a simple one-tap long-term predictor (an LTP, which we will meet later in Chapter 6). The area plot represents the strength of the pitch signal, in per cent, while the black line and crosses mark the detected pitch lag during each analysis frame. This is measured in samples, so it varies inversely to the pitch frequency. Assuming it has been detected correctly, this 'pitch contour' describes the tone of the voice (the perceived frequency), and is in effect the fundamental vocal frequency. It is usually denoted f0 to reflect the fact that it is the fundamental of the higher-order vocal tract resonances F1, F2 and so on. Note that the lower-case notation should be used to distinguish it from formants, although it can be seen written (incorrectly) as F0 in some research papers. It is interesting that, in Figure 3.5, there are some periods when the pitch energy is low, but the frequency is high (e.g. just after 1 s), and there is also some obvious correspondence between the pitch and features in the spectrogram, which may correspond to partitions between words or different phonemes, Incidentally, this plot also demonstrates a common problem with pitch detection algorithms that we will discuss later – pitch doubling, when the detected pitch suddenly increases to twice the correct value (and the opposite problem, which is pitch halving). There are several instances of pitch doubling in the plot, for example around the 1 and 2 s marks. Normally, these would be detected and removed by a post-processing algorithm. This is all discussed more fully in Chapter 6.

Although pitch is very important in causing the formants, and despite the fact that it does itself contain energy, it contributes surprisingly little directly to intelligibility for English and similar European languages [14]. It is, however, a very different matter in a tonal language such as Cantonese, Thai, Vietnamese, Burmese and Mandarin Chinese, which can be totally dependent on slight differences in tone for conveying meaning [15,16]. As an example, in Chinese the single word 'ma' can mean one of five things depending on which tone it is spoken with: mother, horse, scold, question, etc. This is by no means an isolated example: all single Chinese word sounds have multiple meanings that are differentiated only by tone.

3.2.2 Amplitude distribution of speech

The volume or amplitude with which speech is spoken is obviously situational, depending upon the speaker's personality and mood as well as environmental noise, infection and so on. Extremes of level include whispering and shouting, but extreme flatness is equally noticeable: most readers have probably endured monotonous speakers (literally meaning 'single tone' speech).

Feedback from a listener, either a verbal signal like 'speak up please' or a non-verbal one, such as cupping a hand around an ear, can prompt a speaker to alter their vocal characteristics (although it is noticeable that such prompts to the most monotonous speakers – such as falling asleep – seldom elicit the desired alteration in speaking pattern). Despite this obvious variability in speech, it is useful to consider the average speech levels in different environments, as shown in Table 3.1, reproduced from [17], where the sound amplitudes are listed in dB_{SPL}.[3]

Note the wide range in average speech level, and its relationship to location. In a train and in an aeroplane are the only situations listed where a negative signal-to-noise level results (i.e. where the speech level is lower than that of the noise), and that is partly due to the particular shape of the noise spectrum in both environments. It should also be noted that perhaps modern aircraft are a little quieter than they were when this data was obtained. Later, in Chapter 7 we will derive methods of accounting for, and measuring, differences in speech and noise spectra.

Generally speaking, as background noise increases in amplitude by 1 dB, a speaker will raise his or her voice level by 0.5 dB [17]. In low levels of background noise, a male adult speaker can produce 52 dB_{SPL} of speech measured at a distance of 1 m when speaking casually. This rises to about 90 dB_{SPL} when shouting, giving a useful dynamic range of about 48 dB. Typically quoted amplitude figures of 50–82 dB_{SPL} for women

---

[3] The decibel (dB) is a base-10 logarithmic measure of amplitude, with the *SPL* subscript in dB_{SPL} referring to sound pressure level, which is referenced so that 0 dB is the quietest average audible sound at 1 kHz. In terms of measurable pressure, 74 dB_{SPL} is 1 μbar, 1 dyne cm^{-2} or 0.1 Pa in different units.

Table 3.1 Average amplitude of speech in several environments, from [17].[a]

Location	Noise level (dB$_{SPL}$)	Speech level (dB$_{SPL}$)
School	50	71
Home (outside, urban)	61	65
Home (outside, suburban)	48	55
Home (inside, urban)	48	57
Home (inside, suburban)	41	55
Department store	54	58
On a train	74	66
In an aircraft	79	68

[a]These data were originally published in *The Handbook of Hearing and the Effects of Noise*, K. Kryter, Chapter 1, Copyright Elsevier (Academic Press) 1994.

and 53–83 dB$_{SPL}$ for children were probably obtained by a researcher who had little first-hand experience of raising children.

The dynamic range of normal conversational speech is around 30 dB [18], and the mean level for males measured at 1 m is somewhere in the region of 55–60 dBA$_{SPL}$[4] with peaks that extend 12 dB beyond this [14].

3.2.3 The lexical components of speech

Lexical speech (i.e. the content of speech as it is understood and written down by humans) is generally described as a sequence of phonemes that vary in tone, timbre, amplitude and rate. Infobox 3.1: *The structure of speech* on page 63 introduces some of the basic terminology, while Infobox 3.2: *The International Phonetic Alphabet* discusses the way we normally describe individual phonemes.

Although speech amplitude varies according to many factors, phonemes naturally have different average amplitudes depending upon how they are produced – which is how we divide them into different classes. A number of these are shown in Table 3.2, with their relative average amplitudes and ranges. More detail, including phoneme-by-phoneme examples from the original experiments, can be found in [19].

It is quite obvious from the data that vowels are spoken with more power than other phonemes on average, and that the range of intensity for all listed sound classes is rather large, spanning almost 30 dB. Remember that these figures have been averaged over time, so the instantaneous or peak amplitude differences may be even higher.

A useful rule of thumb is that, in normal speech, vowels are approximately 12 dB louder than consonants. This might be surprising bearing in mind that, for English at least, consonants convey a greater share of vocal intelligibility than vowels. I generally

---

[4] The 'A' in dBA refers to the A-weighting curve (discussed in Section 4.2.2), in which a frequency correction is applied to the sound before calculating the average amplitude. The frequency correction is based upon 'hearing curves', and makes the dBA score behave in a similar way to what a human would perceive.

3.2 Characteristics of speech

> **Box 3.1** The structure of speech
>
> A *phoneme* is the smallest structural unit of speech: there may be several of these comprising a single word. Usually we write phonemes between slashes to distinguish them, thus /t/ is the phoneme that ends the word 'cat'. Phonemes often comprise distinctly recognisable *phones* which may vary widely to account for different spoken pronunciations.
>
> Two alternative pronunciations of a phoneme are sometimes the result of a choice between two phones that could be used within that phoneme. In such cases, the alternative phone pair are termed *allophones*. Interestingly, phones which are identical except in their spoken tone can be called *allotones*. Tonal languages such as Mandarin Chinese are replete with allotones, with many phonemes having a choice of tone which can totally change the meaning of a word. Incidentally, this tonal nature of Chinese is profoundly disliked by many Western learners of the language, because their brains lack the automatic means to differentiate between words based on tonal information.
>
> Single or clustered phonemes form units of sound organisation called *syllables* which generally allow a natural rhythm in speaking. Syllables usually contain some form of *initial* sound, followed by a *nucleus* and then a *final*. Both the initial and the final are optional and, if present, are typically consonants, while the syllable nucleus is usually a vowel.
>
> Technically a *vowel* is a sound spoken with an open vocal tract as explained in Section 3.1, whereas a *consonant* is one spoken with a constricted, or partially constricted vocal tract, but, as with many research areas, these definitions which appear clear and unambiguous on paper are blurred substantially in practice.

> **Box 3.2** The International Phonetic Alphabet
>
> The International Phonetic Alphabet (IPA) is a standardised method of describing and transcribing (writing) the various phonemes that make up speech. As defined by the International Phonetic Association, a set of symbols, written using a shorthand notation, describes the basic sound units of words. These symbols can completely describe many different languages using the 107 letters and several diacritical marks that are available in the alphabet [10]. It is beyond the scope of this book to describe the IPA, but researchers working with phonetics would be advised to learn the rudiments of the IPA and apply this notation in their work. This will help to avoid misconceptions, prevent phonemes from being insufficiently specified, and may have the added benefit of making their research appear more scholarly.

ask sceptics who require a demonstration of the relative information carrying content of vowels with respect to consonants to read aloud the following sentence:

The yellow dog had fleas.

Next, I ask them to repeat the sentence having replaced all consonants with a single unchanging phoneme:

Tte tettot tot tat tteat.

Table 3.2 Average amplitude of phonemes by class, also showing amplitude range within each class, measured with respect to the quietest phoneme in English, the voiceless fricative /th/ in 'thought'.

Phoneme class	Example	Amplitude (range), dB
Vowel	c*a*rd	26.0 (4.9)
Glide	*l*uck	21.6 (3.2)
Nasal	*n*ight	17.1 (3.0)
Affricative	*j*ack	14.9 (2.6)
Voiced fricative	a*z*ure	11.5 (2.2)
Voiceless fricative	*sh*ip	10.0 (10.0)
Voiced plosive	*b*ap	9.6 (3.3)
Voiceless plosive	*k*ick	9.5 (3.3)

> **Box 3.3** Speech articulation
>
> Many phonemes are defined by their place, or method of articulation within the vocal tract. Here is a list of some of the more common terms used for vowels and consonants:
>
> - *affricative* – a turbulent airflow fricative following an initial stop. E.g. /ch/ in 'chip'.
> - *diphthong* – a two-part sound consisting of a vowel followed by a glide. E.g. /i/ and /n/ in 'fine'.
> - *fricative* – a very turbulent airflow due to a near closure of the vocal tract. E.g. /sh/ in 'ship'.
> - *glide* – a vowel-like consonant spoken with almost unconstricted vocal tract. E.g. /y/ in 'yacht' (indeed many academics class /y/ as a vowel).
> - *nasal* – a consonant spoken with the vellum lowered, so sound comes through the nasal cavity. E.g. /m/ in 'man'.
> - *plosive* – an explosive release of air caused by the rapid removal of a vocal tract blockage or stop. E.g. /p/ in 'pop'.
>
> Most of the consonant sounds can be either voiced or unvoiced, depending upon whether the glottis is resonating. For example /c/ in 'cap' is unvoiced whereas /g/ in 'gap' is voiced. In true whispers, all sounds, whether lexical consonant or lexical vowel, are unvoiced.

Finally, I ask them to replace all vowels with a single unchanging phoneme and read again:

Tha yallaw dag had flaas.

Apart from utterly humiliating the sceptics by making them sound stupid, it is immediately obvious that although the same-vowel sentence sounds odd, it is still highly intelligible. By contrast the same-consonant sentence is utterly unintelligible. This simple example illustrates that, although vowels are spoken louder than consonants, they tend to convey less intelligibility, in English (in fact the same is true in many other languages, including Mandarin Chinese, although in Chinese the tone is more important than either vowel or consonant sound).

3.2.4 Frequency distribution

The frequency distribution of speech matches the frequency sensitivity of the human ear quite closely. Most of the frequencies involved in speech, and certainly all of those that convey significant intelligibility, lie within the range of frequencies over which the ear is most sensitive. However, within this band (which spans about 300 Hz to 4 kHz) there is a mismatch between the speech frequencies of greatest energy and those of greatest intelligibility. Put another way, the speech frequencies with the greatest concentration of power are not quite the same as those that account for most of the intelligibility [20]. This slight mismatch is hinted at by the vowel/consonant intelligibility difference in the previous section. To examine further, let us now consider both power and intelligibility separately.

Speech energy: Much of the energy in speech is found at low frequencies. around 500 Hz for males and 800 Hz for females. These frequencies are not essential for intelligibility – experiments in which these frequencies are removed can demonstrate that the remaining speech, whilst quiet and abnormal sounding, is still perfectly intelligible. That is to say, the spoken lexical information in the speech remains when the highest energy frequencies are removed. However, the ability to recognise the speaker is severely impaired when those frequencies are removed. In general, around 84% of the energy in speech is located below 1 kHz as shown in Figure 3.6, which has been constructed by combining data from [14] and [21].

Figure 3.6 also shows shaded bands which indicate the ranges where the first three formants usually lie. Note the correspondence between F1 and the band of greatest energy distribution.

Intelligibility: Although this will be discussed more fully in Section 3.4, it is worth noting here that most of the useful information regarding *what* is being said lies above 1 kHz, carried by formants F2 and F3 as mentioned previously. Removal of all speech

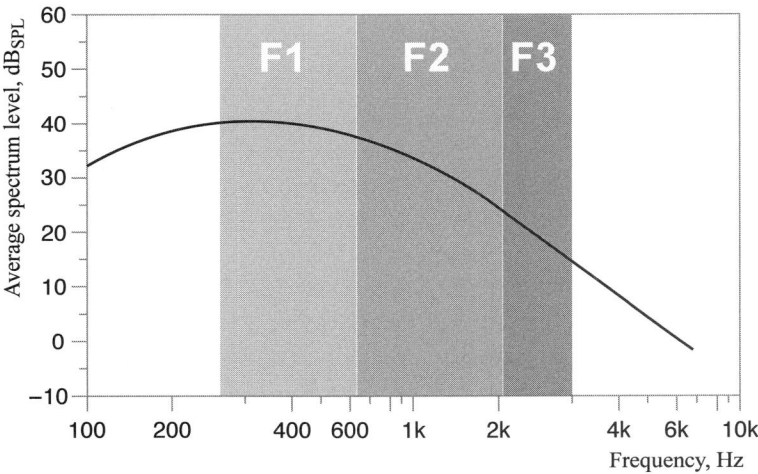

Figure 3.6 Long-time averaged speech power distribution plotted against frequency, with the approximate regions of the first three formants identified through vertical grey bands.

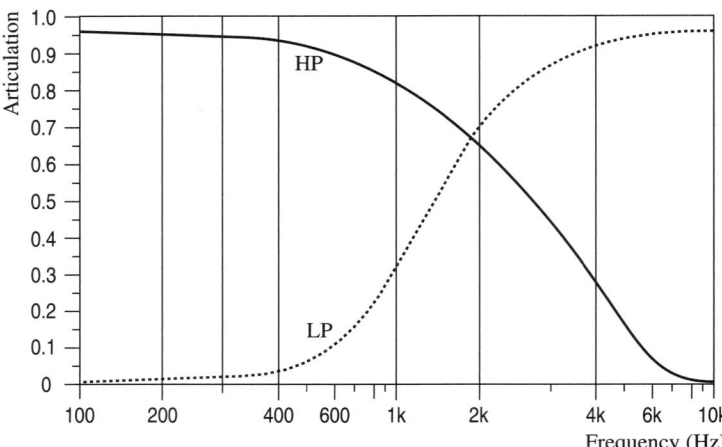

Figure 3.7 Illustration of the effect of limiting speech frequency range on the intelligibility of speech syllables, measured as articulation index.

energy between 800 Hz and 3 kHz would leave a signal that sounded fairly speech-like but which was completely unintelligible [22]. The effect of this is illustrated in Figure 3.7, which plots the articulation index (roughly equivalent to the intelligibility) of speech that has been either low-pass filtered (LP) or high-pass filtered (HP) to remove the frequencies above or below, respectively.

As an example, if speech were low-pass filtered at 1 kHz (i.e. refer to where the LP curve crosses the 1 kHz line), about 28% of speech syllables would be recognisable. If speech were high-pass filtered at 2 kHz (i.e. refer to where the HP curve crosses the 2 kHz line), around 67% would be recognisable. Note that these are approximate curves for average syllables spoken by average speakers without noise. Actual intelligibility would vary quite considerably, and be dependent upon both context and content.

3.2.5 Temporal distribution

Temporally speaking, the major constraint on producing speech is how fast the brain and vocal apparatus can attempt to articulate phonemes or syllables. The various muscles involved in vocal production can only move so fast, as can the muscles controlling the lungs. A further constraint on lung muscle movement is the need for regular lung re-filling required to prevent asphyxia (i.e. to breathe).

Evidence suggests that for normal speech the speed of articulation is relatively independent of the rate of speaking. This means that, when speaking more quickly, most people will use the same length of time to articulate a particular syllable, but will reduce the length of the gaps between syllables and words [12]. When speaking more slowly, the phoneme duration remains similar, but there will be larger gaps between speech components. This near-uniformity in phoneme duration greatly assists speech researchers in the artificial description and modelling of speech. As usual, though, there is wide variation between people, as well as some situational variation (try speaking fast after being woken by a telephone call at 3 am). English is usually spoken at a rate of about 120 to 150 words per minute (WPM) by native speakers, but bursts can exceed 200 WPM. Urban legends state that the late American President John F. Kennedy was able to exceed 300 WPM during some of his public speeches, although it is unlikely that he would be capable of maintaining such a rate for more than a few sentences.

Of all the temporal constraints on speech, the maximum speed at which the muscles are capable of moving is of the most interest to us [4], because it is language independent, independent of context, and may be relatively constant among different speakers. Assuming such a maximum limit means that we can define speech as being almost stationary (i.e. its statistics are constant) over durations that are shorter than the rate of muscle movement. In practice, this means periods of about 20 ms. Thus speech analysis (including short-time Fourier analysis, linear predictive analysis and pitch detection) is usually conducted over windows that span such a duration. These periods are termed pseudo-stationarity (see Section 2.5.1).

One further temporal generalisation is that of syllabic rate, the rate at which syllables are articulated. For most languages average syllabic rate when speaking remains fairly constant, and also does not vary greatly between adult individuals, or between types of speech [14] and situations.

One or more agglomerated phonemes can make up a syllable sound. For humans, it seems that the simplest unit of recognition may be the syllable, whereas the phoneme is a distinction generally made by speech researchers as a convenience in determining a set of basic building blocks for speech sounds. Actual phoneme duration varies from language to language and speaker to speaker. It also differs depending upon the exact phoneme being spoken, since some are naturally longer than others, due in part to the amount of mouth movement needed to form them (i.e. hinting at the maximum speed of muscle movement again). There is evidence to tie phoneme length to word stress – louder and more emphasised phonemes tend to exhibit a longer duration. During research into Mandarin Chinese [23], the author has found evidence that phoneme length is tied to lexical tone for vowels, and particularly so when whispering (in fact, whispered phonemes are almost always longer than spoken – phonated – phonemes, for English too).

3.3 Types of speech

While most human utterances contain speech, the two extremes of shouting and whispering are also very much a part of the normal speaking repertoire for most people (although rarely together). Figure 3.8 plots spectrograms and waveforms of the same sentence having been shouted, spoken and whispered. While many common characteristics are shared between the three types of utterance, it is quite noticeable that amplitude variation, time duration and formant structure vary substantially between the modes. Using normal speech as a baseline, this section will discuss these differences by looking briefly at shouting before considering whispers, and ending with a short description of atypical speech.

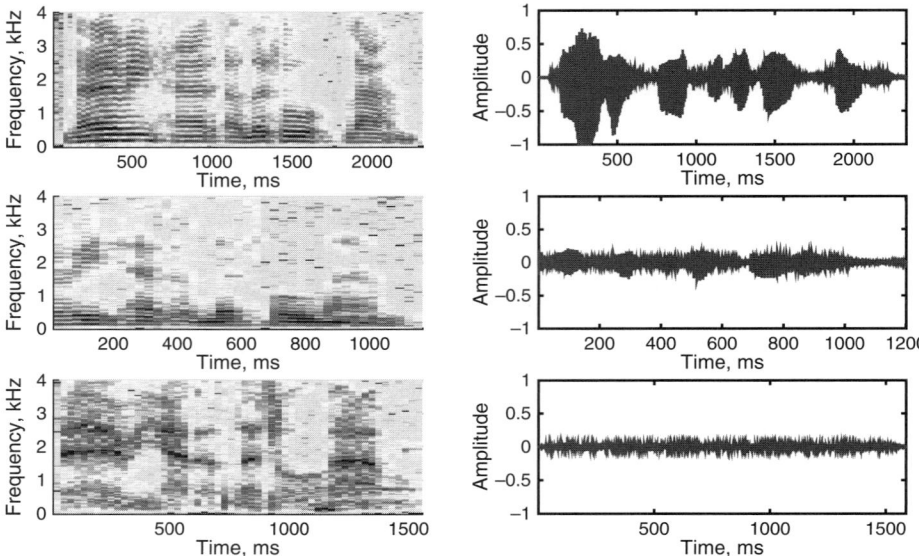

Figure 3.8 Spectrogram (left) and waveform plots (right) of the same sentence shouted (top), spoken (middle) and whispered (bottom).

3.3.1 SHOUTING!

The shouting mechanism is for 'long distance' communications, where speakers adapt their voices to the environment or message. Shouting from one hill across an open valley to a person standing on the opposite hill is natural, but may not be appropriate in crowded urban conditions where it can be socially unacceptable to shout. Shouting after hitting one's thumb with a hammer is also a form of speech adaptation to the situation (as may be the choice of words produced under those circumstances). As speakers, we adapt easily and often, usually without thinking.

The author can relate the experience of standing alone on a grassy mound on a hot summer's day. From the brief details provided, the reader might imagine that this location would be quiet. However, the reality was quite different since he was facing a block of residential flats which had many open windows at that time. Apart from significant sound from a multitude of televisions, there were many overlapping incidences of loud vocal communications within the residences, emanating from some of the open windows. It was a vivid example of an environmental mismatch between source (their mouths) and receiver (my ears), as well as the diversity and flexibility of the human voice.

Shouting, in general, means using a loud voice – produced using increased diaphragm pressure. Apart from the amplitude, the stress pattern of words changes during shouting, as does the duration of phonemes. Pitch, or f0, tends to rise when shouting; however, the formant locations and their relationships are relatively unchanged.

Some phonemes do not lend themselves to being shouted. For example, it is very difficult to shout a /th/ or an /f/ due to their method of production through a constriction in the air path through the mouth. In practice they are therefore sometimes replaced by similar, but more heavily voiced alternatives during shouting. This is another example of an entirely unconscious speech adaptation mechanism.

3.3.2 Whispering

Whispering is another mode of speech, often used for short distance 'private' or quiet communications. In conversations we whisper words for effect to impart a hint of secrecy, while entirely whispered sentences are normally produced when we wish to prevent others from overhearing our speech. This may be due to the sensitive content of the message (e.g. something private), or the sensitive environment in which it is conveyed (e.g. in a library, or during a meeting). Due to the reduced power in whispers, these are often accompanied by a physical shift in posture to bring the whisperer's lips closer to the listener's ear.

As we have seen, normal speech results when air, expelled from the lungs, flows past a taut glottis. The glottis resonates to generate a pitch oscillation which, in turn, resonates through the vocal cavity to create formants. The geometry and tautness of the glottis controls the fundamental frequency and timbre (quality) of the pitch, and both are naturally and unconsciously adjusted during speaking. The isolated pitch excitation which drives speech is similar to an audible buzzing sound. This fills the vocal tract (VT) and nasal cavity before emerging primarily through the mouth (and some through the nasal passages) as speech. Resonances of the pitch fundamental, and their harmonics, are controlled, also largely unconsciously, by the action of vocal tract modulators. Apart from the glottis which we have mentioned, the modulators include the velum, tongue, teeth (jaw position) and lips which adjust to change the resonant cavity of the VT.

Whispering means that all phonemes are spoken unvoiced or unphonated, although there are some variants which we call 'whispering' but which are actually semi-phonated. The main one is 'stage whispers', which are a deliberate attempt to produce voiced and intelligible speech that shares some of the audible characteristics of whispers. Another example is the result of trying to speak loudly or shout when whispering – the result will inevitably include a few phonemes that contain hints of voicing.

Unphonated or whispered utterances, by contrast to phonated or spoken utterances, lack a distinct glottal source of pitch. They are instead driven by a broadband excitation of turbulent exhaled air from the lungs [24]. A simple example is to listen to the spoken (but naturally unvoiced) phoneme /f/ and its voiced near-counterpart /v/. Most of the vocal articulator shapes are, or can be, identical between these two phonemes. The primary difference is the amount of pitch, or glottal modulation, with /f/ being unvoiced and /v/ being voiced (another example would be /p/ and /b/).

Normal whispers are produced when the vocal cords are held open. Air from the lungs rushes through the open glottis and into the back of the throat with a turbulent rushing sound. This sound then fills the vocal tract, forming resonances, before exiting through the mouth and nose as normal. Whispers are also produced by many speakers for medical reasons, for example as a result of having all or part of the glottis removed surgically (this is often as a result of a total or partial laryngectomy operation).

3.3.2.1 Differences between speech and whispers

The open glottis and reduced lung power are the prime differences between these two modes of speaking, and these lead to several characteristic spectral features in whispers.

Firstly, that the formants in whispered phonemes which are normally voiced have much lower energy than their spoken counterparts; they are also much flatter in shape. The flatter, lower energy shape means that they are much less distinct and more easily masked by background noise (which explains the reduced intelligibility of whispers in noise). Whisper formants are also frequency-shifted compared with voiced ones. The precise frequency shift varies with speaker but tends to relate to age and gender. A rule of thumb is that the whispered F1 is 200 Hz higher than the spoken one, F2 is about 150 Hz higher, while F3 and above do not exhibit any consistent amount of shifting [11, 25].

Apart from lack of pitch and formant shape differences, there is anecdotal evidence that whispers are spoken at a slower rate than phonated speech, and that speakers adjust their choice of words – perhaps avoiding the most confused phonemes – to suit the method of speaking.

Most of this discussion pertains to English speech. However, tonal languages such as Mandarin Chinese are a different matter. As we have discussed previously, Chinese relies very heavily upon tone to differentiate between words. Tone in Mandarin is conveyed by pitch contour. It would therefore follow that, since whispers are pitchless and therefore lack tone, whispered Chinese should be virtually unintelligible. While this explanation could be used to explain the basis of the phrase 'Chinese Whispers', it is actually completely incorrect. In fact, whispered Chinese retains a significant degree of intelligibility [23], with the current hypothesis being that whispered Chinese uses a combination of amplitude modulation, formant frequency variation and phoneme duration to convey lexical tone information. In reality, whatever mechanism is being used, whispered Chinese may in fact be more intelligible than whispered English.

3.3.3 Atypical speech

This short introduction would not be complete without mentioning atypical speech, i.e. anything which differs substantially from average or typical speech. Speech impairment leads to the production of atypical speech, and may actually be a lot more common than one would realise. It is also situational, for example anyone speaking after having partaken of excessive alcohol, answering a telephone call when half-asleep at 3 am, or while concentrating hard on another task might expect some form of impairment. But it is also physical: possibly the biggest factor in atypical speech production is age, because it occurs at both ends of the age spectrum, and occurs thanks to both neurological as well as physiological factors. In fact anyone who has worked with either the very young or the very old would probably attest to having had to adapt their ears to occasional unusual speech patterns.

If we imagine for a moment the myriad aspects of voice, we might note a number of opportunities for atypical production, including rate of speaking, volume, pitch level, formant frequencies, incorrect mouth shape for the required phoneme, excessive sibilation (perhaps incorrect tongue position) and airflow or fluency interruptions (e.g. stammer). There are many things that can go wrong – for example, tongue position, lip closure, jaw opening, mouth shape, diaphragm pressure and so on. This list doesn't include how those elements are sequenced, or can be mis-sequenced, when forming phonemes which in turn form words. It doesn't even include the choice of phonemes that make up words, or the choice of words themselves. While we began this chapter by commenting that 'there is nothing special about speech from an audio perspective', we can see that, from a physiological perspective, the act of speaking is a complex and delicate process. This everyday action may in fact be one of the most complex tasks that most human beings are capable of performing.

3.4 Speech understanding

Up to now, this chapter has investigated the production of speech, plus the physical and lexical characteristics of the produced speech. In this section, we will now turn our attention to the understanding of typical speech by humans. We will consider the understanding of normal speech and how the nature of speech structure relates to understanding; however, we will not delve into the related complexities of the human auditory system until Chapter 4.

3.4.1 Speech intelligibility and speech quality

Firstly it is very important to distinguish between quality and intelligibility. At times they are used interchangeably, but their measurement and dependencies are actually very different. In simple terms, quality is a measure of the fidelity or 'correctness' of speech. This includes how well the speech resembles target or reference speech (perhaps compared with our concept of how ideal speech would sound), but it extends beyond that to encompass how 'nice' the speech sounds. Quality is therefore a highly subjective measure which depends upon the ear of the listener. Despite being subjective, it can be approximated using objective measures, and can even be determined with some degree of objective accuracy, as we shall see.

Intelligibility, on the other hand, is a measure of how understandable the speech is. In other words, it concentrates on the information-carrying content of the speech, by referring to the degree of information that is correctly conveyed. Some examples should clarify the difference:

- A string of nonsense syllables, similar to baby speech, spoken by someone with a good speaking voice can sound very pleasant, of extremely high quality, but contain no verbal information, and in fact have no intelligibility at all.

- A recording of speech with a high-frequency buzzing sound in the background will be rated as having low quality even though the words themselves may be perfectly understandable. In this case the intelligibility is high but quality is low.
- When choosing a car audio system, you might test it by tuning to a favourite radio station, or listening to some favourite music. Generally the audio system that sounds nicest (of highest quality) would be the one purchased. It would be unusual to spend time counting the proportion of words spoken by a radio announcer which you can understand before choosing a system – more likely you would choose one that plays music nicely.
- When the military are in a combat situation, it is usually extremely important to understand the speech from a radio, whereas the quality of the sound is almost totally unimportant. In the Second World War, crude speech processing (clipping, or filter-clipping) was applied to radios used in aircraft – making the speech sound shrill and screechy – but significantly improving its intelligibility in a noisy cockpit [13]. This effect can often be heard in films and documentaries of the period.

Despite stressing the difference between quality and intelligibility in this section, it is useful to note that under most circumstances excellent intelligibility implies excellent quality, while very poor intelligibility implies very poor quality. These are the extremes – between these points the relationship between the two is not straightforward.

3.4.2 Measurement of speech quality

Speech quality is normally measured subjectively, in terms of a mean opinion score (MOS). This involves a panel of several listeners, usually tested individually in a soundproofed room, who have the audio recording under evaluation played to them. They will then rate this according to the scale shown in Table 3.3.

The MOS score of a particular recording is the mean of the results reported by all listeners. Obviously the more listeners, the more accurate (and repeatable) the results will be. This test is standardised by the International Telecommunications Union (ITU) in recommendation P.800, widely used in the audio community.

In recent years, several objective quality algorithms have been developed. These are computer programs that will estimate the MOS score of an item of test audio (or

Table 3.3 MOS rating scale for objective assessment of speech quality.

Score	Description	Impairment
5	Excellent	Imperceptible
4	Good	Perceptible but not annoying
3	Fair	Slightly annoying
2	Poor	Annoying
1	Bad	Very annoying

will produce a figure which can be converted to a MOS score). The more common double-ended systems are those that compute their score from two inputs, one of which is a reference (i.e. a clean or good recording), and the other of which is a degraded recording. For example, one could compare an original recording of a concert with the same concert replayed by a television.

Single-ended systems are those that have no reference. They simply consider the test audio to ascertain how much it is degraded compared with an assumed perfect recording. Such systems are much more complex, and likely to be less repeatable (and less correct) than double-ended systems. In fact, most of these measuring programs can be tricked, and at the extremes will track real MOS scores very poorly, but, for normal speech processed in typical ways (especially in the presence of moderate levels of background noise or other analogue effects), have been shown to produce respectable results. They are used primarily because they are cheaper and quicker than forming a panel of listeners to get the MOS. Most importantly they allow automated testing to take place. Some of the more prominent algorithms are:

- PESQ (perceptual evaluation of speech quality);
- PSQM (perceptual speech quality measure);
- MNB (measuring normalised blocks).

Although these are commercially supported algorithms, it has been possible in the past to download working versions from the Internet for non-commercial research use.

We will now present several alternative practical measures that can be used to evaluate the similarity between two speech or audio signals.

3.4.2.1 Mean-squared error

A crude double-ended time-domain measure of the difference in quality between a processed audio vector p and an original audio vector s is the mean-squared error (MSE) E. This is simply calculated on a sample-by-sample basis as the average squared difference between the vectors, both of length N:

$$E = \frac{1}{N} \sum_{i=0}^{N-1} \{s[i] - p[i]\}^2. \tag{3.1}$$

In MATLAB, without using library functions, that would be:

```
mse=mean((s-p).^2)
```

For long recordings, this measure would smear together all features of speech which change over time to compute a single average value. However, it is often more useful to know the mean-squared error on a segment-by-segment basis to see how this varies (think of a typical speech system you are developing: it would be more useful to know that it works very well for voiced speech but not well for unvoiced speech rather than know the overall average). The reader may remember the same argument used previously in Section 2.4 for a similar analysis example.

The *segmental* mean-squared error is this measure of the time-varying MSE over smaller segments. These might be 20 to 30 ms long sequential analysis frames, sometimes with overlap. For a frame size of N samples and no overlap, the jth segment MSE would be:

$$E(j) = \frac{1}{N} \sum_{i=jN}^{(j+1)N-1} \{s[i] - p[i]\}^2. \quad (3.2)$$

Remember that MATLAB indexes arrays from element 1 and not element 0, hence the slight difference in indexing terms between Equation (3.2) and the MATLAB expression:

```
mse(j)=mean(s((j-1)*N+1:j*N)-p((j-1)*N+1:j*N).^2);
```

3.4.2.2 Signal-to-noise ratio

For cases when signals of interest are not really being compared for likeness, but rather one signal is corrupted by another one, the ratios of the signals themselves can be useful, as the signal-to-noise ratio (SNR). Note that SNR is not used to measure the degree of difference between signals, but is simply the base-10 logarithm of the ratio between a wanted signal s and interfering noise vector n, measured in decibels, dB. Like MSE, it is a time-domain measure:

$$\text{SNR} = 20 \log_{10} \left\{ \frac{1}{N} \sum_{i=0}^{N-1} \left(\frac{s}{n}\right) \right\}. \quad (3.3)$$

```
snr=20*log10(s./n)
```

Segmental signal-to-noise ratio (SEGSNR) is a measure of the signal-to-noise ratio of segments of speech, in the same way that we segment other measures to see how things change over time. Like the segmental MSE, segments are typically 20–30 ms in size, perhaps with some overlap. For a frame size of N samples and no overlap, for the jth segment this would be as shown in Equation (3.4):

$$\text{SEGSNR}(j) = 20 \log_{10} \left\{ \frac{1}{N} \sum_{i=jN}^{(j+1)N-1} \left(\frac{s}{n}\right) \right\}. \quad (3.4)$$

```
ssnr(j)=20*log10(s((j-1)*N+1:j*N)./n((j-1)*N+1:j*N))
```

This equation works when we have access to the separate signal and noise arrays. For cases when we only have the original signal s and the noise-corrupted signal $c = s + n$, the equations will be different, such as replacing s/n by $s^2/(c-s)^2$ (see [26]).

3.4.2.3 Spectral distortion

On the basis that hearing is not confined to the time domain, but is mainly frequency domain, the measures discussed above are likely to be only minimally perceptually relevant. Better measures – or at least those more related to a real-world subjective analysis – can be obtained in the frequency domain for many tasks. The primary frequency-domain difference measure is called spectral distortion, shown in Equation (3.7), measured in dB^2, as a comparison between signals $p(t)$ and $s(t)$ for a frame of size N.

First, however, since the equation is in the frequency domain, we convert the time-domain signals to be compared into the frequency domain using the Fourier transform:

$$S(\omega) = \frac{1}{\sqrt{2\pi}} \int_{-\infty}^{+\infty} s(t) e^{-j\omega t} dt, \qquad (3.5)$$

$$P(\omega) = \frac{1}{\sqrt{2\pi}} \int_{-\infty}^{+\infty} p(t) e^{-j\omega t} dt. \qquad (3.6)$$

Only then can we compute the spectral distortion, SD:

$$\text{SD} = \frac{1}{\pi} \int_0^{\pi} (\log(S(\omega)) - \log(P(\omega)))^2 d\omega. \qquad (3.7)$$

In terms of implementation, and performing such analysis in MATLAB, we note firstly that performing an analysis from $-\infty$ to $+\infty$ is an unrealistic expectation, so we would normally choose a segment of N samples of audio to analyse over, then window it and perform a fast Fourier transform to obtain both power spectra, P and S. In discrete sampled versions, we follow the same method that we used in Chapter 2 to visualise signals in the frequency domain:

```
S=fft(s.*hamming(N));
S=20*log10(abs(S(1:N/2)));
P=fft(p.*hamming(N));
P=20*log10(abs(P(1:N/2)));
```

and then proceed with the SD measure without normalisation:

```
SD=mean((S-P).^2);
```

Indeed, SD is a perceptually relevant difference measure for speech and audio; however, it *can* be enhanced further. One way to improve it is by the additional step of

A-weighting the spectra – so that differences in frequency regions that are more audible are weighted more than those in frequency regions that are inaudible. This yields a perceptually weighted spectral distortion, and is used in practical systems that perform high-quality speech and audio signal analysis.

3.4.2.4 LPC cepstral distance

Linear predictive coding (LPC) is one of the staple techniques used in speech analysis, coding and processing. It will be discussed in detail in Chapter 6, but for now we will just make use of it to compute another frequency-domain measure. Around $P = 10$ or 12 order LPC values are usually obtained from one pseudo-stationary speech window, and mainly encode formant information. Here, we will combine LPC with the cepstrum representation of Section 2.6.2.2 to form a perceptually relevant double-ended similarity measure called the LPC cepstral distance [27].

If we denote the LPC coefficients of two signals $s(t)$ and $p(t)$ to be compared as $\mathbf{a}_s$ and $\mathbf{a}_p$, then we first recursively obtain something called the LPC cepstrum coefficients, $Cs(m)$ and $Cp(m)$:

$$C(m) = \mathbf{a}_m + \sum_{k=1}^{m-1} \frac{k}{m}\{C(k)\mathbf{a}_{m-k}\} \quad \text{for } 1 \leq m \leq P. \tag{3.8}$$

Then we can easily compute the LPC cepstral distance (CD) measure as the sum of the squared pair-wise difference between the two [28]:

$$\text{CD} = 10/\log_{10}\sqrt{2\sum_{i=1}^{P}\{Cs(i) - Cp(i)\}^2}. \tag{3.9}$$

In practice, usually only the lower 128 cepstral coefficients (excluding the DC value) are used to compute the distance, and the effect of outliers is often reduced by limiting the CD range to $[0, 10]$ [28]. Infobox 3.4 on page 77 will demonstrate how this and other measures can easily be computed in MATLAB.

3.4.2.5 Log-likelihood ratio

The log-likelihood ratio (LLR) also makes use of the LPCs $\mathbf{a}_s$ and $\mathbf{a}_p$ to compute another frequency-domain measure. This time, however, we need to choose one signal as a reference, while we measure the closeness of the other ('degraded') signal to that reference. We then make use of the autocorrelation matrix of the reference signal $\mathbf{R}_s$ to compute the log-scaled ratio between the two signals, where T indicates matrix transpose [28]:

$$\text{LLR} = \log\left\{\frac{\mathbf{a}_p \mathbf{R}_s \mathbf{a}_p^T}{\mathbf{a}_s \mathbf{R}_s \mathbf{a}_s^T}\right\}. \tag{3.10}$$

A typical recording would consist of many analysis frames, which themselves may be overlapped. Frame size, as usual, should be chosen to ensure that the signal within each frame is pseudo-stationary.

3.4 Speech understanding

Box 3.4 Speech articulation

Perhaps the most cited speech quality evaluation reference is the paper by Hu and Loizou, 'Evaluation of objective quality measures for speech enhancement', which appeared in *IEEE Transactions on Audio, Speech and Language Processing*, vol. 16, no. 1, in January 2008 [28]. This paper reports and compares the main speech quality measures, and provides a good overview of their background and use. More importantly, the late Professor Philip C. Loizou released a MATLAB function, `distance()`, as part of his COLEA speech package, which can compute several speech quality measures (LPC cepstrum distance, weighted cepstrum distance, SNR, IS, linear likelihood ratio, LLR, a weighted likelihood ratio and Klatt's weighted slope distance metric) [26].

`distance()` can be freely downloaded from the MATLAB File Exchange and used as follows (assuming you also downloaded `ilpc()` and `autoc()`). To demonstrate this, we will load (or record) a few seconds of speech, then create a few versions that have been corrupted with random noise:

```
[s,fs]=wavread('speech.wav');    % load a speech
   recording
% now create a few versions with 10 to 80 %AWGN:
p1=s+max(s)*0.1*rand(size(s));
p2=s+max(s)*0.2*rand(size(s));
p4=s+max(s)*0.4*rand(size(s));
p8=s+max(s)*0.8*rand(size(s));
```

Next we can perform the tests. Here we will specify 24 cepstral coefficients and tenth-order LPC when we determine four types of distance measure:

```
m_cep=distance(s,p1',24,'cep','');
m_wcep=distance(s,p1',24,'wcep','');
m_is=distance(s,p1',10,'IS');
m_llr=distance(s,p1',10,'LLR');
%now repeat for p2...p8 if desired
```

Using a standard TIMIT sentence, the scores we get from each of the measures are as follows:

Measure	(s,p1)	(s,p2)	(s,p4)	(s,p8)
m_cep	0.07	0.14	0.23	0.31
m_wcep	5.81	8.70	13.10	15.17
m_is	0.27	0.77	1.57	2.61
m_llr	0.16	0.44	0.84	1.35

It is not meaningful to compare the score from one measure with the score of another, but, when the measure type is fixed, we can see that the more noise is added to the speech, the higher the resulting score becomes.

In use, a set of LPC values and an autocorrelation matrix would be obtained for each windowed analysis frame, and thus one LLR score obtained per frame. The overall distance between the two recordings would then be the mean of all LLR scores; however, it is common to find that some frames are 'outliers', exhibiting much larger scores than others, and thus skewing the results. Thus the highest 5% of scores are often discarded and the remaining LLR scores limited to the range [0, 2] before the overall mean LLR is computed. When reporting an LLR score computed in this way in a paper or report, the fact that outliers were removed should obviously be clearly disclosed. Refer to Infobox 3.4 on page 77 for a MATLAB example of computing LLR.

3.4.2.6 Itakura–Saito distance measure

The widely used Itakura–Saito (IS) measure is computed in a similar way to the LLR, using the same raw input data as follows:

$$\text{IS} = \frac{\sigma_s^2}{\sigma_p^2} \left\{ \frac{\mathbf{a}_p \mathbf{R}_s \mathbf{a}_p^T}{\mathbf{a}_s \mathbf{R}_s \mathbf{a}_s^T} \right\} + \log \left\{ \frac{\sigma_s^2}{\sigma_p^2} \right\} - 1, \qquad (3.11)$$

the main difference is the σ_s^2 and σ_p^2 terms which denote LPC gains from the two signals being compared, obtained from $1/F(e^{j\omega T})$, where $\omega T = 2\pi k/N_r$ for $k = 0, 1, \ldots, (N_r - 1)$, for a frequency resolution of $F_s/2N_r$ Hz at sample frequency F_s computed over an N_r sample segment. Later, in Chapter 6 we will use this method of computing LPC gains to visualise LPC (and line spectral pair, LSP) spectra, but for now we will simply make use of the technique, without too much detail on *how* it works.

As with LLR, an IS score is computed for each analysis window, with the overall score being their mean. IS tends to experience some huge outliers, and thus these are often removed before computing the mean by limiting the score range to [0, 100] [28] (and this range limiting should be clearly disclosed in any research papers or reports which make use of the technique).

An additional issue here is that the IS measure is not symmetrical, i.e. $\text{IS}(a, b) \neq \text{IS}(b, a)$. Thus it is important to determine which signal is the reference and which is the degraded signal when obtaining an IS score. If this cannot be done, then it is better to compute the average score in both directions:

$$\text{IS}_{\text{bidir}}(p, s) = \frac{1}{2} \left\{ \text{IS}(p, s) + \text{IS}(s, p) \right\}. \qquad (3.12)$$

3.4.2.7 Other measures

This is a very active research area, and so new measures are proposed and evaluated from time to time. These could be double-ended but are increasingly single-ended – after all, a human being is capable of listening to a speech recording and judging its quality without needing to listen to a reference. Many of the most recent techniques use 'big-data' approaches to train machine learning systems to try and replicate the human perception of quality.

Already, automated measures of speech quality perform very well in the middle quality regions, especially in cases where quality is degraded primarily through added noise. Systems generally fail at the extremes of very good or very poor quality, as well as when the quality degradation is due to unusual, non-linear or intermittent causes.

3.4.3 Measurement of speech intelligibility

As with speech quality, intelligibility is best measured by a panel of human listeners. This is because it relates to the ability of listeners to correctly identify words, phrases or sentences. An articulation test is similar, but applies to the understanding of individual phonemes (vowels or consonants) in monosyllabic or polysyllabic real or artificial words. Many methods of evaluating intelligibility have been developed over the years, but those standardised by ANSI (in standard S2.3-1989) are the most prominent and popular. Several methods are listed here along with references that can provide more information (unless noted, see [29] for further details):

- diagnostic rhyme test (DRT) [30] – listeners must distinguish between two words which rhyme by their initial phoneme, such as {freak, leak} or {ghost, most};
- modified rhyme test (MRT) – listeners must select one of six words, half differing by their initial phoneme and half by their final phoneme, such as {cap, tap, rap, cat, tan, rat};
- phonetically balanced word lists – presenting listeners with 50 sentences of 20 words each, and asking them to write down the words they hear;
- diagnostic medial consonant test;
- diagnostic alliteration test;
- ICAO spelling alphabet test;
- two-alternative forced choice [31] – a general test category that includes the DRT;
- six-alternative rhyme test [31] – a general test category that includes the MRT;
- four-alternative auditory feature test [30] – asking listeners to select one of four words, chosen to highlight the intelligibility of the given auditory feature;
- consonant–vowel–consonant test [19, 32, 33] – test of vowel syllable sandwiched between two identical consonants, with the recognition of the vowel being the listeners' task, for example {tAt}, {bOb};
- general sentence test [19] – similar to the phonetically balanced word list test, but using self-selected sentences that may be more realistic in content (and in the context of what the test is trying to determine);
- general word test [13] – asking listeners to write down each of a set of (usually 100) spoken words, possibly containing realistic words.

Intelligibility may evidently be tested in terms of phonemes, syllables, words, phrases, sentences, paragraph meaning, or any other arbitrarily grouped, measurable recognition rate. In general, we can say that the smaller the unit tested, the more likely we will be able to relate the results to individual parts of speech. Unfortunately no reliable method has yet been developed for extrapolating between tests, for example using the results of a phoneme test to determine the effectiveness on sentence recognition (although, if the cause of intelligibility loss in a particular system is known, it should be possible to make a fairly good guess).

3.4.4 Contextual information, redundancy and vocabulary size

Everyday experience indicates that contextual information plays an important role in the understanding of speech, often compensating for an extreme lack of original information. For example, the sentence:

'He likes to xxxxx brandy'

can be understood by simple guesswork even though a complete word is missing ('drink'). Of course the missing word could be 'waste', 'buy', 'guzzle', 'wash with' or even 'walk' (assuming a dog of that name). However, we know that 'drink' would probably be quite a safe guess in most cases.

Note that the construction of sentences in English can be such that the importance of missing words is very difficult to predict. It is hard to know in advance whether the start, middle or end of a sentence will be more critical to its overall understanding. For example, the missing word 'stop' differs in both importance and predictability in the two sentences:

'She waited in the long queue at the bus xxxx'

and

'As the car sped towards him he shouted xxxx!'

Contextual information is usually provided by surrounding words and paragraphs which act to constrain the choice of the missing word. On a smaller scale, even the surrounding syllables can help to constrain the choice of a missing or obscured syllable (as certain combinations do not sound comfortable and are thus infrequently juxtaposed in the English language). Vocabulary size reduction also comes into play in allowing a listener to guess a mis-heard or noise-obscured word. This is the effect where speakers – either through deliberate choice, or through impaired speaking ability – use such a small set of adjectives or verbs, and such inflexible grammar construction, that their speech may be accurately guessed. Reduced vocabulary is of clear benefit in noisy environments, when simple and guessable words are preferred. Perhaps this is why speech on the factory floor or building site is often peppered with repeated and easily guessed adjectives.

Contextual information is also naturally provided by speakers in other ways which may be situational, or may be through shared knowledge between two communicating parties. Redundancy, in which information is imparted in more ways than would normally be necessary, has effects similar to contextual constraint. Redundancy may be provided in the form of over-complex sentences in which the role of a single word is limited, by context, to a very small set of choices. In this way, the extra information given in the sentence should enable complete understanding even if some words were lost. A simple example would be the repetition of important words such as 'It is a green car. The green car is now turning.'

Redundancy may also be achieved by the use of longer phrases of words or descriptions, such as the use of 'alpha bravo foxtrot' instead of spelling out 'ABF' (these

3.4 Speech understanding

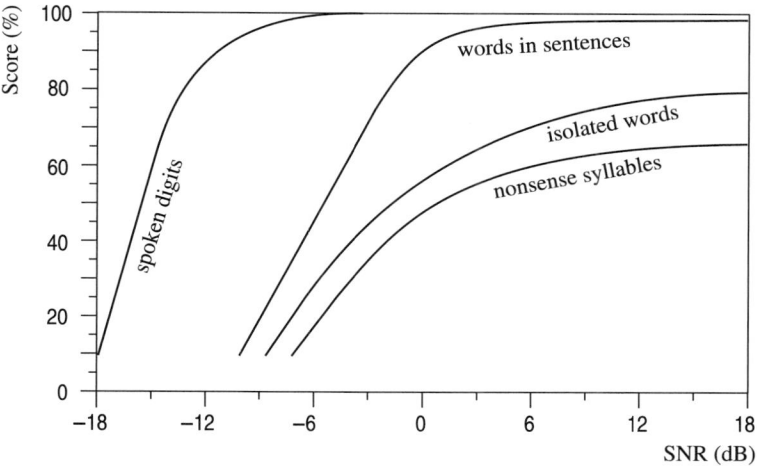

Figure 3.9 Effect of contextual information on the intelligibility of speech.

anti-abbreviations are from the NATO phonetic alphabet). Such redundancy reduces the chance that a short interruption to the audio (perhaps interference during transmission) would obliterate the entire meaning of the word. The same is ensured by asking for or giving repetition (e.g. 'it is a blue, repeat blue car' or 'unit one, confirm the vehicle is headed North West').

A measurement of the effects of contextual information on understanding is extremely difficult to quantify and is, of course, highly subjective. It is also entirely dependent upon the testing method, and the method of information removal necessary to conduct the tests. It goes without saying that context is also language dependent.

Some data is available, however, and it turns out that understanding this may be useful. Figure 3.9 plots the percentage of correctly identified digits, syllables and English words spoken in the presence of the given degree of background noise [19]. Although the shorter digit words are relatively more likely to be corrupted by noise than the predominantly longer unconstrained words, the extremely limited set of possible choices involved (as the listeners knew they were only listening for digits) means that even a corrupted digit may be guessed with some accuracy. It is also interesting to look at the extremes – above about 9 dB SNR for isolated words, −3 dB for digits and +3 dB for words in sentences, the curves become flat, meaning that no further increase in intelligibility is possible even when the speech becomes clearer. Viewed another way, it seems that words gain around 3 to 6 dB of intelligibility just through the context of being in a sentence. Furthermore, isolated words are only about 80% intelligible at the best of times, compared with near 100% for words embedded in a sentence. No wonder humans tend to prefer continuous speech to monosyllabic outbursts.

Vocabulary size can similarly be explored, as shown next in Figure 3.10. Here listeners are given the number of word choices indicated on the graph (i.e. the vocabulary size

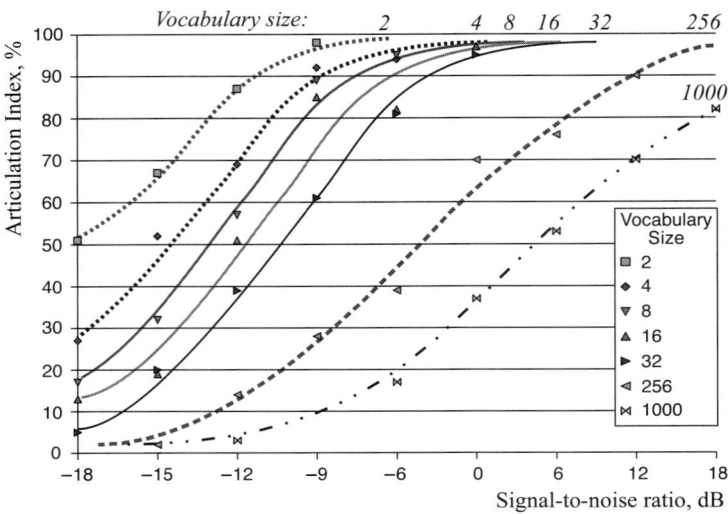

Figure 3.10 Effect of spoken vocabulary size on the intelligibility of speech.

in each test), for identification of each spoken word in the test. This illustrates the large improvement in recognition when vocabulary size is constrained – whether artificially or by context. For example, reducing vocabulary size from 256 to 16 at −9 dB signal-to-noise level results in almost four times as many words being recognised. It should be noted that the articulation index shown here is a measure of the recognition rate of individual phonemes, not words themselves [18], but the same principle would be at play in a word test. The figure was plotted by fitting polynomial curves to tabular data presented in [19], which was itself derived from historic experimental results published in 1951 [34].

3.5 Summary

In this chapter we have studied the physiological source and then the physical fundamentals of the speech signal – its amplitude, frequency distribution, tempo and so on. We then considered the units of speech: phonemes, words and sentences, and how these convey information. Finally, we looked at vocabulary effects and understanding, where we introduced the major topics of speech quality and speech intelligibility (and the testing of these).

We will now leave the subject of speech for a while. In Chapter 4 we move on to the ear and human auditory system. Later, in Chapters 5 and 6, we will then tie the hearing topic back to the characteristics of speech and perceived sound, and then discuss speech handling within communications systems.

Bibliography

- *The Noise Handbook*
  Ed. W. Tempest (Academic Press, 1985)

- *Digital Processing of Speech Signals*
  L. R. Rabiner and R. W. Schafer (Prentice-Hall, 1978)

- *Computer Speech Processing*
  F. Fallside and W. Woods (Prentice-Hall, 1985)

- *Acoustic and Auditory Phonetics*
  K. Johnson (Blackwell Publishing, 2003)

- *The Handbook of Hearing and the Effects of Noise*
  K. Kryter (Academic Press, 1994)

- *The Physics of Speech*
  D. B. Fry (Cambridge University Press, 1979)

- *Speech Intelligibility and Speaker Recognition*
  Ed. M. E. Hawley (Halsted Press/Dowden Hutchinson and Ross, 1977)

- *The Cambridge Encyclopedia of the English Language*
  D. Crystal (Cambridge University Press, 2003)

Questions

Q3.1 Arrange the following types of speech phoneme by their average loudness (loudest first), and give example words for each: nasal, affricative, vowel, voiced fricative, voiceless plosive.

Q3.2 In the context of speech evaluation, what does MOS stand for, and when is it used?

Q3.3 Compute the signal-to-noise ratio (SNR) in MATLAB between arrays `speech` and `noise`.

Q3.4 Compute the spectral distortion (SD) in MATLAB between arrays `speech` and `nspeech=speech+noise`.

Q3.5 The Itakura–Saito distance measure is not symmetrical. What does that mean in practice when I am computing the similarity of two speech arrays named `speechA` and `speechB`?

Q3.6 Which speech intelligibility test uses a two-alternative forced choice arrangement to discern the first syllable of rhyming word pairs?

Q3.7 If a researcher finds that her speech synthesiser consistently creates speech with an MOS of 3.95, how would you describe the quality of the speech that has been produced?

Q3.8 Why is the glottis more important for the production of Mandarin Chinese speech than it is for English speech?

Q3.9 What is the normal range, in words per minute, of spoken English?

Q3.10 Generally speaking, are vowels or consonants more important to the understanding of spoken English? Which component tends to be louder in normal speech?

Q3.11 Explain the Lombard effect using the example of two people in conversation who walk from a quiet balcony into a noisy room.

Q3.12 What is sinewave speech and what does it reveal about the human brain?

Q3.13 What is the difference, in speech production terms, between whispering and shouting?

Q3.14 Contrast a single-ended and a double-ended speech quality evaluation method.

Q3.15 What physiological change occurs during speaking to cause a nasal sound?

4 The human auditory system

A study of human hearing and the biomechanical processes involved in hearing reveals several non-linear steps, or stages, in the perception of sound. Each of these stages contributes to the eventual unequal distribution of subjective features against purely physical ones in human hearing.

Put simply, what we *think* we hear is quite significantly different from the physical sounds that may be present in reality (which in turn differs from what might be recorded onto a computer, given the imperfections of microphones and recording technology). By taking into account the various non-linearities in the hearing process, and some of the basic physical characteristics of the ear, nervous system, and brain, it becomes possible to begin to account for these discrepancies between perception and physical measurements.

Over the years, science and technology has incrementally improved our ability to understand and model the hearing process using purely physical data. One simple example is that of A-law compression (or the similar μ-law used in some regions of the world), where approximately logarithmic amplitude quantisation replaces the linear quantisation of PCM (pulse coded modulation): humans tend to perceive amplitude logarithmically rather than linearly, thus A-law quantisation using 8 bits to represent each sample sounds better than linear PCM quantisation using 8 bits (in truth, it can sound better than speech quantised linearly with 12 bits). It thus achieves a higher degree of subjective speech quality than PCM – for a given bitrate [4].

4.1 Physical processes

A cut-away diagram of the human ear (outer, middle and inner) is shown in Figure 4.1. The outer ear includes the pinna, which filters sound and focuses it into the external auditory canal. Sound then acts upon the eardrum, where it is transmitted and amplified through the middle ear by the three bones, the malleus, incus and stapes, to the oval window, opening on to the cochlea in the inner ear.

The cochlea, as a coiled tube, contains an approximately 35 mm long semi-rigid pair of membranes (basilar and Reissner's) enclosed in a fluid called endolymph [35]. The basilar membrane carries the organs of Corti, each of which contains a number of hair cells arranged in two rows (approximately 3500 inner and 20 000 outer hair cells).

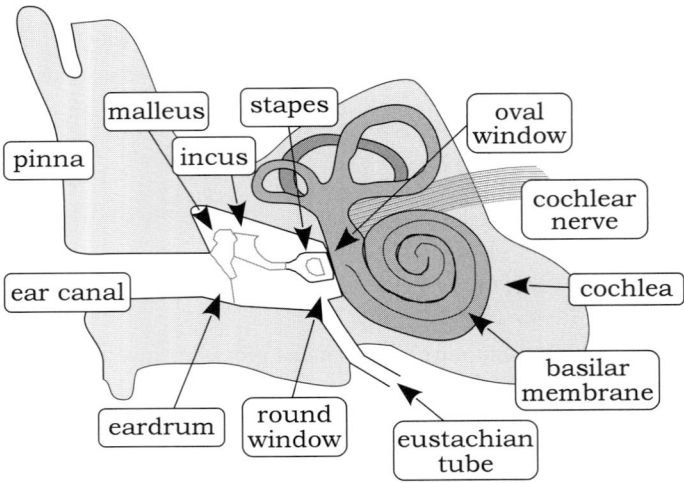

Figure 4.1 Schematic and stylised cut-away diagram of the human ear.

Small electrical impulses are generated by the hair cells when tension is applied to them (usually when they are pulled, so that auditory response occurs during the rarefaction part of an oscillation in the fluid surrounding them). The width and stiffness of the basilar membrane tapers along its length, and thus, when a vibration is applied to the system, the location on the basilar membrane which resonates is dependent upon frequency [21]. This process provides some frequency selectivity to the ear, with active processing by the auditory cortex providing the rest. In fact the auditory cortex processing substantially improves frequency selectivity. Nerves carry electrical impulses from the ear into the brain where extensive unconscious processing occurs [12]. The nature of the processing is not fully understood; however, it is highly non-linear, involves correlation between signals from each ear, and possibly with other senses (such as the feel of vibrations, and sight). Recent years have seen much research using magnetic resonance imaging (MRI) and similar technologies to analyse brain responses to various stimuli, including speech and sounds. Gradually, a brain-map is emerging which reveals locations used within the brain to process sound. What is also being revealed is a large degree of interdependence between brain areas and functions. For example, relating sound and vision (we will meet the McGurk effect later in Section 5.2), as well as sound and movement (unsurprising since the inner ear is also the organ of balance), sound and touch, and even sound and flavour (readers with a taste for more details can refer to some interesting experimental results in [36]).

During the listening process, the physiological construction of the ear means it effectively integrates sound over short time periods as well as processing grossly repetitive patterns in different ways to sporadic or single sounds [12].

Apart from being a passive receptor, active mechanisms also operate in the ear. These include tensioning the muscles operating on the malleus and stapes, which can protect the ear from loud sounds, reducing the firing rate of certain hair cells and reportedly even producing sounds on occasions [12, 37]. Some common evidence for this is the way that audio acuity is reduced by illness or the taking of certain drugs. Earache can be caused by the failure of the inner ear muscles to protect the eardrum [38].

4.2 Perception

The previous section has presented the biological structure of the ear, and looked at how sound (pressure oscillations in the air) is detected and transformed into electrical nerve impulses. In this section, by contrast, we will consider how humans perceive and understand those impulses.

4.2.1 Psychoacoustics

The word 'psychoacoustics' is an abbreviated concatenation of the phrase 'the psychology of acoustics'. In a nutshell, it describes the discrepancy between a purely physical (objective) view of hearing and a subjective view. This discrepancy forms the basis of several auditory illusions – auditory tricks that are analogous to optical illusions – as well as being the basis for a number of computational techniques used to analyse, compress, transform, transmit, store and recreate sound. Psychoacoustics will form the basis of Chapter 5, but for now we will discuss more foundational concepts, including the basis for the subjective perception of sound.

4.2.2 Equal loudness contours

Human subjects do not always judge tones of different frequency, but identical amplitude, to be equal in loudness [18, 37]. Put another way, a 1 kHz signal at 40 dB$_{SPL}$ would sound louder than a 15 kHz signal at exactly the same amplitude, and both would sound louder than a 40 Hz sound of the same amplitude.

After long and careful experimentation, scientists created a set of 'equal loudness contours' in the 1940s that model this behaviour, shown plotted in Figure 4.2. The contours are widely known today as Fletcher–Munson curves, after their discoverers [39]. Several contours are shown in Figure 4.2, identified in phons, where the n phon contour follows the frequency amplitudes that sound equally loud to a subject as an n dB$_{SPL}$ tone at a reference frequency of 1 kHz. The lowest curve (the dashed line) indicates the threshold of hearing.

The effect of equal loudness should be relatively easy to test in MATLAB; however, one should be aware that, even when creating two different sounds of the same amplitude inside MATLAB, the technology used to reproduce them might change their amplitudes. In fact, all loudspeakers and headphones have a non-linear frequency response, meaning that some frequencies will be converted from an electrical signal to a mechanical vibration with higher efficiencies than others. Put another way, it means that a 1 kHz tone at a 50% amplitude level might create a 40 dB$_{SPL}$ sound, but a 2 kHz tone at exactly the same amplitude might create a slightly louder or slightly quieter tone, depending upon the speaker technology, construction and quality. Even if we were to measure

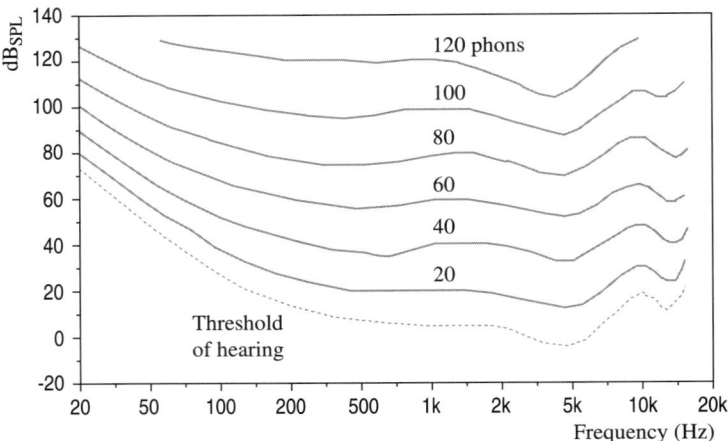

Figure 4.2 Equal-loudness contours (re-plotted from Fletcher–Munson data).

the sounds to check their amplitude, we would be faced with a non-linear microphone frequency response! In fact we would need to use a calibrated reference microphone before we could begin to be certain of the real amplitudes being produced. Similarly, we could use a calibrated reference speaker to produce sounds of known amplitude. However, all of these devices are expensive, delicate, and require periodic re-calibration (which is itself a costly service).

Fortunately, we do not need to produce absolutely correct amplitudes for our experiment, but rather sounds with relatively correct amplitudes and frequencies.[1] A good pair of headphones will have a relatively flat frequency response, sometimes given to purchasers as a plot, or specified in a similar way to '20 Hz to 20 kHz $\pm$ 3 dB'. Such a specification means that two equal-amplitude tones of different frequency within the given range will be converted to sounds with amplitudes of no more than 3 dB difference from each other. In fact, the 3 dB measure usually applies to the extremes at either end of the frequency range. Within the middle span of frequencies, a high-quality set of headphones or loudspeakers should be flat to within 1 to 1.5 dB. Assuming such equipment is available, we can use the `tonegen()` function from Section 2.7.1 to create and listen to three two second long tones, all of equal amplitude:

```
lo= tonegen(250, 441000, 2);
mi= tonegen(1200, 441000, 2);
hi= tonegen(11000, 441000, 2);
soundsc(lo, 441000);
soundsc(mi, 441000);
soundsc(hi, 441000);
```

---

[1] Absolutely correct means that the frequencies and amplitudes are located at the specified values, to within some known tolerance (for example 1 kHz $\pm$ 20 Hz, 40 dB$_{SPL}$ $\pm$ 0.05 dB). Relatively correct means that they may be absolutely incorrect, but the relative amplitude and frequency difference between them is constrained to a known tolerance.

> **Box 4.1** Weighting
>
> For speech and hearing purposes, voice power, background noise and other sound levels are usually measured in dBA, where the signal is A-weighted before being quantified. This is the application of a frequency weighting based upon the 40-phon equal loudness contour to the signal. Thus all frequency components in the signal are weighted so that they make a contribution to the overall figure dependent upon their perceived loudness, rather than upon their actual intensity.
>
> Although this scheme appears reasonable, when applied to speech, it takes no account of the importance of different frequencies to quality or intelligibility – only to their hearing perception. Additionally it does not account for absolute loudness, since the 40-phon curve really only describes the perception by a mythical average human of a 1 kHz signal at 40 dB$_{SPL}$.
>
> More concerning is the fact that these historical and pioneering data were gathered based on single-tone signals rather than complex sounds, but are almost always applied to complex sounds. Still, in the absence of better methods, and scientific data indicating otherwise, A-weighting can be considered a reasonable approximation to many situations and applications outside the boundary of its known validity.
>
> A-weighting has been ratified by the International Standards Organisation (ISO), as have ISO B- and C-weighting based on the shapes of the 70- and 100-phon curves respectively.
>
> As an aside, dBA is commonly used for level measurements of 'nuisance' noise. However, this works only if the audio frequencies being measured fit to the 40-phon curve. Infrasonic frequencies (2–32 Hz) and ultrasonic frequencies (> 20 kHz) both fall almost entirely outside the measurement domain. An example is given in [21], where householders near New York's JFK airport complained about excessive rumble from passing aeroplanes. This prompted tests for noise pollution which indicated that very little low-frequency noise was present. The rumble sound came from much higher-frequency harmonics inducing a low-frequency perceived sound (we will see how this happens in Section 5.1.1). On an A-weighted scale the noise level measured would be low, whereas the sound experienced by the householders was loud.

Based on the loudness curves, the 1.2 kHz middle tone should be perceived as being louder than either the 250 Hz lower tone or the very high 11 kHz sound (which may not even be audible at all for some people). We have used a 44.1 kHz sample rate in this case in order to faithfully reproduce the 11 kHz sinewave (although we could have used any sample rate above 22 kHz, for example 24 kHz).

It should be noted that the equal loudness contours are derived from averaged tests from average listeners, and apply to single tones only: there may be substantial person-to-person hearing differences, especially in cases of people who listened to too much rock music when young (i.e. suffering from hearing damage). Despite this, the curves are very useful at describing listening behaviour, and are generally considered to be quite accurate average responses. They are often applied to describe complex sounds that are not just simple continuous tones.

The equal loudness contours as plotted in Figure 4.2 are the result of a number of factors in the human hearing system, one of which is the filtering (frequency selectivity) introduced by the pinna: *orthotelephonic gain*. The frequency distribution impinging on the eardrum differs when inner-ear headphones are used as opposed to loudspeakers, as the pinna provides around 5 dB gain at 2 kHz, 10 dB gain at 4 kHz and 2 dB gain at 8 kHz [40]. The filtering effect of the pinna below 500 Hz is negligible [21].

4.2.3 Combination tones

When hearing response is stimulated with two pure tones, active processing produces components with differing frequencies and amplitudes compared to the stimulating tones [12]. Thus listeners may hear additional sounds apart from the two tones which have been applied. For example, given two tones at frequencies $f1$ and $f2$ with $f2 > f1$, listeners may perceive additional tones at a frequency of $2f1 - f2$, caused by non-linear hearing processes, as well as tones at a frequency of $f2 - f1$, caused by linear hearing processes. On occasion, tones of $f2/f1 < 1.5$ will also be heard [41]. In fact, these effects have long been exploited by musicians to induce musical harmony. We will now consider the $2f1 - f2$ and $f2 - f1$ effects in turn.

4.2.3.1 $2f1 - f2$ tone induction

As mentioned above, some listeners will discern a tone of frequency $2f1 - f2$ when sounds of frequency $f1$ and $f2$ (with $f2 > f1$) are presented simultaneously. This can work across a wide range of amplitudes and frequencies, and is likely to be produced as a consequence of active mechanisms within the inner ear. Although the effect is not always easy to isolate, we will use MATLAB to attempt to recreate an excellent demonstration that was published by Kaoru Ashihara [42], and which should be audible for most listeners.

In this demonstration, tone $f1$ is of fixed frequency – we will use 1.8 kHz since a higher frequency seems to sound better – while tone $f2$ will sweep slowly upwards in frequency from 2.0 to 2.2 kHz. In this case we will use a quadratic chirp to accentuate the perceived generated tone. Here we prepare the signals for left and right loudspeakers:

```
fs=44100;  %sample frequency
f1=tonegen(1800,fs,2);
f2=chirp([1:fs*2]/fs,2000,2,2200,'q');
left=[f1,zeros(1,fs),f1];  %repeat sounds twice
right=[f2,zeros(1,fs),f2];
```

We will use two separate loudspeakers (one for each tone) to ensure that any combination tones that are produced are from our own ears – not from interference of the tones within a loudspeaker. Since these tones need to be combined within the ear, the demonstration requires external loudspeakers rather than headphones or earphones.

First we should listen to the two signals in isolation, ensuring that they are easily audible by themselves:

```
soundsc([left;],fs)
%wait for this to finish, then
soundsc([;right],fs)
```

These should be clear and pure tones when heard separately. One tone is flat, the other rises in frequency. Now we will listen to them together – one from the right loudspeaker and one from the left:

```
soundsc([left;right],fs)
```

You will hear a constant tone, and one that rises in frequency, but can you also hear a tone that *falls* in frequency? If so, this is the combination tone.[2] To explain, we can compute the combination tone frequency at the beginning of the chirp: $f1 = 1800\,\text{Hz}$ and $f2 = 2000\,\text{Hz}$, so $2f1 - f2 = 1600\,\text{Hz}$. At the end of the chirp, $f1 = 1800\,\text{Hz}$ and $f2 = 2200\,\text{Hz}$, so $2f1 - f2 = 1400\,\text{Hz}$. So the combination tone sweeps downwards in frequency from 1.6 kHz to 1.4 kHz.

4.2.3.2 $f2 - f1$ tone induction

A similar effect is found when listening to complex harmonic structures with the fundamental frequency removed, which are perceived as having a frequency equal to the fundamental, despite its absence.

For example, a series of tones at 200, 400, 600, 800 Hz, ... evokes a pitch sensation of 200 Hz (as may be expected since this is the fundamental frequency in the set); however, removal of the 200 Hz frequency does not affect the overall perceived pitch. It does, however, affect the timbre or quality of the sound experienced. The perceived tone, when not actually present, is termed the residue [12].

The effect is very easy to demonstrate using MATLAB. First we create four pure sinusoidal tones of duration two seconds at an 8 kHz sample rate. We will use a fundamental of 196 Hz (G_3 on the musical scale) and its multiples:

```
t1=tonegen(196, 8000, 2);
t2=tonegen(196*2, 8000, 2);
t3=tonegen(196*3, 8000, 2);
t4=tonegen(196*4, 8000, 2);
```

Next we will simply play back these tones, starting with a pure G_3. Remember the pitch of this tone, then we will play back the complex note both with and without the fundamental:

```
soundsc(t1, 8000);
soundsc(t1+t2+t3+t4, 8000);
soundsc(t2+t3+t4, 8000);
```

---

[2] Note that this effect may not be audible by all listeners in all situations – many factors including listening environment, loudspeaker and amplifier quality and the characteristics of the listener's ear can all affect its audibility. If the effect is not audible at first, try changing the frequencies and adjusting the loudspeaker volume.

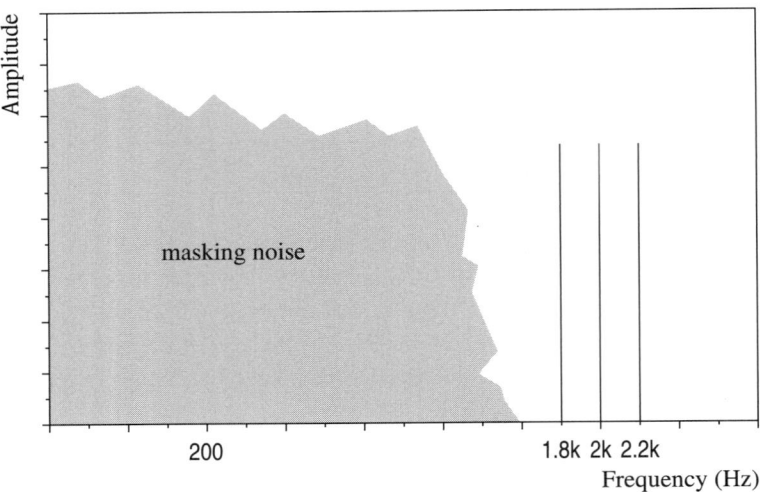

Figure 4.3 An artificial arrangement causing 200 Hz to be heard, but where the 200 Hz tone, if present, would not be discernible due to heavy masking noise.

It should be noticeable that, although the quality of the sound differs somewhat, the musical note being replayed is the same in each case.

Musically speaking, the fundamental frequency is the perceived note being played. Harmonics and other tones may be present, with timbre or quality changing greatly depending on the proportions. This is one of the defining characteristics of different musical instruments. Even when low-frequency noise masks the perceived note (the residue frequency) such that it cannot be heard, the frequency of the residue is still perceived [43]. This is illustrated in Figure 4.3, where the tones shown would evoke a response (i.e. it would be heard) at 200 Hz, a frequency that if really present would be swamped by noise and therefore could not actually be heard.

We find that both arithmetic and geometric relationships of single tones are discerned by the ear as being composite in some way, something we will explore further in Chapter 5.

4.2.4 Phase locking

As noted previously, neural excitation by the hair cells only occurs at the rarefaction part of the sound wave, at approximately fixed phase, although cells can vary considerably as to their individual phase positions. The average firing frequency for all cells will be of the correct frequency and correct phase. Some cells only fire every two, four or six cycles but this does not alter the overall firing rate [38]. Due to this cycle averaging process that the ear seems to use to distinguish between frequencies, it is possible for the ear to become attuned to a particular frequency, with hair cells firing in time with a rarefaction and not recovering until a short time later. Another rarefaction in this recovery period may be missed, the result being that a louder than usual amplitude will be required in order for a second tone (causing the additional rarefaction) to be heard [12]. This phase locking, as it is called, may well be part of the explanation for why a single tone will suppress tones of similar frequency and lower amplitude, part of the simultaneous masking phenomenon discussed in Section 4.2.9.

4.2.5 Signal processing

Cells in the auditory cortex are excited by a variety of acoustic signals; however, some cells do not respond to single tones, but require more complex sounds [38] to evoke a response. Natural processing in the human brain can detect tone start, tone stop, tone pulse, frequency slide up, frequency slide down, amplitude variation and noise burst conditions. One experiment has even determined the position of a group of brain cells that specialises in detecting 'kissing' sounds [12]. One wonders at the experimental methods employed.

4.2.6 Temporal integration

The ear's response with respect to time is highly non-linear. For a tone duration below about 200 ms, the intensity required for detection increases with decreasing duration, being linearly proportional to the duration multiplied by the intensity required for detection of a constant tone of that frequency. Tones of longer duration – above about 500 ms – are detected irrespective of their duration, complexity and pattern [12].

In a similar way, periods of silence introduced into a constant tone are detectable to an extent which is dependent upon duration, up to a duration exceeding about 200 ms. This effect will be explored a little further, and demonstrated, in Chapter 5.

4.2.7 Post-stimulatory auditory fatigue

After an abnormally loud sound the ear's response is reduced during a recovery period, after which it returns to normal [21]. This is termed *temporary threshold shift* (TTS) [12]. The degree of TTS depends upon the intensity of the fatiguing stimulus, its duration, its frequency and the recovery interval. It is frequency specific, in that the TTS effect is distributed symmetrically about the frequency of the fatiguing tone and is limited to its immediate neighbourhood, but the actual frequency spread of the TTS curve is related to the absolute frequency. It is also related to the amplitude of the tone and to the logarithm of the duration of the fatiguing tone (although the middle ear reflex muscle action reduces TTS for low frequencies, and a tone duration of over five minutes will produce no appreciable increase in TTS). When the fatiguing noise is broadband, TTS occurs mostly between 4 and 6 kHz, begins immediately and may still be noticeable up to 16 hours after the noise onset [19]. Tones louder than about 110 or 120 dB_{SPL} can cause permanent hearing loss, but TTS is most prominent for amplitudes of 90 to 100 dB_{SPL}. Caution is advised when dealing with loud sounds as the resulting permanent hearing damage is not an uncommon consequence. This can be selective (i.e. confined to a few frequency regions) or apply across all audible frequencies. As with many adverse health effects, continuous or frequent exposure to loud sounds may be more dangerous than sporadic exposure.

4.2.8 Auditory adaptation

The response of a subject to steady-state tones will decline to a minimum over time (i.e. they become less noticeable when they do not vary), although an amplitude of about 30 dB$_{SPL}$ is needed to trigger the effect. It is worth mentioning that this effect appears to be highly subjective, with some subjects reporting that continuous tones completely disappear, whilst others experience only a 3 dB or smaller reduction. This applies to pure, unchanging, tones, but, for broadband or complex noise, the auditory system is unable to adapt to the same degree. The literature also reports that high-frequency tones are easier to adapt to than low-frequency tones [12]. Despite the reports concerning broadband noise, it appears anecdotally that, on long aeroplane journeys, an initial high level of cabin noise (which is subjectively quite broadband) can become almost unnoticed by the end of a flight. One hopes that this is an example of auditory adaptation, rather than hearing damage.

4.2.9 Masking

Masking deserves a chapter by itself, and indeed will form a large part of the basis for psychoacoustic modelling in Chapter 5, but for now a brief introduction is sufficient. Masking in general is defined by the American standards agency as 'the process by which the threshold of audibility for one sound is raised by the presence of another sound' and 'the amount by which the threshold of audibility of sound is raised by the presence of another sound'. The frequency selectivity of the basilar membrane may be considered similar to the functionality provided by a bank of bandpass filters with a threshold of audibility in each filter being dependent upon the noise energy falling within its passband [22]. The filters each have similarly shaped responses with bandwidths approximately 100 Hz up to frequencies of about 1 kHz. Above this, bandwidth increases in a linear fashion with frequency up to a 3 kHz bandwidth at 10 kHz. Each 'filter' in the imaginary array is termed a critical-band filter [9]. We will return to these filters later in Section 4.3.2 when we attempt to numerically model them.

The effect of the critical-band filters is that, for a given tone having fixed frequency and amplitude, the sensitivity of the ear to tones of similar frequency is reduced. This is illustrated in Figure 4.4, which plots the frequency response of an example tone, and indicates the area within which this tone might induce masking. While listening to the tone shown, any new tone which is introduced within the identified masking area will be inaudible.

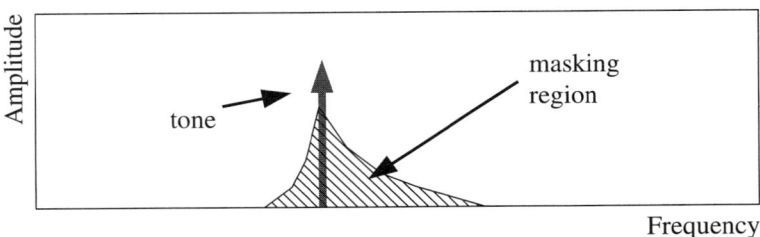

Figure 4.4 Illustration of masking effect due to a single tone.

4.2 Perception

In general, once a louder tone has 'occupied' the sensors of one critical-band filter, the same filter is less sensitive to other coincident sounds. Many researchers have created logical models of this masking process, ranging in complexity and accuracy from very simple to highly complex. Some established models can be found in the literature, including the following: [9, 19, 44, 45], and in fact we will implement and discuss several of these in Chapter 5.

In the meantime, as an example to assist in understanding the masking effect, let us create two pure tones in MATLAB, again using the `tonegen()` function from Section 2.7.1. We will ensure that the lower-frequency tone is only 10% of the amplitude of the louder one so that it is (hopefully) masked by it:

```
lo=0.1*tonegen(800, 8000, 2);
hi=tonegen(880, 8000, 2);
```

Next we will use `sound()` to replay the audio instead of `soundsc()` so that we can appreciate the differences in amplitude of the two tones (since `soundsc()` would scale them both so they have the same amplitude). We will first listen to both tones in isolation, then we will listen to the two tones mixed together:

```
% first adjust volume
sound(hi/2, 8000);
% that was the louder sound
sound(lo/2, 8000);
% that was the quieter sound - now together:
sound((lo+hi)/2, 8000);
```

Both of the individual tones can be heard when played alone, although the lower-frequency tone is clearly quieter. However, when replayed together the result is a slightly high tone exhibiting a very slight warble. The low tone should be inaudible – masked by the louder tone.

One further interesting related point is that, for sounds whose bandwidth falls entirely within one critical band, the intensity of that sound is independent of its bandwidth. However, for sounds with bandwidth greater than one critical band, the intensity depends strongly on the proportion of the sound's bandwidth falling within each critical band. Thus, in general, a complex sound having components in more critical bands sounds louder to a listener than a complex sound with components in fewer critical bands does [12].

4.2.10 Co-modulation masking release

Although the amount of masking relies upon noise power, modulated noise (even when randomly modulated) whose bandwidth extends across more than one critical band induces less masking as its bandwidth increases [37]. This is really an extension of the point mentioned above, for sounds.

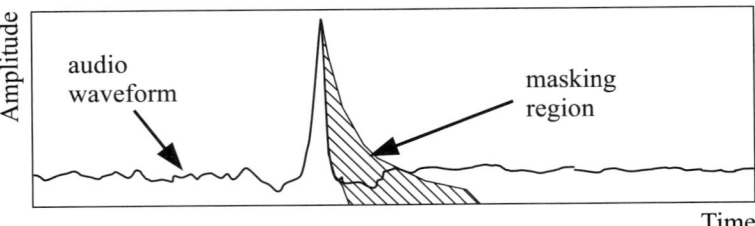

Figure 4.5 Illustration of post-stimulatory masking in the time domain caused by a momentary loud sound.

This phenomenon indicates that correlation between critical bands can be effectively utilised by the auditory system to enhance its performance (in this case, to reduce the effect of masking noise).

4.2.11 Non-simultaneous masking

Non-simultaneous or temporal masking can be subdivided into pre-stimulatory (backward) and post-stimulatory (forward) masking. In the former, masking effects occur *before* the onset of the masking tone, and in the latter, masking effects occur after the onset of the masking tone.

Figure 4.5 provides an illustration of the effects of post-stimulatory masking. It plots the time-domain waveform of some audio, and shows a masking region immediately following a loud sound, during which other sounds will not be audible.

Pre-stimulatory masking is notoriously difficult to measure, as it is highly dependent on the subject, and also on any previous exposure the subject has had to the test situation (which may cause the measurable effects to disappear completely) [12]. Normal practice in auditory testing is to 'warm up' with some dummy or calibration tests prior to the procedure beginning, which is not really possible when testing for pre-stimulatory masking.

Effects may occur for signals lasting up to a few hundred milliseconds, and persist for a similar length of time, with the rate of recovery (rate of reduction of masking effect) being higher for louder sounds. The duration of the masking tone determines the degree of masking up to about 20 ms, where the dependence levels out, and obviously the masking effect has a frequency dependence based on the frequency of the masking tone [12, 46].

4.2.12 Frequency discrimination

Frequency discrimination by humans is dependent upon absolute frequency, and to some extent amplitude; however, an approximate figure would be around 2 Hz for a 65 dB$_{SPL}$ signal at 1 kHz. Thus a tone of 1002 Hz can just be distinguished from a 1000 Hz tone by an average listener.

4.2 Perception

Frequency discrimination is related to pitch perception, which decreases with increasing amplitude for tones below about 2 kHz, but increases with increasing amplitude for tones above about 4 kHz. Unless the subject is one of the 1% of the population capable of *absolute pitch* or *perfect pitch* then a pitch above 2.5 kHz, even when presented with a reference, cannot be discriminated. It is worth noting that, due to this, tones of frequency greater than 5 kHz cannot evoke a sensation of melody [12].

We can easily demonstrate frequency discrimination using MATLAB with our trusty `tonegen()` function. We simply need to create and replay slightly different pure sinusoidal tones and see if there is any noticeable difference. In this case, let us use 1000 Hz and 1002 Hz tones of two seconds duration at 8 kHz:

```
t1=tonegen(1000, 8000, 2);
t2=tonegen(1002, 8000, 2);
%listen to both, 1/4 second pause between
soundsc([t1,zeros(1,1000),t2], 8000);
```

It is likely that very few (if any) listeners would be able to note the difference between `t1` and `t2`. If someone claims to be able to, just try several tests with `t1` and `t2` in random sequence and see how many they get correct. Over many repetitions, a score significantly greater than guesswork (i.e. 50% correct) would indicate that the tones can be discriminated. This is a test of absolute tonal acuity.

On the other hand, if we remove the pause, allowing us to directly compare `t1` with `t2`, the difference is still subtle but many listeners will now be able to hear the change between tones:

```
soundsc([t1,t2], 8000);
```

This has now become a test of relative tonal acuity, since there is a reference (the first tone) with which to compare the second tone.

4.2.13 Binaural masking

The human brain can correlate the signals received by two ears to provide a processing gain. This is measured by presenting a signal of noise plus tone to both ears, and adjusting the tone level until it is just masked by the noise. From this situation a change is made, such as inverting the tone phase to one ear, and the tone amplitude is again adjusted until it is just barely audible. The difference in amplitude of the tone in the second case reveals the processing gain. Table 4.1 (constructed from data presented in [12, 19]) summarises the results under different conditions.

When the signal is speech, intelligibility is reduced in antiphasic (uncorrelated noise, out of phase speech) conditions over phasic conditions (uncorrelated noise, speech in phase) as revealed in Table 4.2.

Table 4.1 Binaural masking conditions and features (constructed from data presented in [17]).

Noise condition	Signal condition	Processing gain
In phase	In phase	0 dB (reference condition)
Uncorrelated	Out of phase	3 dB
Uncorrelated	In phase	4 dB
Out of phase	In phase	6–13 dB
In phase	Out of phase	9–15 dB

Table 4.2 Binaural masking conditions and features (constructed from data presented in [17]).

		Percentage of words correctly identified		
		Phasic	Random	Antiphasic
Stereo	Phasic	18	27	35
	Antiphasic	43	27	16
Mono	Right ear	30	13	20
	Left ear	18	8	15

The effects of binaural masking are greatest at lower frequencies, but depend upon the frequency distribution of both the test signal and the noise.

Note that *binaural* is a term used by auditory professionals and researchers. It means 'pertaining to both ears', and is frequently confused with *stereo*. Stereo actually refers to sound recorded for multiple loudspeakers, which is designed to be replayed several metres in front of a listener, and generally enjoyed by both ears. Users of headphones are thus exposed to a significantly different sound field to those listening with loudspeakers – a fact which is not normally exploited by audio equipment manufacturers.

4.2.13.1 Two ears and intelligibility

Inspired by the ideas of binaural masking applied to speech, the author performed a series of experiments to determine how the angle of incidence of speech affects intelligibility [47]. The idea was to replay speech in a noisy environment to listeners from different horizontal directions with respect to their head position, and evaluate whether there is any relationship between the angle and the intelligibility. Apart from being an interesting study in its own right, this has application in speech playback applications in noisy environments such as hands-free communications systems in vehicles, or in smart homes.

Figure 4.6 illustrates the angles of incidence tested for speech, including from straight ahead, from the right, from the left and from a slowly moving source located $\pm 30°$ in front of the listener. The angle was adjusted in each case using MATLAB code to adjust the phase and amplitude (we will develop this later in Section 10.2), and the intelligibility evaluated through a DRT test (see Section 3.4.3). Figure 4.7 illustrates the evaluation process, where recorded word lists are randomised and then played back

4.2 Perception

Table 4.3 Intelligibility reduction of speech by type when replayed from different positions.

	Voicing	Nasal	Sustain	Sibilate	Grave	Compact	*Mean*
Left[1]	21.3%	0.6%	5.0%	6.4%	9.1%	16.3%	*6.6%*
Right[1]	18.0%	0.0%	12.4%	20.0%	5.9%	16.4%	*12.1%*
Front[1]	2.8%	2.9%	8.1%	1.1%	10.4%	12.9%	*5.4%*
Front[2]	2.5%	0.5%	6.6%	9.5%	15.2%	3.8%	*4.0%*

[1] in AWGN, [2] in multi-speaker babble

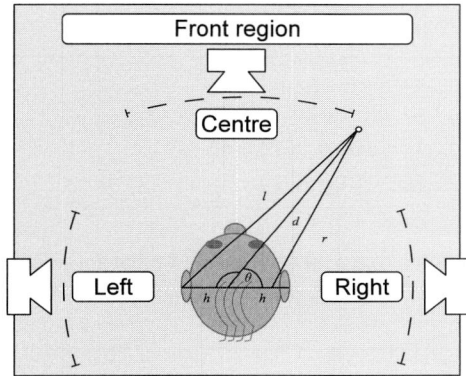

Figure 4.6 Diagram of a listener hearing speech from different directions.

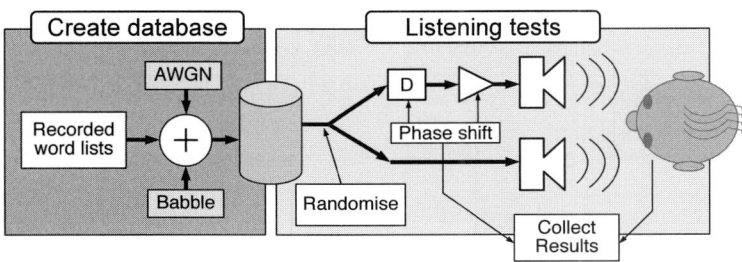

Figure 4.7 The evaluation process for binaural speech listening tests.

through two speakers to change stereo position to simulate the different directionalities of the speech source. Background noise was added to reduce the average intelligibility to around 75% before commencing the DRT. AWGN was used in all cases, apart from one test which used multi-speaker babble noise. Results from seven normal-hearing listeners are presented in Table 4.3, in terms of intelligibility difference compared with the centre position for the various DRT classes.

Interestingly, all cases showed that speech intelligibility was degraded, particularly from the right and left positions. The lowest degradation was for nasal sounds, perhaps because of their lower basic frequency, perception of which by humans is known to be less sensitive to positioning (due to the longer wavelength of the lower-frequency sounds).

4.2.14 Mistuning of harmonics

Two complex musical tones are perceived as separate when they have different fundamental frequencies; however, the hearing process is capable of dealing with slight mistuning of certain components, so almost-equal fundamentals can sometimes be perceived as being equal.

Again, MATLAB can be used to demonstrate this effect. The following section of code creates a complex musical chord, in this case an A_4, plus the notes a third and an eighth of an octave above this (refer to Infobox 2.6: *Musical notes* on page 45 for an explanation of the one-twelfth power used below):

```
note=440;
t1=tonegen(note, 8000, 1);
t2=tonegen(note*2^(3/12), 8000, 1);
t3=tonegen(note*2^(8/12), 8000, 1);
```

When replayed, the combined notes make an obvious and pleasant-sounding musical chord:

```
soundsc(t1+t2+t3, 8000);
```

Now we will mistune the centre note in the chord by a factor of 5% and then listen to the result:

```
m2=tonegen(note*1.05*2^(3/12), 8000, 1);
soundsc(t1+m2+t3, 8000);
```

The resulting sound should still be perceived as a fairly pleasant-sounding musical chord, but with a different quality to the correctly tuned chord. However, it is musically still compatible as a chord (to the relief of piano tuners everywhere). Try repeating the test with a larger degree of mistuning – the pleasant sound should begin to deteriorate quite quickly as the mistuning becomes worse.

In general, a slight mistuning of one harmonic will result in a perception of reduced amplitude, until the degree of mistuning becomes such that the harmonic is perceived as a tone in its own right. Again, the effects depend on duration, amplitude and absolute frequency (as well as person-to-person differences), but a rule of thumb is that 400 ms long tones must be mistuned by over 3% for them to be heard separately [48]. Note that this is not the same effect at all as the beat frequency caused by two interfering tones.

4.2.15 The precedence effect

The *Haas* or precedence effect ensures that, if similar versions of a sound reach an observer at slightly delayed times, then the observer will hear the first signal but suppress the subsequent versions [21]. This effect, only acting on signals reaching the ear within 50 ms of each other, explains why we can still understand speech in an environment containing multiple echoes (such as a small room). The first sound to reach the observer will be heard in preference to further sounds even if the secondary sounds are up to 10 dB louder.

Once echoes reach an observer with time delays of more than about 65 ms (corresponding to a distance of approximately 20 m in air), they will be perceived as being distinct echoes. For this reason, echoes are only noticeable in large rooms – they are present in small rooms but we are unable to hear them.

MATLAB can be used to construct a demo of the Haas effect. First record a word or two of clear speech into MATLAB – store it in array called `audio`, then use the following code to listen to the speech replayed with progressively longer unattenuated echoes. Assume that `Fs` is set to the sample rate used to record the speech, and that the echo starts at 10 ms, increasing in steps of 20 ms up to 100 ms:

```
audio=reshape(audio,1,length(audio));
for echo=0.01:0.020:0.1
    pad=zeros(1,fix(Fs*echo));
    input('Press any key to hear next echo');
    soundsc([audio,pad]+[pad,audio],Fs);
end
```

Note that the `reshape()` function is used to ensure that the audio vector is $(1 \times N)$ rather than $(N \times 1)$ in dimension, so it can be combined with the variable sized padding.

One useful rule of thumb when working with speech is that echoes of 100 ms or longer in two-way conversational speech, such as during a telephone call, will become annoying and distracting to callers. This is a major quality factor for telephony systems, and presents a hard-to-meet target for IP-based Internet telephony systems such as Skype which use digital processing and relatively slow or congested communications networks.

4.2.16 Speech perception

Evidence exists to indicate that the human aural system processes speech in a completely different way from other non-similar sounds [12]. One experiment which uses sinewave speech to demonstrate this was discussed in the introduction to Chapter 3. Whatever the exact cause, certain brain mechanisms cause a listener to perceive two sentences of speech as being similar when the physical differences between them may be very large (for example, same speaker, same words, different acoustic situation). Two examples of one speaker saying the same words in two different environments would be judged by listeners as similar, irrespective of possibly major physical differences in amplitude, timing, pitch and so on. Speech can be detected and understood when the noise-to-signal power is such that single tones would be inaudible.

MATLAB can be used to demonstrate that speech is processed in a different way to general sounds. One fairly common example is to see how easy it is to understand piecewise backwards speech. First of all, record (or load) some speech – just a sentence should be sufficient. Let us assume that the speech array is named speech, and was recorded with sample rate Fs.

We can now slice the speech into 0.1 s blocks, reversing the content of each block completely, before replaying the sequence of reversed blocks:

```
soundsc(speech,Fs)
ls=length(speech);
ws=Fs*0.1;
s2=[];
for i=1:floor(ls/ws)
   s2=[s2;speech(i*ws:-1:1+(i-1)*ws)];
end
%now listen to the reversed segments
soundsc(s2,Fs)
```

Most interestingly, although every part of the waveform has been reversed, the speech should still be intelligible. This is because it has been divided into roughly phoneme-sized blocks, and, while each phoneme is replayed backwards within the block, the overall sequence of phonemes is preserved (Figure 4.8).

The speech certainly sounds strange – its quality has degraded substantially – but the intelligibility remains high. Try experimenting with different block sizes. You may find that blocks smaller than about 100 ms do not degrade intelligibility, but, as blocks

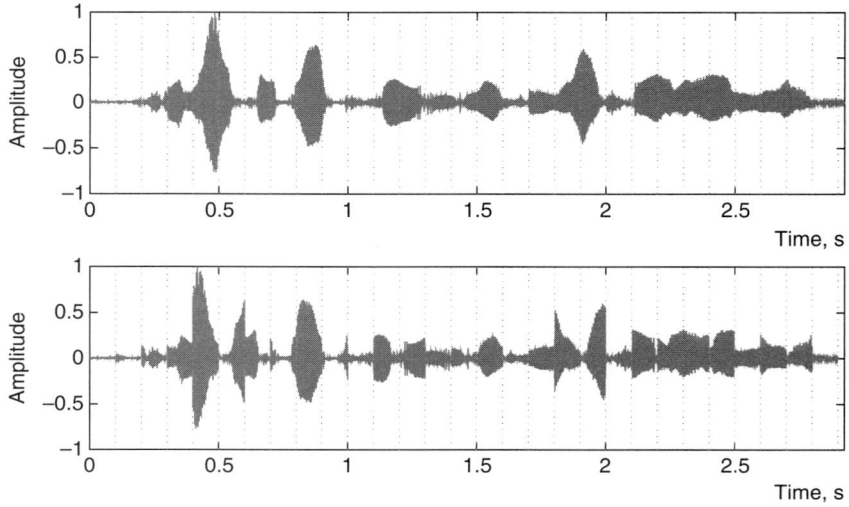

Figure 4.8 A speech waveform (top) and a version with reversed 100 ms segments.

become bigger than 100 ms, intelligibility decreases. In fact, this provides very good evidence that speech is processed in roughly phoneme-sized units, and that within those units it is primarily the mean spectral content and mean amplitudes that convey lexical information, but between units the sequencing is based in the time domain.

Of course, we could reverse the entire recording of speech:

```
soundsc(speech(1s:-1:1),Fs)
```

Unless you are very fortunate, this will be unintelligible – however, conspiracy theories abound regarding secret 'backwards messages' embedded in popular music from The Beatles onwards. If so, it is highly likely that The Beatles failed to anticipate the ease with which speech recordings would be able to be reversed (also listened to, and processed in many other ways) using tools like MATLAB.

Despite the evidence suggesting that speech processing in the brain is separate from sound processing, speech perception still suffers from simultaneous and non-simultaneous masking, binaural masking and auditory adaptation (indeed this may be more pronounced: monotonic speech can lead rapidly to drowsiness or complete lack of attention). Speech perception interlinks speech production with the psychological aspects of the communication channel, the subject matter and the state of mind of the recipient [21].

4.3 Amplitude and frequency models

Attempts to classify and regularise the human hearing system have been made for many years. Most of these have revolved around the analysis of frequency and amplitude effects, and in particular the measurements of loudness and frequency selectivity discussed below.

4.3.1 Loudness

Backtracking from the equal loudness contours of Section 4.2.2, it is fairly evident that humans do not measure the concept of loudness in the same way as a physical measurement. From our discussions previously we have seen that people interpret amplitude approximately logarithmically.

In fact, several researchers indicate that perceived loudness is proportional to physical amplitude raised to the power of 0.3 [12]. It is likely that there are upper and lower bounds where the relationship no longer holds true, nevertheless across the main portion of the hearing range the relationship is defined by a constant multiplier that varies from person to person. The measure of loudness is the *sone*, although this is rarely used in current literature. One sone is the loudness of a 1 kHz tone at 40 dB$_{SPL}$, with every 10 dB increase in amplitude increasing the loudness by one sone (i.e. a 50 dB$_{SPL}$ 1 kHz tone has a loudness of two sones).

4.3.2 The Bark scale

Since the hearing process is often considered to derive from some set of bandpass filters, and the resultant processing within these, researchers attempted to identify and characterise these filters [9, 12, 22]. Known as *critical-band filters*, each has similar shape but different bandwidth, centre frequency and amplitude weighting. The different amplitude weightings contribute to the sone loudness scale (Section 4.3.1), as well as the equal loudness contours of Section 4.2.2 – both of which originally described the loudness of single tones only.

More generally, the loudness and frequency selectivity of sounds depend on the bands within which they fall. In particular, this mechanism can be used to explain the masking effects of Section 4.2.9: a loud tone in one critical band will occupy that band and prevent a quieter tone in the same band from being heard. Moving the quieter tone gradually away from the louder tone until it enters a neighbouring critical band will result in it becoming audible once more. The *Bark*[3] scale is one way to express this relationship [49].

The first few lower critical bands are shown in Table 4.4, where they are numbered according to the Bark scale. The scale is arranged so that a unit change in Bark corresponds to a perceived unit change in frequency effect by listeners. It is therefore a psychoacoustically relevant frequency scale, but as always we need to remember that this table describes the mythical 'standard human'. Real-life listeners will all have slightly different hearing characteristics.

Table 4.4 Critical bands and corresponding centre frequencies.

Critical band (Bark)	Lower cutoff frequency (Hz)
1	100
2	204
3	313
4	430
5	560
6	705
7	870
8	1059
9	1278
10	1532
11	1828
12	2176
13	2584
14	3065
15	3630

---

[3] Since the scale is named after a person, Heinrich Barkhausen, for his early work on loudness perception, the name Bark should be capitalised to distinguish it from the sound a dog makes.

Denoting a Bark unit as Ω, and the angular frequency as ω, then Hermansky [50] defines the Bark in the following way:

$$\Omega(\omega) = 6 \log\left(\omega/1200\pi + \sqrt{(\omega/1200\pi)^2 + 1}\right). \tag{4.1}$$

Simple MATLAB functions to convert bidirectionally between frequencies in Hz and Bark are given below:

```
function [bark]=f2bark(hz)
   cn=2*hz/1200;
   bark=6*log(cn+(cn^2+1)^0.5);
```

```
function [hz]=bark2f(bark)
   hz=600*sinh(bark/6);
```

One final note of caution is to beware of alternative definitions of the Bark. To the author's knowledge, there are at least three separate definitions of the simple mapping between Hz and Bark in use by research authors worldwide. Exactly which one is in use is not particularly important, since all are relative, all map to real frequencies with a similar shaped representation, and all are approximate. However, it is critically important to be consistent and not combine or confuse the different definitions. We will make use of the Bark mapping again later to form equal-Bark frequency regions for evaluating the perception of sounds, as an alternative to the possibly more common mel frequency scale shown in the following section.

4.3.3 The mel frequency scale

Like the Bark scale discussed in Section 4.3.2, the mel frequency scale is often used today to account for the frequency selectivity within the human auditory system. It attempts to model the non-linearity in a subjective way, derived by assessing pitch perception. In experiments originally published in 1937 [51], a small group of musicians was used to create an equal-difference pitch scale, called the mel scale (an abbreviation of the word 'melody'). The expert listeners were asked to fix tones an equal perceived distance apart, to form a musical scale which might be considered these days to relate to the size of critical bands. Many others have refined this work since, most notably by making use of just-noticeable difference (JND) in pitch, which may be more reliable and universal.

As with the Bark scale, there are also quite a few alternative methods around to convert between Hz and mel scale, although in general they all agree at least that 1 kHz equates to 1000 mels. A simple mapping is that attributed to the popular hidden Markov toolkit (HTK), derived from original research by O'Shaughnessy [52], which computes the mel frequency, m, from the frequency in Hz, f, as follows:

Table 4.5 Mapping between frequency in Hz and mel scales from Equations (4.2) and (4.3).

Frequency (Hz)	mel scale (Eq. 4.2)	mel scale (Eq. 4.3)
100	150.5	137.5
200	283.2	263.0
400	509.4	485.4
600	697.7	678.1
800	858.9	848.0
1000	1000.0	1000.0
1400	1238.1	1263.0
1800	1434.6	1485.4
2000	1521.4	1585.0
2500	1712.8	1807.4
3000	1876.5	2000.0
4000	2146.1	2321.9
6000	2545.6	2807.4
10000	3073.2	3459.4
12000	3266.3	3700.4

$$m = 2595 \times \log_{10}(1 + f/700). \tag{4.2}$$

Table 4.5 shows a few mel-to-frequency mappings using this equation, as well as those from another popular definition attributed to G. Fant [53], in Equation (4.3):

$$m = (1000/\log(2)) \times \log(1 + f/1000). \tag{4.3}$$

Fant's equation is interesting because it does not depend upon the type of logarithm used, and so it could be computed using an extremely computationally efficient base-2 logarithm if desired.

Simple MATLAB functions for computing the HTK version of the equation (which is probably the most frequently used) are given below:

```
function [mel]=f2mel(hz)
   mel=2595 * log10(1+hz/700);
```

```
function [hz]=mel2f(mel)
   hz=700*(10^(mel/2595)-1);
```

The mel scale is very commonly used to create a mel-filterbank for mapping sounds to critical bands defined by the mel scale. This will be explored (along with a Bark version) later.

4.4 Summary

In this chapter we studied human hearing, beginning with the physical structure of the ear, and the processes occurring within it. We then discussed several counter-intuitive aspects of hearing, where the human auditory system fails to behave in the predictable way that an examination of the physics of sound would suggest. This departure from a purely physical interpretation of sound, named psychoacoustics, will be explored further in the next chapter, which will in turn set the scene for Chapter 6 where we begin to make use of many of the features we have discussed for speech analysis and communications topics.

Bibliography

- *An Introduction to the Psychology of Hearing*
  B. C. J. Moore (Academic Press, 4th edition 1997)
  This book is highly recommended as an introduction to hearing research, in any of its editions. It is a well-written and engaging scholarly work, covering almost all aspects of hearing which may relate to computer handling of speech and music. It does not discuss any processing topics, and mentions few computer-related practicalities, since it concentrates on human experience, psychology and physiology.

- *The Handbook of Hearing and the Effects of Noise*
  K. D. Kryter (Academic Press, 1994)

- *Hearing (Handbook of Perception and Cognition)*
  B. C. J. Moore (Academic Press, 2nd edition 1995)

- *Speech Intelligibility and Speaker Recognition*
  Ed. M. E. Hawley (Halsted Press/Dowden Hutchinson and Ross, 1977)

- *The Psychophysics of Speech Perception*
  Ed. M. E. H. Schouten (NATO Science Series D, Martinus Nijhoff Publishers, Springer, 1987)

- *Auditory Scene Analysis: theory and demos*
  A large number of downloadable ASA demonstration recordings with explanations, and a fair amount of background information, is available at http://webpages.mcgill.ca/staff/Group2/abregm1/web/asa.htm

- *Auditory Scene Analysis*
  A. S. Bregman (MIT Press, 1990)

Questions

Q4.1 Which sounds loudest to the 'average' human, a 1 kHz tone at 40 dB$_{SPL}$ or a 10 kHz tone of the same amplitude?

Q4.2 Explain how the principle of frequency masking can make some sounds seem to disappear.

Q4.3 Can we hear a quiet stereo tone replayed in noise better when the same noise is heard in both ears, or when different noise is heard in the two ears?

Q4.4 Roughly how much frequency mis-tuning of notes can the average listener perceive, 3%, 13% or 23%?

Q4.5 Using either the Bark formula or the MATLAB `f2bark()` function, what would 440 Hz (A_4) and 493.9 Hz (B_4) correspond to in Barks? Repeat the calculation using the mel formula or the `f2mel()` function to find the same note on the mel scale.

Q4.6 Name the type of short-term reduction in hearing that listeners may experience after being suddenly exposed to a very loud sound.

Q4.7 Repeat the experiment of Section 4.2.16 where speech blocks of size 100 ms are reversed, but this time try flipping the amplitude (i.e. multiply each block by the constant value -1). Does this appreciably change the intelligibility of the speech?

Q4.8 An audio sound meter contains a switch with three positions marked 'A-weight', 'C-weight' and 'no-weight'. Which setting would you choose when (a) measuring loud rock music in order to complain to your city council, (b) comparing the sound produced by a robin and a blackbird to decide which has the strongest singing voice?

Q4.9 What is orthotelephonic gain, and how does that (among many other factors) change what you would hear between a live concert and a recording of the event you listen to later?

Q4.10 If I listen simultaneously to two pure tones of frequencies 100 Hz and 120 Hz, I might hear another sound being produced. If I do hear something, what would its likely frequency be?

5 Psychoacoustics

If there is one topic that has most deeply impacted audio and speech research over the past two decades or so, it is psychoacoustics (by contrast, the deepest impact in audio and speech *engineering* has probably been big data – explored separately in Chapter 8). We now know that perceived sounds and speech owe just as much to psychology as they do to physiology. The state and activity of the human brain and nervous system have a profound influence on the characteristics of speech and sounds that are perceived by human listeners.

It is definitely beyond the scope of this book to delve into too much detail concerning the psychological reasons underpinning psychoacoustics – and indeed much of that detail remains to be discovered – but we will discuss, demonstrate and uncover many interesting and useful psychoacoustic phenomena in this chapter. Extensive experiments by cross-disciplinary researchers over the past two decades have allowed computational models to be developed to begin to describe the effects of psychoacoustics. While these models vary in complexity and accuracy, and continue to increase in quality and usefulness, they have already found applications in many areas of daily life. The following sections will overview many of the effects that the models can (or could) describe. We will take a fascinating look at auditory scene analysis (which includes a number of auditory-illusion style demonstrations), before building and applying our own phsychoacoustic models.

5.1 Psychoacoustic processing

The use of psychoacoustic criteria to improve communications systems, or rather to target the available resources towards more subjectively important areas, is now common. Many telephone communications systems use A-law compression. Around 1990, Philips and Sony respectively produced the DCC (digital compact cassette) and the MiniDisc formats which both make extensive use of equal-loudness contours and masking information to compress high-quality audio [54]. Whilst neither of these was a runaway market success, they introduced psychoacoustics to the music industry, and paved the way for solid state music players such as the Creative Technologies Zen micro, Apple iPod and various devices from other innovative companies.

Most of these devices use the popular MP3 compression format, although more recent formats, such as Ogg Vorbis, MP4 and various proprietary alternatives such as WMA also exist (refer to Infobox 2.2 on page 15 for descriptions of these). It is important to note that all of these music compression methods have something in common: they all use psychoacoustic criteria to reduce audio file sizes whilst maintaining quality. All take account of masking thresholds and consider harmonic relationships between sounds, and some also exploit binaural masking effects.

In the remainder of this section, several promising psychoacoustic techniques with potential for use in speech and music compression are discussed. Not all of these have yet been exploited commercially, and few are in mainstream research, but all hold future promise.

5.1.1 Tone induction

Knowledge of the residue effect (Section 4.2.3.2) allows one to induce a low-frequency tone by the addition of higher-frequency tones. Only three or so high-frequency tones are required, and the technique is useful where it would otherwise be necessary to directly add an extremely loud low-frequency tone in order for it to be heard above a low-frequency noise. One application of this is in the construction of an artificial speech formant (formants apparently being regarded by Klatt, a leading audio and speech researcher, as the most important aspect of speech recognition [12]). Despite sounding like a good idea, this application of tone induction is yet to be substantiated. Critical to its success is the location within the auditory cortex of the mechanisms causing the residue effect. If these mechanisms occur posterior to (after) the mechanisms of speech recognition, then formant induction is unlikely to be successful, and the listener would simply hear speech degraded by high-frequency tones, but if they lie anterior to (before) the specialised speech circuitry in the brain, then there is potential for them to strengthen speech recognition in noise. Another application is in the induction of bass notes using small speaker cones. Generally, for low-frequency bass notes to be reproduced, a large loudspeaker cone is needed. However, such items are expensive, unwieldy and power hungry – so why not use higher-frequency sounds to induce the low-frequency tones? This method could potentially be used to allow smaller speakers to produce the kind of sound normally requiring larger, more costly, speakers.

5.1.2 Sound strengthening

As stated in Section 4.2.3.2, adding harmonics to a complex tone, or adding geometrically related tones, does not change the perceived frequency of a sound, but does change the amplitude (firstly by allocating more signal power to the sound, and secondly by spreading the components into more critical bands). This relationship allows the 'strengthening' of a sound by the addition of related frequency components without causing undue concentration of energy at specific frequencies, as would be the case with direct amplification of those fundamental frequencies to achieve a perceptually similar result.

This technique may allow a speech formant obscured by localised noise to be perceptually strengthened [32].

5.1.3 Temporal masking release

Post-stimulatory temporal masking (see Section 4.2.11), causing sounds occurring just after a loud tone or pulse to be perceived as being less loud, may be negated by detecting such events and applying an initial increase, followed by a decaying level of amplification after it. A knowledge of the exact amplitude at the ear of a listener is probably a necessary requirement for this processing. In a similar way for much louder sounds, allowances may be made for TTS (temporary threshold shift – see Section 4.2.7).

5.1.4 Masking and two-tone suppression

If a wanted tone lies close to a masking tone of higher amplitude, it is known that the wanted tone will be masked (see Section 4.2.9). To alleviate this situation, the wanted tone could be amplified, strengthened by the addition of harmonics, or simply shifted slightly. If the wanted tone is harmonically related to other tones, we know that the shift may not be more than about 3% [12]. Techniques already exist for predicting the effects of masking for the purpose of enhancement, many of which have been reported to perform fairly well [20, 31, 46, 49, 55].

5.1.5 Use of correlated noise

Remember that modulated noise, present in more than one critical band, allows the auditory system to use a correlation process to reduce its effects (refer to Section 4.2.10). In this way, we can conjecture that a frequency shifted, reduced amplitude version of that noise induced elsewhere in an unimportant part of the frequency spectrum should reduce its effect. Does that mean that it is possible for a speech system to deliberately *increase* the amount of background noise (by spreading it wider in frequency) and yet improve intelligibility as a result? Only detailed experimentation would tell us for sure.

5.1.6 Binaural masking

As noted in Section 4.2.13, judicious control of the signal fed to each ear of a listener may help to improve the perceived signal-to-noise level. For speech systems contaminated with additive white Gaussian noise, it appears that applying out-of-phase speech to each ear, whilst subject to the same noise field, can reduce the effect of the noise.

5.1.7 Summary

All of the techniques mentioned in this section are aspects of the human auditory system that might conceivably be exploited for future processing, compression, analysis or synthesis systems. However, much research and experimentation remains before some of these techniques are fully understood or modelled. By contrast, the following section relates a number of psychoacoustic principles that are better understood and can be quite easily demonstrated using MATLAB.

5.2 Auditory scene analysis

Auditory scene analysis (ASA) [56] describes the psychoacoustic and psychological processing by the human auditory system that it uses to deal with and interpret complex mixtures of sound. This topic is clearly the result of some type of analytical process occurring within the brain.

As an illustration, consider an environment when a person talking, a dog barking and a police siren are heard concurrently. Each of these sounds arrives at the ears of a listener simultaneously, and they are thoroughly mixed together. A computer, analysing this sound mixture, would have great trouble in determining that there were three separate sound sources, and, even if it could determine this, may not readily be able to 'tune in' to each of those sounds in the way a human can.

This tuning-in process comes into play when participating in a conversation in a crowded room filled with multi-speaker babble. Tens or even hundreds of separate conversations may be simultaneously in progress, and yet most people will be able to tune out much of the surrounding babble and conduct a conversation with the person next to them. This is true even when the particular speech they are listening to is of lower amplitude than the interfering sounds.

Imagine yourself in such a situation. You are straining to hear the person opposite you who is barely at the edge of audibility, and yet somehow a conversation is possible in all that noise. It is true that occasionally a particular remark or name, or perhaps an overloud laugh from a nearby speaker, may capture your attention, thus intruding on your conversation and prompting a request to repeat something. You would probably turn your head directly to face them (refer to the experiment in Section 4.2.13), as well as leaning closer. However, somehow, you are able to tune in to the particular wanted conversation while ignoring louder conversations occurring around you and other background noises.

Sometimes visual clues help to maintain that focus – if you can see the mouth of the person speaking, the movement and shape of the mouth are interpreted by the human visual system, providing clues to the auditory system (another reason to turn and face the speaker). This was illustrated in a famous experiment in which listeners were played /b/ and /m/ sounds whilst simultaneously watching videos of people saying /b/ or /m/.

When the correct video was played simultaneously with a replayed sound, listeners correctly identified the spoken phoneme. However, when the video of /b/ was replayed alongside the audio for /m/, most listeners reported that they had heard /b/. The image of the mouth closure at the start of the /b/ sound was sufficient to unconsciously convince listeners that they had heard /b/ rather than /m/. This is known as the McGurk effect, and is an example of the human visual system assisting (and indeed overriding) the human auditory system (HAS).

Interested readers may wish to refer to an article in *New Scientist* of 22 September 2007, 'Mind tricks: six ways to explore your brain', in which a McGurk effect experiment and several similar experiments are described.

All of these are illustrations of the way in which the HAS deals with, or interprets, sounds. In order to begin characterising these effects, researchers have identified several

5.2 Auditory scene analysis

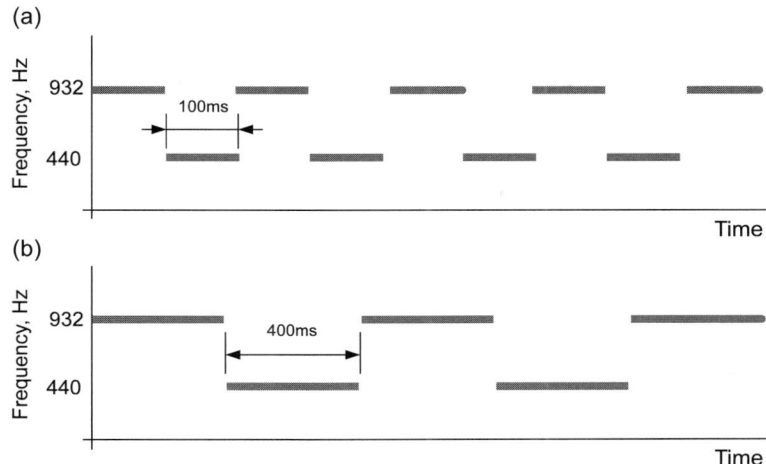

Figure 5.1 Illustration of the feature of proximity. Two sequences of alternating tones are shown. In plot (a) the absolute time difference between like tones is short, thus they exhibit closer proximity compared with the larger time difference between like tones in plot (b).

basic principles [56, 57] which relate complex sounds and sequences to perception, and which we will consider in the following subsections.

5.2.1 Proximity

Sounds of similar proximity are those that are close in terms of amplitude, pitch, duration, timbre and so on. There is definitely a tendency for the brain to classify similar sounds as belonging to the same parts of an auditory scene.

For example, when listening to a recording of a wind quartet, the brain would probably be correct in assuming that each oboe-like sound played in a sequence originates from the same oboe. If, in fact, two oboes are playing, then that may only become evident when both play different notes simultaneously.

We shall use MATLAB code to generate two sound sequences, each consisting of two alternating tones. The first sequence plays the tones slowly, the other sequence plays them quickly. The sequences are illustrated in Figure 5.1 showing a closer proximity in the upper plot than in the lower, due to the increased gap between similar frequencies in the latter. A rather simple MATLAB demonstration, using the `tonegen` function of Section 2.7.1, can illustrate this point:

```
ss=0.1;      %short sound length in seconds
ls=0.4;      %long sound length in seconds
Fs=8000;     %sample rate
short_a=tonegen(440,Fs,ss); %musical note A4
short_b=tonegen(932,Fs,ss); %musical note B5
long_a=tonegen(440,Fs,ls);
```

```
long_b=tonegen(932,Fs,ls);
%put alternating sounds into single matrix
short_mat=[short_a, short_b];
long_mat=[long_a, long_b];
%Repeat this matrix several times
long=repmat(long_mat,1,3);
short=repmat(short_mat,1,12);
```

The two tone sequences reside in vectors long and short respectively. Let us begin by listening to the long duration tones:

```
soundsc(long, Fs);
```

Most likely this is heard as a single sequence consisting of two tones. However, when listening to the faster sequence of short duration tones, it is more likely that listeners notice two streams of sound – one high frequency, and one low frequency, but both are pulsed:

```
soundsc(short, Fs);
```

To explain, both sequences exhibit no gaps between tones, thus there is close temporal proximity between an A_4 and the following B_5. Although there is not a great pitch proximity between the A_4 and the B_5 (which is slightly over an octave higher), in the slower sequence, the close temporal proximity dominates, leading the brain to associate the tones together into one slow string.

In the faster sequence, however, the proximity of the similar notes is increased – they are now only 100 ms apart – and thus the proximity of one A_4 to the next is increased over the slow sequence. Likewise, each B_5 is nearer to the next one. In this example, the close tonal proximity dominates, and thus two separate streams are heard. One is of beeping A_4 notes and one is of beeping B_5 notes.

5.2.2 Closure

The closure principle allows the brain to fill any auditory 'gaps' to match its expectations regarding continuity, origin and so on. This is also known as auditory induction since our brains do not just *deduce* the presence of missing audio, but actually *induce* a replacement to cover the missing gaps. It seems our brains dislike sounds that do not fit into an ongoing scene interpretation, to the extent that additional material will be created to bring the user perceived audio in line with brain expectations. The same process is involved with images, and is the basis for many optical illusions, from impossible three-dimensional shapes to discovering hidden faces within abstract images.

In audio terms, this process of filling in expected information happens commonly during speech communications, especially when carried out in the presence of interfering

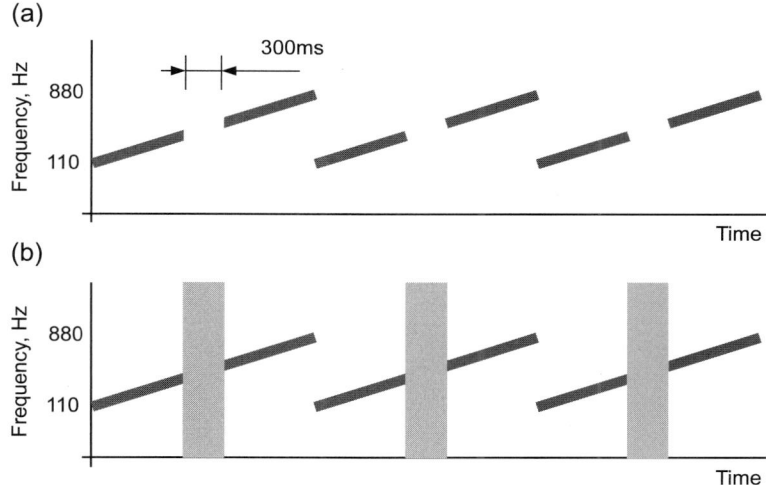

Figure 5.2 Illustration of the feature of closure using steadily rising tonal frequencies. In the first sequence (a), these are interrupted by gaps, whereas in the second sequence (b), the rising tones are interrupted by white noise.

noise. At a higher language level, this would be one of the reasons why context is so important to speech intelligibility (discussed in Section 3.4.4).

In MATLAB, we can demonstrate this effect by again using the freqgen function of Section 2.7.4 to create a sound consisting of rising tones, and then to insert either gaps or noise portions into the sound. The pattern created is shown in Figure 5.2, using the following MATLAB code:

```
gs=0.30;      %gap/noise length in seconds
ls=1.50;      %length of sound in seconds
Fs=8000;      %sample rate
fr=110;
to=880;
gap_len=Fs*gs;
au_len=Fs*ls;
gap=zeros(1,gap_len);
noise=rand(1,gap_len);
%Make a steadily rising note
note_f=[fr:(to-fr)/(au_len-1):to];
au=freqgen(note_f,Fs);
au_gap=au;
au_noise=au;
%Now put the gaps 1/2 way up
au_gap(au_len/2+1:au_len/2+gap_len)=gap;
au_noise(au_len/2+1:au_len/2+gap_len)=noise;
%Now repeat several times
```

```
au_gap=repmat(au_gap,1,3);
au_noise=repmat(au_noise,1,3);
```

Replaying the noisy version first, listeners will hear a steadily rising sinewave, moving from 110 Hz to 880 Hz. During each of the rising periods, there will be a period of loud white noise. The interpretation by most listeners is that the sinewave is continuous, but parts of it cannot be heard because of the noise. Try and decide whether you think the sinewave is continuous:

```
soundsc(au_noise, Fs);
```

By contrast, when replacing the noise periods by a gap of silence, the discontinuity in the sinewave is immediately obvious:

```
soundsc(au_gap, Fs);
```

By and large, humans are fairly sensitive to discontinuities – whether audio, visual (edges) or even tactile – and thus the fact that two discontinuities can be hidden by white noise of the same average amplitude is significant. As a final check, we can compare how audible the sinewave would really be if it were replayed simultaneously with the noise instead of being replaced by the noise:

```
soundsc(repmat(au,1,3)+au_noise, Fs)
```

The answer is that the sinewave is clearly audible even in the presence of that noise. The HAS interpreted the first playback as being a continuous sinewave, even when the sinewave was not audible, but would have been were it actually present. This is clearly an illustration of the brain filling in gaps in its interpretation of the events which it witnesses.

Interestingly, GSM and several other speech coders introduce a sound known as *comfort noise*, a pseudo-random white noise sound, at times when silence would otherwise be present. This is done to improve the perception of listeners, based on observations which showed that listeners were discomforted by periods of total silence during a voice communications session. Perhaps, in line with our experiment to test the effects of closure using periods of noise, another advantage of comfort noise is the ability to 'blind' the HAS to otherwise disturbing and obvious waveform discontinuities in a speech coder output.

5.2.3 Common fate

When groups of tones or noises start, stop or fluctuate together, they are more likely to be interpreted as being part of a combined sound or at least having a common source.

5.2 Auditory scene analysis

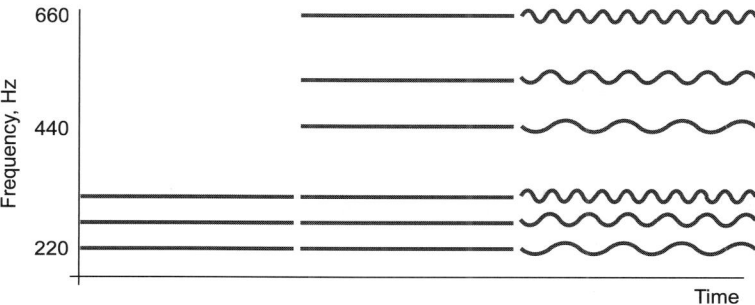

Figure 5.3 Illustration of common fate by comparing three chords. The first generates a pleasing note from three related sinewaves of different frequency. The second adds a first harmonic to each of the three fundamentals. The third modulates each of the three tones plus their first harmonic with unrelated modulating frequencies.

As an example, a human voice actually consists of many component tones which are modulated by lung power, vocal tract shape, lip closure and so on, to create speech. It is likely that the ability to cluster sounds together by 'common fate' is the mechanism by which we interpret many sounds heard simultaneously as being speech from a single mouth. In noisy environments, or in the case of multi-speaker babble, common fate may well be the dominant mechanism by which we can separate out the sounds of particular individuals who are speaking simultaneously.

The same effect, and outcome, can be heard in an orchestra, where several musical instruments, even playing the same fundamental note simultaneously, can be interpreted by the HAS as being played separately (and the more different the instrument is, the more unique is the modulation, and thus the easier they are to distinguish; for example a violin and a bassoon are easier to distinguish than two bassoons playing together).

Let us attempt to demonstrate this effect in MATLAB by creating three sets of combined audio notes for playback, as illustrated in Figure 5.3. First of all we will define three sinewave frequencies of a, b and c, related by a power of $2^{(1/12)}$ so that they sound pleasant (see Infobox 2.6 on page 45 which describes the frequency relationship of musical notes). We will then use the tonegen function of Section 2.7.1 to create three sinewaves with these frequencies, and also three sinewaves of double the frequency:

```
dur=1.2;
Fs=8000;
a=220;
b=a*2^(3/12);
c=a*2^(7/12);
sa=tonegen(a, Fs, dur);
sb=tonegen(b, Fs, dur);
sc=tonegen(c, Fs, dur);
sa2=tonegen(a*2, Fs, dur);
sb2=tonegen(b*2, Fs, dur);
sc2=tonegen(c*2, Fs, dur);
```

Next, two sounds are defined which in turn mix together these notes:

```
sound1=sa+sb+sc;
sound2=sound1+sa2+sb2+sc2;
```

Now, three different modulation patterns are created, again using the `tonegen` function, but at much lower frequencies of 7, 27 and 51 Hz, chosen arbitrarily to not be harmonically related either to the original tones or to each other. These are then used to modulate the various original tones:

```
mod1=tonegen(7, Fs, dur);
mod2=tonegen(27, Fs, dur);
mod3=tonegen(51, Fs, dur);
am=mod1.*(sa+sa2);
bm=mod2.*(sb+sb2);
cm=mod3.*(sc+sc2);
```

Finally, a short gap is defined for the third sound to accentuate the common starting point of three sound components, which are combined into `sound3` and finally placed into a composite vector for replaying:

```
gap=zeros(1,Fs*0.05);
sound3=[am,gap,gap]+[gap,bm,gap]+[gap,gap,cm];
soundsc([sound1,sound2,sound3], Fs)
```

A listener exposed to the sounds created should hear, in turn, three segments of audio lasting approximately 1.2 s each. Firstly a pleasing musical chord consisting of three harmonics is heard. Then this musical chord is augmented with some harmonically related higher frequencies. For both of these segments, the perception is of a single musical chord being produced. However, the final segment is rather discordant, and appears to consist of three separate audible components.

Because each of the three harmonic notes from the second segment is now modulated differently, the brain no longer considers them as being from the same source, but rather from different sources. It thus separates them in its perceptual space.

To really show that this interpretation is not simply a side-effect of the small gap that was inserted, the reader is encouraged to repeat the experiment, but modulate bm and cm in the same way as am, namely:

```
am=mod1.*(sa+sa2);
bm=mod1.*(sb+sb2);
cm=mod1.*(sc+sc2);
```

With all subsequent steps repeated as previously, the listener will hear, instead of a discordant mixture of three notes, a warbling single chord. This is to be expected since now each of the three harmonic notes shares a common fate (a common modulation).

5.2.4 Good continuation

Generally, sounds in nature do not start, stop or switch frequency instantaneously. There will normally be some amplitude attack at the beginning (ramping up from zero) and decay (ramping down to zero) at the end of a sound. Similarly, sounds in nature normally do not just switch frequency instantaneously, they slide either quickly or slowly.

Both of these properties are due to the physical methods of creation of such sounds. Changes in frequency are caused by changes in the shape or geometry of the producing object, and such changes would be driven by muscles in humans and animals, which move relatively slowly and gradually in audio terms, so that at least some intermediate frequencies will be present between two extremes. Similarly, sounds are conveyed by the movement of air, and air movements can be sharp but not completely instantaneous. They must first build up, and later die away.

So, when presented with complex sets of sound, the human brain tends to classify sounds that are connected in some way by these gradual changes as coming from a single source. By contrast, sounds which do not blend together are more likely to be regarded by the brain as coming from separate sources. Researchers have termed this effect the good continuation of sounds, although it could perhaps better be described as being the *connectedness* of individual sounds.

Musical instruments are similar. In particular, their notes always exhibit some form of amplitude attack and decay; however, notes do not always glide into one another.

Computers provide us with the ability to generate almost any arbitrary set of sounds, with whatever features we may require. Computer-generated sounds can start, stop and switch frequency instantaneously if we wish, or could be tailored to show some connectedness.

In the following MATLAB code, we will aim to demonstrate the effect of the good continuation property by constructing two sequences of sound. The first, shown in Figure 5.4(a), has two alternating notes joined by a sliding frequency, or *glissando*, while the second, in plot (b), presents the same two notes without any glissando.

Listeners of both sequences should consider that the first could conceivably be produced by a single source which slides upwards and downwards in frequency. The second, by contrast, is more likely to be interpreted as two separate, single frequency sound sources that alternate in producing a sound.

In MATLAB, first we will produce the two note sequences, sa and sb, using the freqgen() function, with the latter sequence exhibiting the glissando:

```
Fs=8000;
n1=832;
n2=350;
```

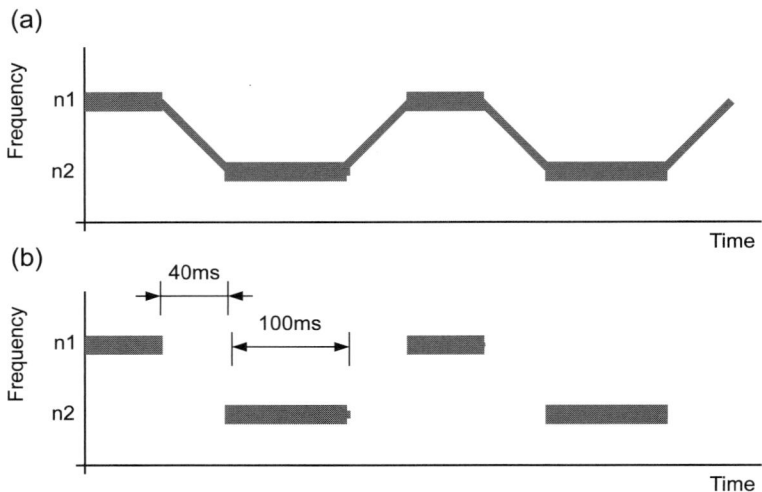

Figure 5.4 Illustration of the feature of good continuation between notes in plot (a) which is missing in plot (b).

```
d1=0.1*Fs;
dm=0.04*Fs;
d2=0.1*Fs;
a=[n1*ones(1,d1), zeros(1,dm), n2*ones(1,d2), zeros
    (1,dm)];
b=[n1*ones(1,d1), n1-[1:dm]*(n1-n2)/(dm), n2*ones(1,d2),
    n2+[1:dm]*(n1-n2)/(dm)];
sa=freqgen(a,Fs);
sb=freqgen(b,Fs);
```

We could plot the two frequency tracks, a and b, and see that they are very similar to the plots of Figure 5.4. Next we will perform some amplitude shaping by amplifying the notes at the expense of the glissando. This helps to present the glissando as a secondary feature rather than as the focus of the sounds:

```
amp=0.4+sign(a)/2;
sa=sa.*amp;
sb=sb.*amp;
```

Finally, we can replay several repetitions of the discontinuous sound:

```
soundsc(repmat(sa,1,8))
```

and follow that with the connected sounds:

```
soundsc(repmat(sb,1,8))
```

5.2.5 Summary

This section has delved deeper into the area of psychoacoustics through several examples. Hopefully these served as interesting illustrations of the ways in which the human auditory system perceives sound in ways that are not purely representative of the basic physical signals, but rather bring in higher brain functions to begin interpreting audio.

In the next section, we will begin to see how we can use some of the aspects of psychoacoustics. In particular, we will develop a working psychoacoustic model. While this might appear complex at first glance, it actually barely scratches the surface of what is possible with psychoacoustics.

5.3 Psychoacoustic modelling

Remember back in Section 4.2.1 where we claimed that this marriage of the art of psychology and the science of acoustics was important in forming a link between the purely physical domain of sound and the experience of a listener? In this section we will examine further to see why and how that happens in practice.

It follows that a recording of a physical sound wave – which is a physical representation of the audio – contains elements which are very relevant to a listener and elements which are not. At one extreme, part of the recorded sound may be inaudible to a listener (e.g. too high or low in frequency, or too quiet). Equally, it is possible that a recording of sound does not contain some of the original audible features. This may be one reason why many listeners would prefer to spend a sum of money listening to live music rather than an equivalent sum to purchase a compact disc (CD) which allows them to listen to the music again and again, or download an MP3 version which they may be able to listen to for a period of time.

Psychoacoustics as a study is predominantly computer-based these days: it rarely considers information which has not been recorded to computer, but is often used to identify parts of recorded information which are not useful to listeners.

One use of psychoacoustics is illustrated simply in Figure 5.5 showing a time-domain waveform with something like a sudden loud pop occurring part way through. A listening human ear might suffer from post-stimulatory masking (see Section 4.2.11) if the pop is a very loud sound: quieter sounds following soon after this will tend to be inaudible.

Figure 5.5(a) shows a detected masking region (with sound falling in the shaded region being inaudible). Any speech processing system capable of detecting this region could practically strip out all audio within that region, as in Figure 5.5(b), resulting

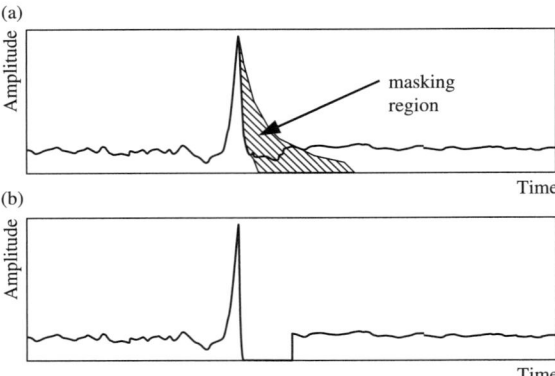

Figure 5.5 Illustration of post-stimulatory masking in the time domain detected (a) and inaudible information stripped out (b).

in something that sounds identical to the original, but could be smaller in its storage requirements (or transmission bandwidth). In real life, the pop sound could be the musical drum beat, part of the rhythm, as well as many numerous smaller instances where musical notes are played. Throughout a typical piece of music, there are instances where sounds might be inaudible. Each instance is an opportunity to reduce the bandwidth/storage requirements of that piece of audio.

For speech compression systems, it is an enormous advantage to identify and exploit such regions: why waste energy and time coding, or waste data space in storing, or waste bandwidth in transmitting information which is inaudible? The popular MP3 format automatically detects many such inaudible regions in original music, and strips them out, compressing only the remaining audio. This mechanism allows very-low-bitrate representation of music (and indeed speech): every compressed bit is fairly sure to contribute to overall sound quality.

Readers should, at this point, understand the primary application of psychoacoustics: to obtain some advantage in coding, compression, analysis or understanding sound by determining the difference between what is physically present and what is actually heard by a listener. We have illustrated this with a post-stimulatory masking example – but be aware that most of the other hearing characteristics described in Chapter 4 could potentially lead to some kind of psychoacoustic effect. They are most definitely not all used in practice, but at least the potential exists.

Psychoacoustics became a 'hot' topic during the 1990s, and has since been applied to almost every sphere in electronic audio systems. A comprehensive list of systems using psychoacoustics-based algorithms would be extremely long, but the following few categories probably encompass the vast majority:

- compression of high-fidelity audio;[1]
- compression of speech;

---

[1] Many purists maintain that high-fidelity audio *can't* be compressed, and others argue that only analogue systems like the long play (LP) record represent the best fidelity. Here we take the view that better things are possible with digital processing.

- audio steganography: data hiding, or audio watermarking [58];
- active noise cancellation systems;
- speech intelligibility improvement in noise [59];
- automatic speech recognition systems (see Chapter 9);
- machine hearing (see Chapter 8);
- speech and sound synthesis (see Chapter 10).

Despite the profusion of applications, relatively few fundamental psychoacoustic models or modelling techniques are used. Most of the applications employ a subset of a model predicting masking effect, an equal-loudness pre-emphasis, and an appreciation of frequency discrimination. For example, the highly popular MFCCs (mel-frequency cepstral coefficients, see Section 5.5) have been applied by researchers across all of the areas listed above. Unfortunately, we should note that most current masking models do not cater for temporal masking, which we presented as an example above (Figure 5.5).

By contrast, simultaneous frequency masking is very well represented in masking models. This type of masking was described and illustrated back in Chapter 4, specifically in Section 4.2.9 and in Figure 4.4. Simultaneous frequency is well catered for because it is relatively easily modelled by computer. Models relate to single tones, of given frequency and power, which cause nearby tones of lower power to be inaudible. Referring back to Figure 4.4 on page 94, the shaded area showed the extent of the masking effect for the signal of given frequency response: in essence it is a modified threshold of audibility similar to that of the equal-loudness contours of Section 4.2.2. The difference in audibility caused by the presence of the tone is the masking effect.

Computational models exist for this masking effect due to tones. Of note is that much of the historical auditory data used to derive the computerised models was obtained under well-defined and controlled conditions, with artificial signals such as white noise and sinewaves used (for example, the mel scale introduced in Section 4.3.3 was based on the perception of trained musicians). While it is beyond reasonable doubt that the models describe those scenarios very well, it has not been established with the same confidence that the models accurately describe complex real sounds. In fact there is even some doubt that they can be applied to compound sounds at all [12]. Despite the doubts, these models are used in practice, and assume that complex sounds can be broken down into a set of tones, each of which results in a masking effect, with the overall masking effect from the sound being the summation of the separate contributions. When calculating the overall effect, it is possible to introduce non-linear 'corrections' to the model to compensate for the fact that the effect is not a straightforward summation. We should also note that masking in the real world depends on the exact amplitude of the sound as it reaches an average ear, whereas most applications (such as portable music players) cannot control or even sense such information – they do not know where the listening ears are located, or what kind of loudspeaker or headphones is being used. Instead they use the *relative* amplitude of sounds, and assume an approximate listening level (as well as an approximately average listener).

Perhaps most useful in psychoacoustic modelling is a critical-band model of the ear. The loudest tone in each critical band is audible, and the masking effect is the weighted sum of all sound and noise components within that band. Whilst this does not account for several auditory factors, it does model the overall situation quite well – especially when there is a clear distinction between wanted signal and interfering noise. We will first consider a few alternative computational methods to account for critical bands, before applying them in two basic psychoacoustic models

5.3.1 Critical-band modelling

Experimental results indicate that a sound in one critical band will spread its influence into neighbouring critical bands. Given a loud sound at a particular frequency, its influence tends to diminish with distance in frequency, thus it will tend to mask nearby sounds more than it does distant sounds. So we can say that each sound, note or tone, influences other sounds in accordance with their relative amplitudes and frequencies. The band within which the influence spreads is the critical band. Due to the complexities of the cochlea and brain processing, the effects (and hence the size of the critical band) are non-linear and also scale with *absolute* amplitude and frequency. Many models exist to attempt to quantify some or all of the effects of these critical bands – known by many names, including critical-band spreading functions, noise masking curves and lateral inhibition functions.

5.3.1.1 Critical-band spreading functions

Some approximations that have been empirically formulated by various authors are plotted in Figure 5.6 on a Bark scale. Of the curves shown, only the model of Cheng and O'Shaughnessy [44, 60] attempts to account for lateral inhibition [12], where a sound in one band de-sensitises neighbouring frequencies. Jayant *et al.* [61], Virag [45] and Sen and Holmes [62] report very similar curves, differing mainly in the upper frequency region. However, Sen and Holmes introduce corrections for both absolute frequency (the plot shows frequency relative to critical-band centre frequency) and absolute power, in an attempt to improve accuracy. For the plot shown, a centre frequency of 1 kHz and absolute power of $70\,\mathrm{dB}_{SPL}$ were assumed. The former authors do not account for absolute position in frequency or amplitude.

The Hermansky function curve [50] has a flat-topped response that not only approximates the bandwidth of each critical band, but accounts for spreading with a dependence on absolute frequency and amplitude in a fashion similar to the model from Sen [62]. Most importantly the flat top of the function eases the computational load of the model and helps to avoid situations where frequency aliasing occurs (i.e., where frequency sample points not aligning precisely may adversely affect its amplitude response).

The models in Figure 5.6 are all supposed to be applied on a per-band basis. The centre frequency of each critical band is used as a reference, and sound falling within that critical band will have a masking effect on other sounds as represented by the function. For example, using the Hermansky model to account for one main sound component, a second sound falling within the critical band that is plotted would be masked. In this way, if the second sound were 0.8 Bark lower in frequency, then it would be masked if its amplitude relative to the main sound were 0.2 or lower. If the

5.3 Psychoacoustic modelling

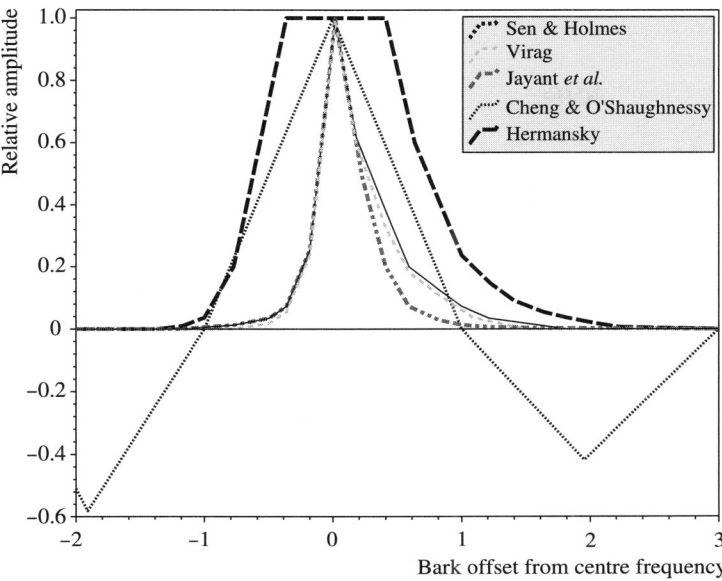

Figure 5.6 Comparison of several critical-band spreading functions from various authors. See text for references.

amplitude of the second sound were increased, it would become audible. Likewise, if it had a relative amplitude of 0.4 then it would need to be at least 1 Bark above the main sound to be audible, or 0.6 Barks below.

The Hermansky model is computed by first warping into the frequency domain, so instead of having units in Hz, we use a Bark scale (Section 4.3.2). In other words, the spectral index is represented in Barks. Then the effect of the critical-band filters can be calculated using a Bark-domain spreading function $\Psi(\Omega_\delta)$ defined as follows [50]:

$$\Psi(\Omega_\delta) = \begin{cases} 0 & \text{for } \Omega_\delta < -1.3 \\ 10^{2.5(\Omega_\delta + 0.5)} & \text{for } -1.3 \leq \Omega_\delta \leq -0.5 \\ 1 & \text{for } -0.5 < \Omega_\delta < 0.5 \\ 10^{-1.0(\Omega_\delta - 0.5)} & \text{for } 0.5 \leq \Omega_\delta \leq 2.5 \\ 0 & \text{for } \Omega_\delta \geq 2.5. \end{cases} \quad (5.1)$$

In this definition, $\Omega_\delta = \Omega_c - \Omega_i$ is the Bark-domain difference between the critical-band centre frequency Ω_c and the frequency of interest Ω_i. In other words, to plot the curve in Figure 5.6, Ω_δ simply sweeps from -2 to $+3$.

We have already defined two MATLAB functions to convert between Hz and Bark x-scales in either direction, $\Omega = f2bark(f)$ and $f = bark2f(\Omega)$ in Section 4.3.2, named f2bark() and bark2f(). Along with those functions, the Hermansky model is not difficult to compute in MATLAB. Function spread_hz(hz_array, hz_c) in Listing 5.1 implements the spreading function (quantised into discrete frequency bands) from a single tone, using the Hermansky model. hz_array is a discrete array of frequencies in Hz, while `hz_c` is the frequency for which the spreading effect is to be computed.

Listing 5.1 spread_hz.m

```
function band=spread_hz(hz_array,hz_c)
    %hz_array is an array of Hz frequencies
    %hz_c is the current centre frequency in Hz
barkc=f2bark(hz_c);
band=zeros(size(hz_array));

for hi=1:length(hz_array)
    barki=f2bark(hz_array(hi));
    barkd=barki-barkc;
    if barkd >= -2.5 & barkd <=-0.5
        band(hi)=10^(1*(barkd+0.5));
    elseif barkd > -0.5 & barkd <0.5
        band(hi)=1;
    elseif barkd >= 0.5 & barkd <=1.3
        band(hi)=10^-(2.5*(barkd-0.5));
    end
end
```

As an example, we could create a 512-element array of frequencies from 0 to 4 kHz, and then compute the spreading function due to a frequency at 1.2 kHz as follows:

```
F=[0:512]*4000/512;
spread=spread_hz(F, 1200);
plot(F,spread,'o-')
```

The filters discussed above and plotted in Figure 5.6 all used the Bark frequency scale (from Section 4.3.2) to attempt to map sounds into a perceptually relevant scale. However, we know that there are alternatives, and in particular the mel frequency scale (see Section 4.3.3) is perhaps more common these days – having been applied to almost all domains of speech and hearing research. Thus it is no surprise that computing a spreading function using the mel frequency scale is also popular.

In fact, the most common mel spreading function uses a triangular shape (in the mel domain), where each triangular area is 50% overlapped with its neighbours. An example of one triangle is plotted in Figure 5.7 to model the spreading effect of a 2.2759 kHz tone (later, we will plot a bank of filters). Each triangular filter $\mathcal{M}_k(m)$ is computed as follows:

5.3 Psychoacoustic modelling

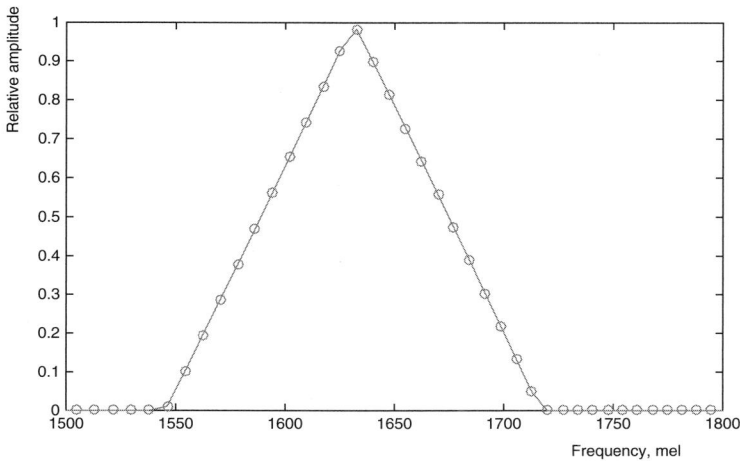

Figure 5.7 The triangular mel-based spreading function.

$$\mathcal{M}_m(f) = \begin{cases} 0 & \text{for} \quad f < F(m-1) \\ \frac{f - F(m-1)}{F(m) - F(m-1)} & \text{for} \quad F(m-1) \leq f \leq F(m) \\ \frac{F(m+1) - f}{F(m+1) - F(m)} & \text{for} \quad F(m) < f \leq F(m+1) \\ 0 & \text{for} \quad f > F(m+1), \end{cases} \quad (5.2)$$

where $F(m)$ is the frequency in Hertz of the mth filter, evaluated at frequency f. As with the Hermansky filter, f is swept across the frequency range to evaluate the frequency response of each of the m bands.

Function `spread_mel(hz_points,hz_c,hz_size,hz_max)` in Listing 5.2 computes the mel spreading function in this way. While `hz_points` is a discrete array of frequencies in Hz, it should be spaced in mel frequencies. Unlike the Hermansky critical-band function, `hz_c` is defined here as the frequency **index** for which we are computing the masking (rather than the actual frequency). `hz_size` defines the resolution of the output masking array, and `hz_max` is the upper frequency limit to evaluate over (normally set to the Nyquist frequency).

Listing 5.2 spread_mel.m

```
function band=spread_mel(hz_points,hz_c,hz_size,hz_max)
%hz_array is an array spaced in Hz
%hz_c is the current index
band=zeros(1, hz_size);
hz1=hz_points(max(1,hz_c-1));                    %start
hz2=hz_points(hz_c);                             %middle
hz3=hz_points(min(length(hz_points),hz_c+1));    %end
%-----
for hi=1:hz_size
    hz=hi*hz_max/hz_size;
```

```
         if hz > hz3
             band(hi)=0;
         elseif hz>=hz2
             band(hi)=(hz3-hz)/(hz3-hz2);
         elseif hz>=hz1
             band(hi)=(hz-hz1)/(hz2-hz1);
         else
             band(hi)=0;
         end
end
```

As an example, we could create a 256-element array of frequencies spaced equally on the mel scale from 0 to 4 kHz and then compute the spreading function for a frequency at 1.2 kHz as follows:

```
mmax=f2mel(4000);
melarray=[0:mmax/255:mmax]; %256 elements
hzarray=mel2f(melarray);
[idx idx]=min(abs(hzarray-1200));
spread=spread_mel(hzarray,idx,100,4000);
plot(f2mel([1:100]*4000/100),spread,'o-')
```

We will later use both the mel and the Bark spreading functions for computing perceptual models. However, we first apply them to create Bark and mel (respectively) filterbank representations – which are also integral parts of the perceptual models.

5.3.1.2 Critical-band filterbanks

The spreading functions explored above were used to model the effect of single critical bands in Figures 5.7 and 5.6. Normally, they are used to help represent sound in a perceptual way, as a vector that corresponds to critical-band excitations.

To explain this further, remember that a single frame of speech or sound can always be represented in a number of alternative ways. For example, a single frame of recorded audio starts as a vector of samples extracted from a longer recording:

```
[s,fs]=wavread('my_speech.wav');
seg=my_speech(1001:1256);
```

We could then take a Fourier transform (or fast Fourier transform), to represent the sound in the frequency domain:

```
S=abs(fft(seg));
```

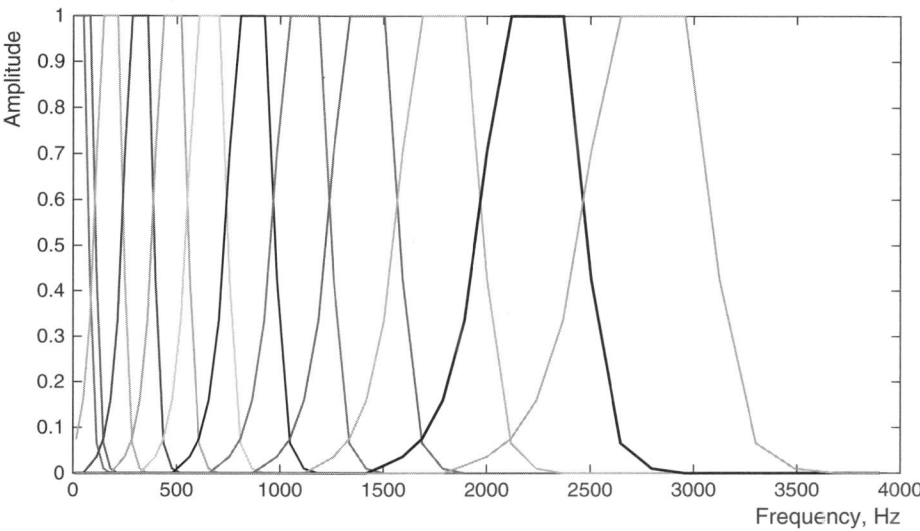

Figure 5.8 A plot of 12 critical-band spreading functions, based on Equation (5.1), over a 48-element array.

In these examples, both `seg` and `S` are vectors that represent the same information (namely the content of a segment of speech); however, the information is stored in the time domain and frequency domain respectively and can highlight different features of the audio. The frequency domain is more useful for some applications, whereas others might best use a time-domain representation, but both are *physical* in nature. The idea of a critical-band filterbank is to provide a perceptually relevant representation of the same information, i.e. a vector of information that better represents what that sound would mean to a listener (in theory, at least).

To compute this perceptual vector, we start with a frequency-domain representation, and then compute which frequencies contribute to which critical bands. Although other alternative methods do exist, the critical bands are normally computed either in the Bark domain, using something like the Hermansky function (Figure 5.6), or in the mel domain using the triangular function (Figure 5.7). To obtain N filterbank elements, we first create N overlapping critical bands and then convolve the power spectral elements into the respective bands, summing their contributions on a per-band basis.

Plots of the Bark and mel spreading functions used are shown in Figures 5.8 and 5.9 respectively.

A segment of pseudo-stationary speech sampled at 8 kHz would normally be represented with between 26 and 40 bands in the respective filterbanks. In the remainder of this chapter we will use exactly this approach to develop two simple but usable psychoacoustic models that have been applied, and tested, in many applications.

5.3.2 Psychoacoustic model computation

Perceptual models based on critical-band spreading functions involve several stages of auditory processing. While there are a number of uses for perceptual models, we will

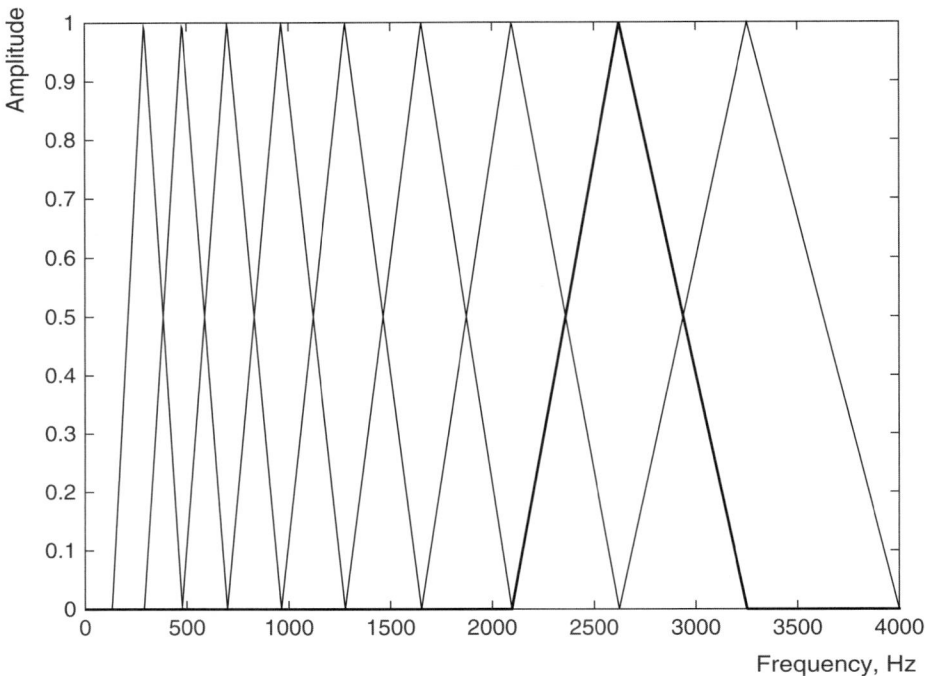

Figure 5.9 A plot of nine mel frequency critical-band spreading functions.

assume here that we wish to examine a particular audio signal to determine a threshold of masking (audibility) due to the sound contained within it. Since there is very little likelihood, as we have mentioned, that the absolute sound levels recorded on computer are the same as those that would impinge on the ear of a listener, it must be stressed that the conclusions drawn by using models such as these should be treated as being approximations.

The processing stages of most simple psychoacoustic models are as follows:

1. spectral analysis;
2. critical-band warping and spreading function convolution;
3. intensity–loudness conversion/pre-emphasis.

We will look at these steps before presenting two alternative complete models based on Bark and mel frequency representations respectively.

5.3.3 Step 1: Spectral analysis

Since the simultaneous masking effect is defined in terms of frequency difference, the first step is to select a frame of audio to analyse, window it, and convert to frequency domain. If $H(n)$ implements an N-point Hamming window, the spectral representation $S(k)$ of the speech segment $s(n)$ would thus be:

5.3 Psychoacoustic modelling

$$S(k) = \sum_{n=0}^{N-1} H(n)s(n)e^{-2\pi jkn/N} \quad \text{for} \quad k = 0, 1, \ldots, \{N-1\}. \tag{5.3}$$

Starting with an example frame of speech, seg of length 256 in MATLAB, we would thus do the following:

```
S=fft(seg.*hamming(1,256));
S=S(1:128);
```

5.3.4 Step 2: Critical-band warping and spreading function convolution

This uses critical-band filterbanks, based on the spreading functions already derived above in Section 5.3.1.2. The filterbanks are convolved with the spectral representation of the signal being analysed to compute per-bank responses. Since this uses slightly different methods for the Bark and mel-domain models, we will present the complete models below in Sections 5.4 and 5.5 respectively.

5.3.5 Step 3: Intensity–loudness conversion/pre-emphasis.

The base frequency selectivity of human hearing is generally approximated by the A-weighting curve of Section 4.2.2, which itself has been nicely approximated in the

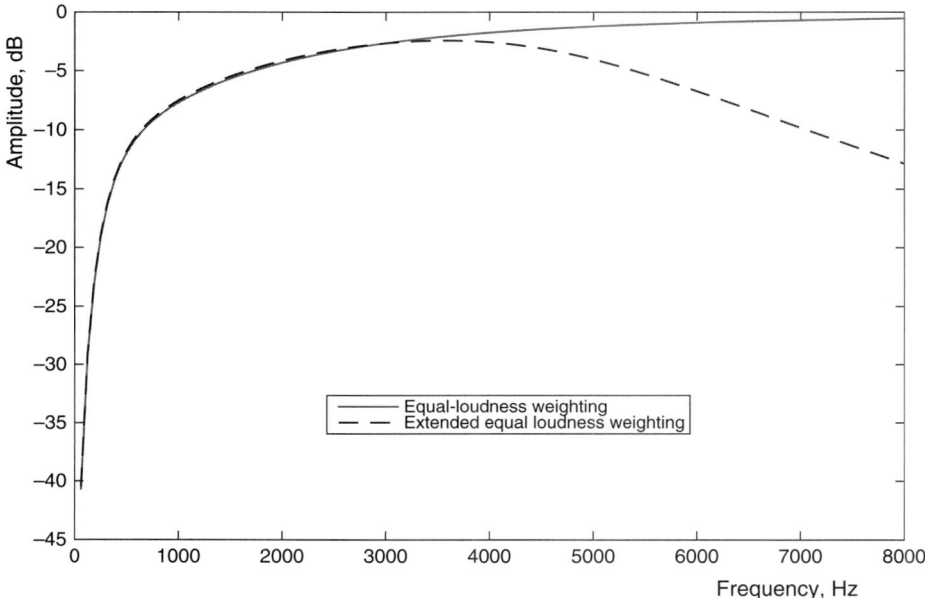

Figure 5.10 The shape of the equal-loudness emphasis curves of Equations (5.5) and (5.4).

frequency domain by several authors. Hermansky [50] used the following in his model, where $\omega = 2\pi f$:

$$E(\omega) = \frac{\omega^4(\omega^2 + 56.8 \times 10^6)}{(\omega^2 + 6.3 \times 10^6)^2(\omega^2 + 0.38 \times 10^9)}. \tag{5.4}$$

It should also be noted from the original source that Equation (5.4) is only accurate up to about 5 kHz (which is beyond what we require). However, if higher frequencies are to be used then a further term should be added to better match the upper frequency region of the A-weighting curve (note the addition of a 10^{27} scaling over Equation (5.4) to normalise the two responses):

$$E(\omega) = \frac{\omega^4(\omega^2 + 56.8 \times 10^6)10^{27}}{(\omega^2 + 6.3 \times 10^6)^2(\omega^2 + 0.38 \times 10^9)(\omega^6 + 9.58 \times 10^{26})}. \tag{5.5}$$

Listing 5.3 equal_loudness_preemph.m

```
function eql=equal_loudness_preemph(hz,varargin)
%hz is frequency in Hz
%if second argument given, use extended form
w=hz*2*pi; %convert to abs angular freq.
w2=w.^2;   %squared version
if ~isempty(varargin)
    eql=1e27*((568e5+w2).*(w2.^2))./(((63e5+w2).^2)
        .*(38e7+w2).*((w.^6)+958e24));
else
    eql=((568e5+w2).*(w2.^2))./(((63e5+w2).^2).*(38e7+w
        .^2));
end
```

This is followed by an intensity–loudness conversion to allow the computed bands to relate the almost arbitrary units of the model to a perceived loudness scale, based on the power law of hearing. This is included here for completeness, but is not needed when the model is used to compare two signals directly (i.e. the relative rather than the absolute difference is required). The final perceptual response, $\mathcal{P}(\omega)$, is thus the equal-loudness

$$\mathcal{P}(\omega) = \{E(\omega)\Theta(\omega)\}^{0.33}. \tag{5.6}$$

For the mel-domain model, a logarithmic conversion is performed instead before convolving the spreading functions to derive MFCCs.

5.4 Hermansky-style model

The Hermansky-style perceptual model first defines an equal-spaced array of Bark frequencies (each of which will be the centre point of one of the individual spreading

5.4 Hermansky-style model

functions). Next it creates the spreading functions, and finally it convolves these with the power spectrum, summing the contribution within each critical band.

The equal-loudness pre–emphasis and intensity–loudness conversions come next, to create a Bark-domain perceptual response.

Given a speech segment, seg, sampled at Fs Hz, the function percep_model() in Listing 5.4 computes the N-point spectral response:

Listing 5.4 percept_model.m

```
1 function [x, xf]=percept_model(seg, N, Fs)
2 % Map audio frame seg to N-point 0:(Fs/2)Hz percept
     model. N typically 40 at Fs = 8 kHz
3 b_low=0;                %Bark span lower limit
4 b_top=f2bark(Fs/2);     %Bark span upper limit
5 bdiv=(b_top-b_low)/N;   %Bark resolution
6 %Define an array of centre frequencies
7 xb=b_low+bdiv/2:bdiv:b_top-bdiv/2;
8 xf=bark2f(xb);   %Convert to Hz
9 S=abs(fft(seg));
10 S=S(1:length(S)/2);
11 F=[1:length(S)]*(Fs/2)/length(S);
12 x=zeros(1,N);
13 for xi=1:N
14    bark=xb(xi);
15    hz=bark2f(bark);
16    %compute spreading function
17    spr=spread_hz(F,hz);
18    %compute summed influence
19 x_sum=sum(spr.*S')*equal_loudness_preemph(hz);
20    %intensity loudness power law
21    x(xi)=(x_sum)^0.33;
22 end
```

In practice, the response would be computed on a frame-by-frame basis (probably using overlapping frames) across an entire audio recording, as illustrated in Figure 5.11, plotted using the MATLAB script from Listing 5.5.

Listing 5.5 Implement_Hermansky.m

```
1 %%This is for testing the Hermansky model
2 [s,fs]=wavread('b0022.wav');
3 Ws=1024;
4 Ol=512;
```

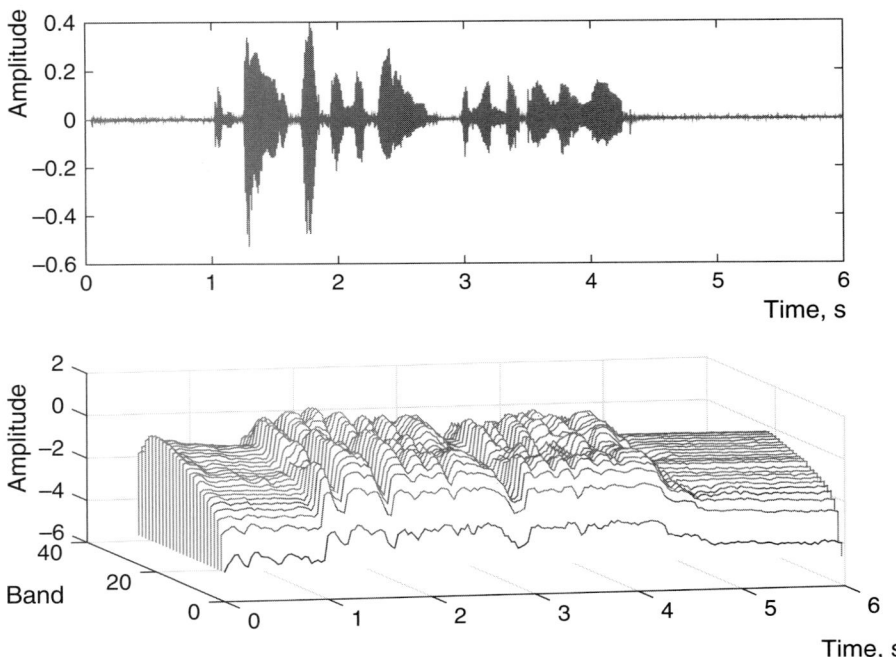

Figure 5.11 Analysis of a speech recording using 26 Bark-based critical-band filters, following the Hermansky method.

```
L=floor((length(s)-Ol)/Ol);
N=26;
bs=zeros(N,L);
for n=1:L
    seg=s(1+(n-1)*Ol:Ws+(n-1)*Ol);
    [x,xf]=percept_model(seg.*hamming(1,Ws),N,fs);
    bs(:,n)=x;
end
waterfall([1:L]*length(s)/(L*fs),[1:26],log(bs))
xlabel('Time, s   '); ylabel('Amplitude')
ylabel('Band    ');    zlabel('Amplitude')
```

5.5 MFCC model

The MFCC model similarly defines an equal-spaced array of mel frequencies which will become centre frequencies of triangular spreading functions (i.e. a mel filterbank). This is convolved with the speech segment power spectrum, then the contribution of each filter is summed to yield a power vector. Intensity–loudness conversion is performed

5.5 MFCC model

by transforming the power vector to the logarithmic domain, before a discrete cosine transform is taken to yield the final MFCCs.

Given a speech segment, seg, sampled at Fs Hz, the function mfcc_model() in Listing 5.6 computes N MFCCs, over M critical bands:

Listing 5.6 mfcc_model.m

```
function cc=mfcc_model(seg, N, M, Fs)
% Do FFT of audio frame seg, map to M MFCCs
% from 0 Hz to Fs/2 Hz, using N filterbanks
% typical values N=26,M=12,Fs=8000,seg~20ms
m_low=0;              %mel span lower limit
m_top=f2mel(Fs/2);    %mel span upper limit
mdiv=(m_top-m_low)/(N-1); %mel resolution
%Define an array of centre frequencies
xm=m_low:mdiv:m_top;
%Convert this to Hz frequencies
xf=mel2f(xm);
%Quantise to the FFT resolution
xq = floor((length(seg)/2 + 1)*xf/(Fs/2));
%Take the FFT of the speech...
S=fft(seg);
S=abs(2*(S.*S)/length(S));
S=S(1:length(S)/2);
F=[1:length(S)]*(Fs/2)/length(S);
%Compute the mel filterbanks.m
x1=zeros(1,N);
for xi=1:N
    band=spread_mel(xf,xi,length(S),Fs/2);
    x1(xi)=sum(band.*S');
end
x=log(x1);
%Convert to MFCC using loop (could use matrix)
cc=zeros(1,M);
for xc=1:M
    cc(xc)=sqrt(2/N)*sum(x.*cos(pi*xc*([1:N]-0.5)/N));
end
```

Again, this would be computed for every frame in a recording (probably overlapped by at least 50%). The frame-to-frame difference in MFCCs is often computed – this is called the 'delta' – because it can yield useful information on the stationarity of the signal. This is often accompanied by the frame-to-frame difference in delta (Δ), called the 'delta-delta' ($\Delta\Delta$) and, in speech recognition at least, the combined MFCC,

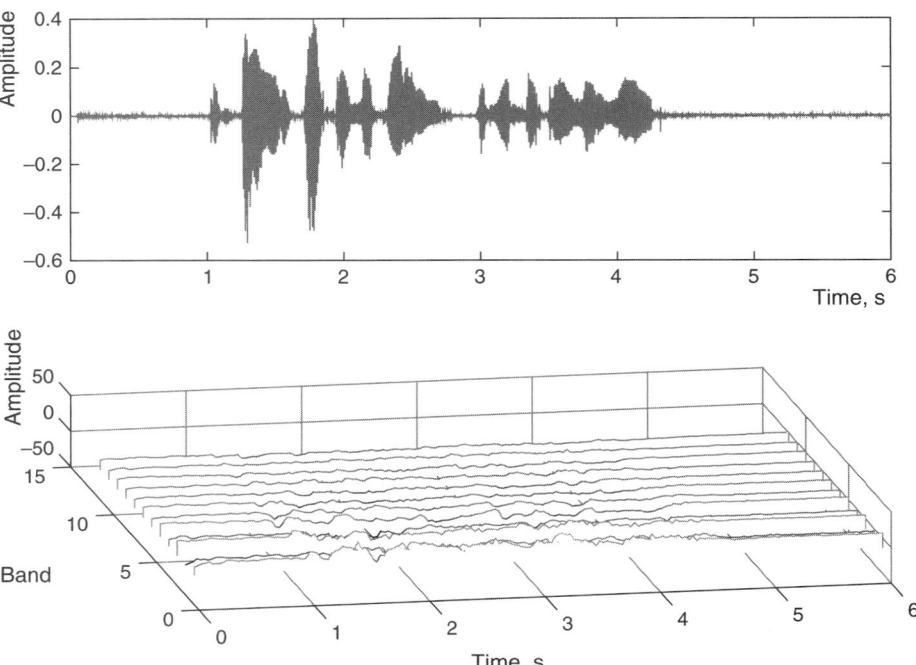

Figure 5.12 Analysis of a speech recording using 12 mel-based critical-band filters, following the MFCC method.

delta and delta-delta of each frame is augmented by its context, which means the same information from neighbouring frames.

Figure 5.12 plots the frame-by-frame variation in the first 12 MFCCs using the MAT-LAB script from Listing 5.7.

Listing 5.7 Implement_MFCC.m

```
%%This is for testing the MFCC model

[s,fs]=wavread('b0022.wav');
Ws=1024;
Ol=512;
L=floor((length(s)-Ol)/Ol);
N=12;

ccs=zeros(N,L);
for n=1:L
    seg=s(1+(n-1)*Ol:Ws+(n-1)*Ol);
    ccs(:,n)=mfcc_model(seg.*hamming(1,Ws),40,N,fs);
end
```

```
14  waterfall([1:L]*length(s)/(L*fs),[1:N],ccs)
15  xlabel('Time, s')
16  ylabel('Amplitude')
17  ylabel('Band')
18  zlabel('Amplitude')
```

5.6 Masking effect of speech

Most computational masking models have been derived for a situation where a single loud tone masks a quieter tone. Some models have been developed beyond that: for cases of noise masking tones, or even tones masking noise. Unfortunately though, evidence to suggest that the models are accurate for any and every combination of sounds is weak. Despite this the models *have* been applied, apparently successfully, to generalised audio, most notably in MP3 players.

A method of application in such generalised scenarios is shown in Figure 5.13, where a particular sound spectrum is analysed across a set of critical-band regions. The sound falling within each critical band is totalled, and its masking contribution within that band determined. Then, for each band, the effect of masking spread from immediate neighbouring bands is factored in, assuming it is additive to the masking originating from the band itself. It is unusual to consider masking spread from bands beyond the immediate neighbours.

The result of such an analysis is that each critical band will have an individual masking level. Sounds within that band that are above the masking level will be audible, and sounds below the masking level will not be audible.

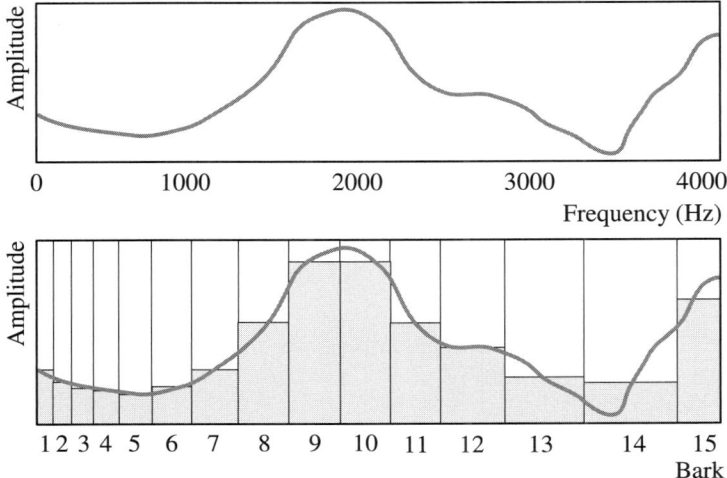

Figure 5.13 In-band masking level derived and plotted (bottom) from an example spectrum (top) for 15 critical bands of constant Bark width.

For application to speech, one relatively useful method of determining audibility is to look at the formant frequencies, and consider these as independent tones, most important being formants F2 and F3 which contribute more to intelligibility than the others (see Section 3.2.4). For segments of speech with no formants, the audibility of the overall spectrum (perhaps within the range 250 Hz to 2 kHz) can be determined. If the overall spectrum is inaudible, or F2 and F3 are inaudible, then it is likely that the speech itself will be unintelligible. It is possible, however, that elements of the speech, whilst unintelligible, will be audible. This underlines the fact that the ability of the human brain and hearing system to extract signal from noise is quite awesome at times.

5.7 Summary

This chapter has developed, discussed, demonstrated and implemented psychoacoustic techniques that are foundational to the world of advanced audio and speech processing. A number of themes developed here will recur in later chapters, particularly when we begin to consider audio coding, as well as speech recognition, language and speaker identification, and even the emerging topic of machine hearing.

Bibliography

- *An Introduction to the Psychology of Hearing*
  B. C. J. Moore (Academic Press, 4th edition 1997)

- *Hearing (Handbook of Perception and Cognition)*
  B. C. J. Moore (Academic Press, 2nd edition 1995)

- *Advances in Speech, Hearing and Language Processing*
  Ed. W. A. Ainsworth (JAI Press, 1990)

- *Psychoacoustics: Facts and Models*
  Eds. H. Fastl and E. Zwicker (Springer, 3rd edition 2006)

- *Acoustics and Psychoacoustics*
  D. Howard and J. Angus (Focal Press, 3rd edition 2006)

- *Music, Cognition and Computerized Sound: An Introduction to Psychoacoustics*
  P. R. Cook (MIT Press, 2001)

- *The Acoustics of Speech Communication: Fundamentals, Speech Perception Theory, and Technology*
  J. M. Pickett (Allyn and Bacon, 1998)

Questions

Q5.1 What is the McGurk effect, and what does it tell us about the interrelationship between the human auditory and visual systems?

Q5.2 What is a Δ and what is a $\Delta\Delta$ when related to MFCC parameters?

Q5.3 If there are 12 MFCC parameters $c[t]_1, \ldots, c[t]_{12}$ for current analysis frame t, how would $\Delta c[t]_4$ be computed?

Q5.4 How might the principle of proximity assist a listener in a highly resonant noisy environment to identify the source of a sound?

Q5.5 Explain the similarity between a speech coder introducing comfort noise and the principle of closure.

Q5.6 How can the principle of good continuation apply to a musical instrument in an orchestra, such as a solo violin, that is playing against a backdrop of many other violins?

Q5.7 Do current psychoacoustic models most commonly make use of simultaneous masking or of temporal masking?

Q5.8 In Figures 5.8 and 5.9, why do you think the triangle and flat-topped regions are plotted overlapping?

6 Speech communications

Chapters 1 to 4 covered the foundations of speech signal processing including the characteristics of audio signals, methods of handling and processing them, the human speech production mechanism and the human auditory system. Chapter 5 then looked in more detail at psychoacoustics – the difference between what a human perceives and what is actually physically present. This chapter will now build upon these foundations as we embark on an exploration of the handling of speech in more depth, in particular in the coding of speech for communications purposes.

The chapter will consider typical speech processing in terms of speech coding and compression (rather than in terms of speech classification and recognition, which we will describe separately in later chapters). We will first consider the important topic of quantisation, which assumes speech to be a general audio waveform (i.e. the technique does not incorporate any specialist knowledge of the characteristics of speech).

Knowledge of speech features and characteristics allows parameterisation of the speech signal, in particular the important source filter model. Perhaps the pinnacle of achievement in these approaches is the CELP (codebook excited linear prediction) speech compression technique, which will be discussed in the final section.

6.1 Quantisation

As mentioned at the beginning of Chapter 1, audio samples need to be quantised in some way during the conversion from analogue quantities to their representations on computer. In effect, the quantisation process acts to reduce the amount of information stored: the fewer bits used to quantise the signal, the less audio information is preserved.

Most real-world systems are bandwidth (rate) or size constrained, such as an MP3 player only being able to store 4 or 8 Gbyte of audio, or a Bluetooth connected speaker only being able to replay sound at 44.1 kHz in 16 bits because this results in the maximum bandwidth audio signal that Bluetooth wireless can convey.

Manufacturers of MP3 devices may quote how many songs their devices can store, or how many hours of audio they can contain – these are both considered more customer-friendly than specifying memory capacity in Gbytes – however, it is the memory capacity in Gbytes that tends to influence the cost of the device. It is therefore also evident that a method of reducing the size of audio recordings is important, since it allows more songs to be stored on a device with smaller memory capacity. This process is generally known as audio compression, and may be applied (preferably using different methods or algorithms), to both speech and general audio such as music. The degree of compression is defined as the ratio between the compressed and original data sizes, in bits. There is a general trade-off between the degree of compression and the quality (or intelligibility) of the compressed signal, but it is possible to use some very clever techniques to achieve both – at the expense of computational complexity.

6.1.1 Pulse coded modulation

Pulse coded modulation (PCM) is the format delivered by most analogue-to-digital converters (ADCs) and the format of choice for representing audio on a computer. Sound is stored as a vector of samples, with each sample usually represented as a single 16-bit value (one notable exception is in MATLAB, where audio data is often stored in floating point format). In theory, samples are supposed to be related to the analogue amplitude of the audio waves travelling through the air, with the timing between samples being determined by the sample rate. This process is illustrated in Figure 6.1, where a waveform is plotted on a grid of lines at regular time intervals and amplitudes The time intervals are drawn at every 0.125 ms (which means a $1/0.000\,125 = 8\,\text{kHz}$ sample rate), and the amplitude might be the voltage from a microphone (in millivolts), or some other quantity that describes the sound amplitude. The exact units are unimportant at this stage, as long as they relate linearly to the actual sound amplitude. In the scheme shown, the analogue waveform will be quantised by choosing the nearest whole amplitude at each time instant. The result is the dark solid 'square' version. Within a computer, the quantised sound would be stored as a vector of amplitudes, one per time instant. Thus, in the example shown, the vector would simply be $\{1, 0, 1, 2, 3, 6, 8, 9, 5, 1, -2, -3, -1, 1, 3, 6, 8\}$. However, this would normally be stored in binary, often 8 bit for

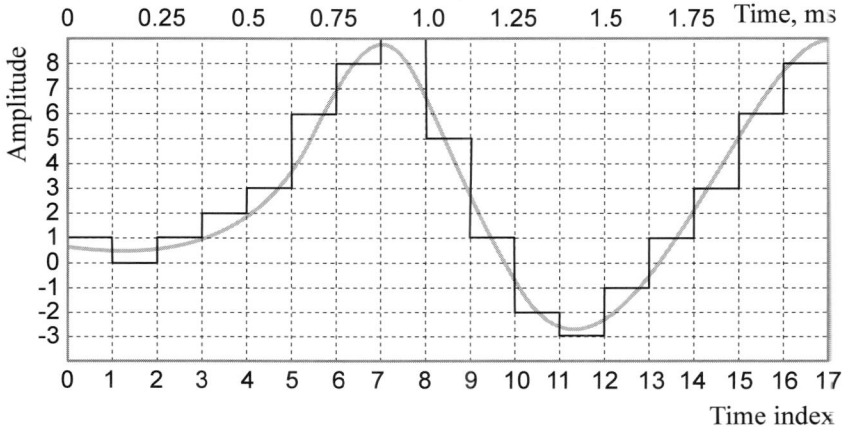

Figure 6.1 Illustration of an audio waveform being pulse code modulated by quantising the amplitude of the signal to the nearest integer at sampling points spaced regularly in time.

low-quality speech but typically 16 bit for music and high-quality speech. In 8-bit two's complement binary, the vector would then become:

{00000001, 00000000, 00000001, 00000010, 00000011,
00000110, 00001000, 00001001, 00000101, 00000001,
11111101, 11111100, ...}

It is quite obvious from Figure 6.1 that there is a fairly significant difference between the original and quantised waveforms. This difference is called the **quantisation error**. It can be reduced (i.e. we could get a better 'fit') by having a denser grid, with more amplitude and time lines. This means using a higher sampling frequency and a higher-precision amplitude representation, making the time lines and the amplitude lines closer together, respectively. The trade-off would then be that the signal consumes more memory (along with bandwidth, processing time, power and so on). Since the trade-off is linear, doubling the sample frequency doubles the storage requirement, and doubling the number of bits used to encode the signal would again double the storage requirement).

Most discrete digital signal processing relies upon PCM samples being available; however, there has recently been a resurgence of interest in non-uniform sampling representations – these are vectors that have a higher sampling frequency around regions of the signal representing important information, or perhaps containing higher-frequency components. Perhaps in time such techniques will find their way into mainstream algorithms, but for now the difficulty in understanding the theory of such systems, and working around the practical difficulties involved in their execution, has limited their usefulness.

Box 6.1 Speech coding objectives

Speech compression, or codec, systems are classified according to what they compress: speech or general audio, how well they compress the content and how well they perform in terms of quality or intelligibility (which were differentiated and measured in Section 3.4.1). To aid in this classification, there is a general agreement on terms used to describe the quality of speech handled by each method. The table below lists the more common terms and describes them in terms of sample rate, bandwidth, approximate dynamic range and mean opinion score (MOS – see Section 3.4.2). All figures given are approximate guides to the typical characteristics of such systems:

Name	Sample rate	Bandwidth	Dynamic range	MOS
Synthetic quality	–	–	48 dB	2.5–3.5
Comms quality	7200 Hz	200–2000 Hz	56 dB	3.5–4.0
Toll quality	8000 Hz	200–3200 Hz	64 dB	4.0
Network quality	16 000 Hz	20–7000 Hz	80 dB	4.0–4.5

Toll quality refers to 'telephone audio', based on the analogue telephone network, but often brought into the realm of digital measurements. For analogue systems a signal-to-noise ratio of 30 dB and 200 Hz to 3.2 kHz bandwidth, measured at the 3 dB points, would be typical.

6.1.2 Delta modulation

The 'delta' refers to amplitude difference, measured from one sampling instant to the next. In this simple scheme, the difference is limited to +1 or −1. In operation, the system maintains an accumulator that starts at zero. At every sample position, this accumulator either steps up by one or steps down by one, whichever best approximates the analogue signal. It can never maintain exactly the same value. The decision to step up or down at each sample input is made by comparing the current accumulator value with the desired waveform amplitude at the same time (and the decision is 'are we too low? then step up, otherwise step down'). This process is illustrated in Figure 6.2. It should be noted that this method would typically have a far higher sample rate than for PCM audio, perhaps 16 or 20 times faster. However, it is much simpler in terms of the encoding performed at each sampling instant – which only requires a 1 or 0 (up or down) decision.

The problem with delta modulation is that the quantisation error depends on the stepsize. This means that, in order to represent audio as accurately as possible, the step height should be small. However, a small step means in turn that more of them are needed to reach up to larger waveform peaks. In Figure 6.2, when rising up to the first peak and dropping down after it, quite a large gap develops between the analogue waveform and the step values – this is because 'delta mod' can only increase by a single step at a time, but the gradient of the waveform has exceeded this. Such a limit on the gradient is termed the *slew rate* or sometimes *slope overload*. The output bitstream which represents the waveform in the delta modulation format is shown below the waveform as a continuous binary stream.

In order to support a higher slew rate without increasing the stepsize, it might appear that it is necessary to sample more quickly, increasing the number of bits per second.

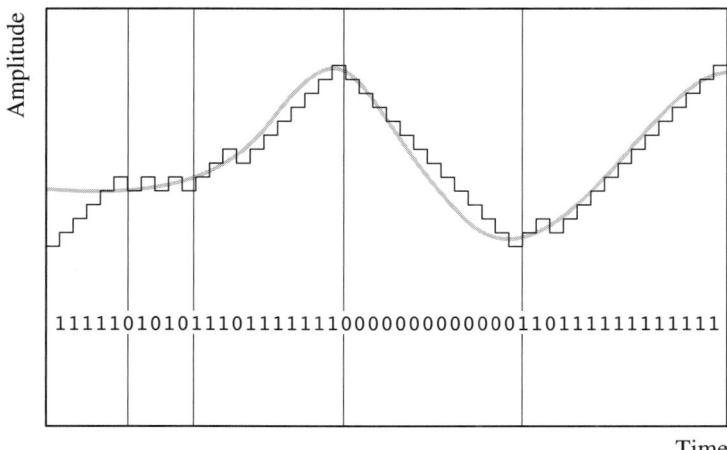

Figure 6.2 Illustration of an audio waveform being represented using delta modulation, showing the encoded vector below the analogue and quantised waveforms. In the encoded vector, a 1 indicates a stepwise increase in amplitude while a 0 indicates a stepwise decrease in amplitude. There is one element in the vector at each sampling instant.

The same trade-off then occurs here between the amount of data used for the representation and the quantisation noise introduced by it (just like with the PCM format). However, with some clever thinking it is possible to sidestep the trade-off, achieving the aim of reducing both quantisation error and data rate.

This clever thinking is the driving force behind audio (and speech) compression. This has led to the development of methods able to reduce the storage or transmission requirements of audio while maintaining the quality and limiting the quantisation error. The remainder of this chapter will gradually introduce more and more capable (and complex) ways of doing this.

6.1.3 Adaptive delta modulation

In an attempt to maintain the beneficial features (and simplicity) of delta modulation, while overcoming the slew rate limitations, designers came up with several methods to vary the step height based on the past history of the audio. These adapt the quantisation level so it can be small when dealing with slowly changing waveforms, but bigger when dealing with rapid waveform changes. The technique is also known by the mouthful *continuously variable slope delta modulation*, abbreviated CVSDM or CSVD, and used as a speech compression method in some older radio systems.

In the most basic form, such a system relies upon some fixed rules to change stepsize, such as the following artificial example:

'If the past n values were the same then double the step height, otherwise halve it.'

Upper and lower limits are applied to the step height changes so they do not become either too small or too large to be useful. Some systems might even adapt their stepsize *rules* over time, in addition to adapting the stepsize itself. In other techniques, step heights gradually increase or decrease, rather than being doubled or halved.

The technique – with stepsize doubling and halving according to the example rule above – is illustrated in Figure 6.3, for $n = 3$. Thus the step height can be seen to double following three successive moves in the same direction. Several step height reductions are similarly illustrated in the figure. The bitstream resulting from this process, which is a quantised representation of the original waveform, is shown across the bottom of the plot.

The analogue waveforms in Figures 6.3 and 6.2 are identical, as are the sampling instants. Both methods use the same amount of digital data to represent the waveform; however, the adaptive delta-mod technique is slightly better at tracking the underlying waveform. If we compute quantisation error as the area between the analogue and quantised waveforms, then the adaptive method error would be seen to be slightly lower – notwithstanding a few places where its stepsize was probably increased a little too much.

6.1.4 ADPCM

Adaptive differential pulse coded modulation (ADPCM) is a method of applying the adaptive delta modulation technique to PCM samples (instead of to a delta-modulated bitstream). The word 'differential' indicates that the system calculates a difference

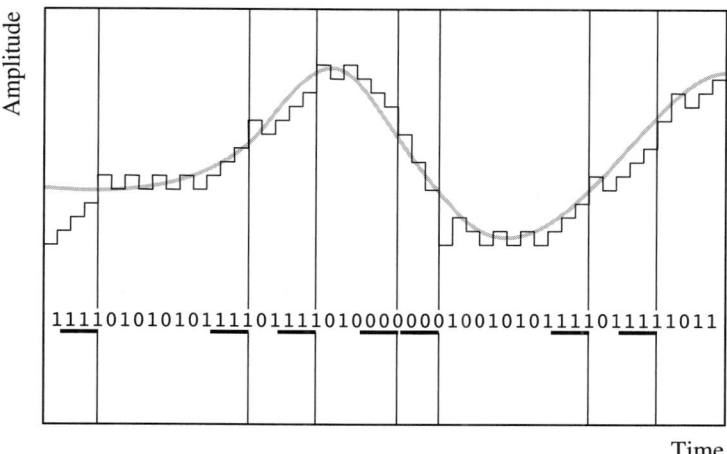

Figure 6.3 Illustration of an audio waveform being represented using adaptive delta modulation, where a 1 indicates a stepwise increase in signal amplitude and a 0 indicates a stepwise decrease in signal amplitude. Repetition of three similar sample values triggers a doubling of stepsize, otherwise the stepsize is halved, until it reaches the minimum (which is the predominant stepsize used throughout).

between sample instants, and 'adaptive' indicates that stepsizes change or adapt based on past history.

As with delta modulation, there is an accumulator that starts at zero. At the first sample instant a difference is calculated between the accumulator and the waveform being coded. This difference is quantised and then added to the accumulator. The next waveform sample value is then compared with the new accumulator value, and a difference calculated, quantised and then added to the accumulator. Evidently the difference can be either positive or negative, and it is thus the quantised difference value that is transmitted at each sample instant.

The adaptive nature of the system is brought into play during the quantisation stage, by adjusting the size of the quantisation steps. Typically, a 16-bit sample vector would be encoded to 3-, 4- or 5-bit ADPCM values, one per sample.

As an example, consider the artificial PCM sample stream in Figure 6.4, with each sample being quantised to 16 levels.[1] At each sampled time instant the value of the quantised level represents the waveform amplitude. In this case the standard PCM sample vector would therefore be {07 07 07 08 09 12 15 15 13 08 04 03 07 12 13 14 14}.

With *differential* PCM encoding, the difference between neighbouring samples is stored, rather than their absolute values. Assume we start with an accumulator of zero (i.e. the starting point is zero), then the differences for the example sample vector would be {07 00 00 01 01 03 03 00 −02 −05 −04 −01 04 05 01 01 00}. Apart from the first sample, we can see that the differential vector values are

---

[1] Just 16 levels are used here for clarity. In reality a sample stream would be quantised to 65 536 levels in a 16-bit system.

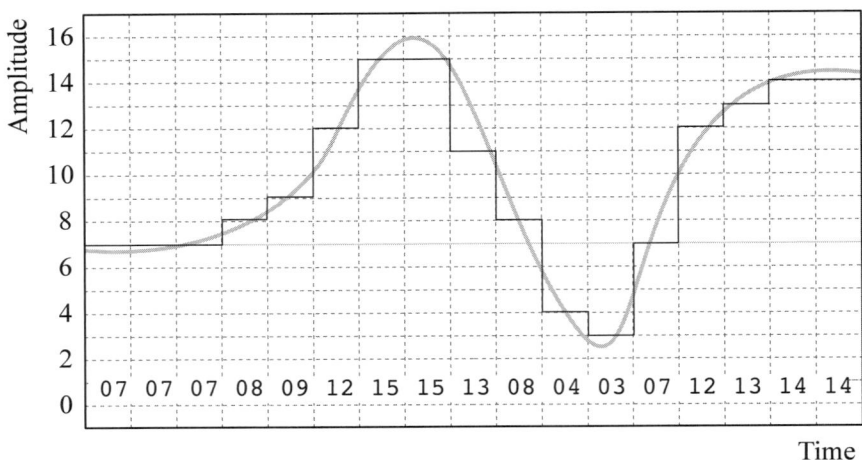

Figure 6.4 Illustration of an audio waveform being quantised to 16 levels of PCM.

much smaller in value than the elements in the original PCM vector. Thus, fewer bits could be used to represent them.

ADPCM uses this methodology but takes it one step further (please excuse the pun) by changing the quantisation stepsize at each sampling instant based on past history. For an example, the same waveform is coded, using exactly the same number of bits per sample, but using adaptive quantisation, in Figure 6.5. Starting with the same initial quantisation levels, the rule used here is that if the sample value is in the middle four quantisation levels then, for the next sample instant, the quantisation stepsize is halved, otherwise it is doubled. This allows the quantisation to 'zoom in' on areas of the waveform where only small sample changes occur, but 'zoom out' sufficiently to capture large changes in amplitude when needed.

In this case the adaptively quantised PCM sample vector would be {07 08 09 10 10 11 11 09 07 05 04 05 14 12 08 10 08} and once we have used differential coding on this it would become {07 01 01 01 00 01 00 −02 −02 −02 −01 01 09 −02 −04 02 −02}. The illustration in Figure 6.5 shows how the quantisation step zooms in on slowly changing waveform amplitudes, and then zooms out to capture large changes in amplitude. This reduces quantisation error, yet still manages to achieve high slew rate (except where a flat waveform is followed by big changes and conversely when large waveform features are followed immediately by much smaller ones).

In reality, the adaptation rule would consider the last few sample values instead of looking only at the last one as the example above did. Always, the *next* sample value is predicted at the encoder based on past history, and then compared with the *actual* value, with the difference between the two being encoded. This is shown in Figure 6.6, where simplified block diagrams of the encoder and decoder, containing the major blocks of quantiser, predictor and de-quantiser, are compared. In practice, ADPCM is actually fairly good at tracking a waveform – especially where the waveform values

6.1 Quantisation

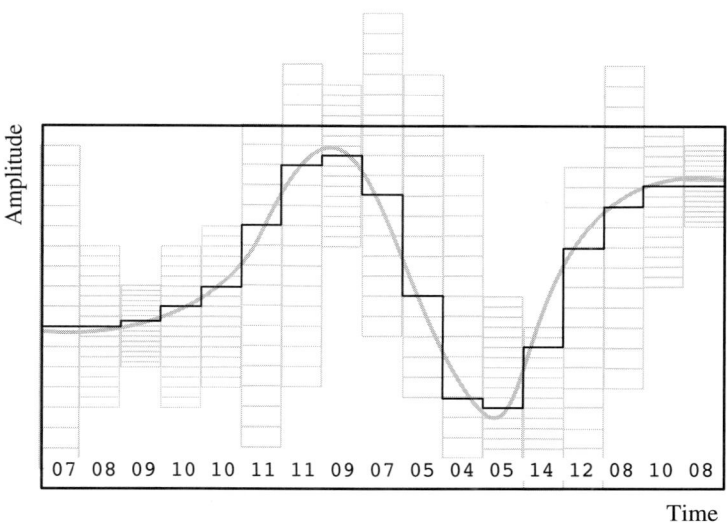

07 08 09 10 10 11 11 09 07 05 04 05 14 12 08 10 08

Figure 6.5 Illustration of an audio waveform being quantised to 16 step levels, with the step sizes being adapted based on the previous sample.

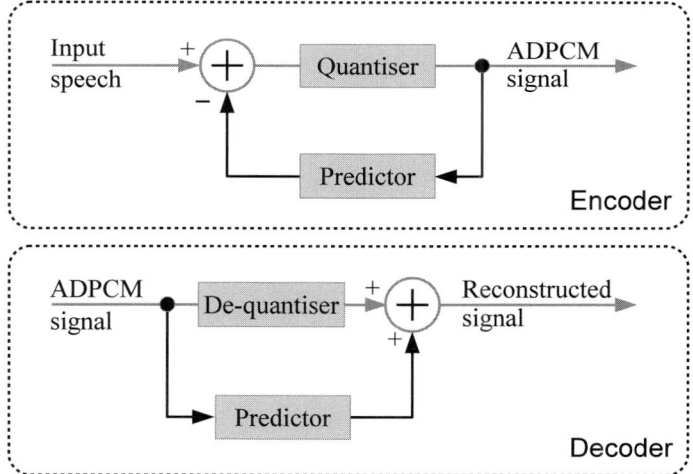

Figure 6.6 A block diagram of an ADPCM encoder (top) and decoder (bottom), showing the predictor in both, fed with identical information (i.e. the ADPCM signal). Both the predictor and the quantiser would usually be adaptive.

can be reasonably accurately predicted, which happens to be the case with many of the useful real-world signals such as music and speech (Figure 6.5).

6.1.5 SB-ADPCM

The sub-band adaptive differential pulse coded modulation (SB-ADPCM) coder, introduced through ITU standard G.722, includes two separate ADPCM coding units. Each

> **Box 6.2** PCM-based speech coding standards
>
> There exists a plethora of different speech coding techniques based around PCM. The main standardisation body in this area is the International Telecommunications Union (ITU) since the primary driver in voice communications worldwide has been telecommunications. In the past the ITU was known as the Comité consultatif international téléphonique et télégraphique, or CCITT. Even the English version of the name is a mouthful: 'International Telegraph and Telephone Consultative Committee', hence the move to the far simpler title ITU.
>
> Several of the more common ITU standards, all beginning with the prefix G, are shown in the following table:
>
> | Name | Description |
> | --- | --- |
> | G.711 | 8 kHz sampling A-law and μ-law compression |
> | G.721 | 32 kbits/s ADPCM standard (replaced by G.726) |
> | G.723 | 24 and 40 kbits/s ADPCM (replaced by G.726) |
> | G.722 | 64 kbits/s SB-ADPCM sampled at 16 kHz |
> | G.726 | 24, 32 and 40 kbits/s ADPCM sampled at 8 kHz |
>
> Some other related ITU speech standards are shown on page 190.

unit operates on half of the frequency range (which are 0–4 kHz and 4–8 kHz respectively, since this is considered a 'wideband' coder). The bit weighting and quantisation in each band differ by a factor of 4 in favour of the lower band, which is thus capable of better audio fidelity, particularly since it conveys frequencies that are more important to speech (see Section 3.2.4).

Remember the simultaneous masking property of the human auditory system (Section 4.2.9)? This states that loud tones will mask nearby, but quieter, frequencies. ADPCM tends to work in a similar way by matching its quantisation step to the loudest sounds, while condemning quieter sounds to be lost in the quantisation noise. Since this behaviour is similar to the human auditory system, it is not a great disadvantage – except when the quiet sound is far away in frequency from the loud sound. In humans, masking is mainly confined to narrow critical bands, but in standard ADPCM the masking effect occurs across the frequency band. Thus ADPCM has no concept of frequency bands and would probably lose the quieter sound even if it were far enough away in frequency that it could be heard by a human. For this reason, SB-ADPCM, being able to simultaneously code one very loud and one very quiet sound, as long as they are in different frequency ranges, is perceived as having much higher quality than the ADPCM equivalent.

6.2 Parameterisation

Coding techniques that follow, or try to predict, a waveform shape tend to be relatively simple and consequently achieve limited results. These techniques typically assume very little about the waveform being coded – except perhaps maximum extents and slew rate. There is a trade-off between coding quality and bitrate, and, although our clever compression techniques can gain some room to manoeuvre, they are unable to achieve high-fidelity speech coding at very low bitrates (or indeed general audio and music coding at low bitrate).

Instead of coding the physical speech waveform directly, researchers use their knowledge of the way speech is produced, or the way sound is heard, in order to parameterise the signal in a way that relates to its content, method of production and way of being perceived. These parameters are values that are chosen to represent important aspects of the speech signal, transmitted from coder to decoder, where they are used to recreate a similar (but not identical) waveform. The process of breaking down speech into these parameters at the coder is called analysis, while reconstruction at the decoder is called synthesis.

Apart from the likelihood of the transmitted parameters requiring fewer bits to represent than a directly coded waveform, parameterisation can hold two other benefits. Firstly, if the parameters are chosen to be particularly relevant to the underlying sound (i.e. a better match to the speech signal) then the difference between the original and coded–decoded signal can be reduced, leading to better fidelity. Secondly, the method of quantising the parameters themselves – or rather the number of bits assigned to each parameter – can be carefully adjusted to improve quality. In simpler terms, when given a 'pool' of bits that are allowed to represent the parameters being transmitted from encoder to decoder, it is possible to 'spend' more bits on parameters that are more important to overall quality. In the more advanced speech coding algorithms, parameters are chosen that match the component signals within the speech, or that match the important aspects of the human auditory system, or perhaps cater a little to both speech production and speech understanding.

Figure 6.7 shows the process used in a great many modern speech coding systems, where the original speech signal is split into components that describe the overall gain or amplitude of the speech vector being coded (G), the pitch information (P), a representation of the vocal tract resonances (VT) and lung excitation (EX). Each of these parameters is determined from input speech through an analysis process, quantised and then conveyed from encoder to decoder. At the decoder the parameters are used together in a synthesis process to recreate speech.

In the sections that follow, many speech coding techniques will be introduced and discussed – but all will relate back to this parameterisation of the speech signal into different components that are relevant in some way to the human speech production mechanism.

6.2.1 Linear prediction

Linear prediction has been the mainstay of speech communications technology, and has been applied to speech coding since at least 1976 [63]. It relies upon several characteristics of speech derived from the fact that speech is produced by a human muscular system. These muscles act to shape speech sounds through their movement, which is

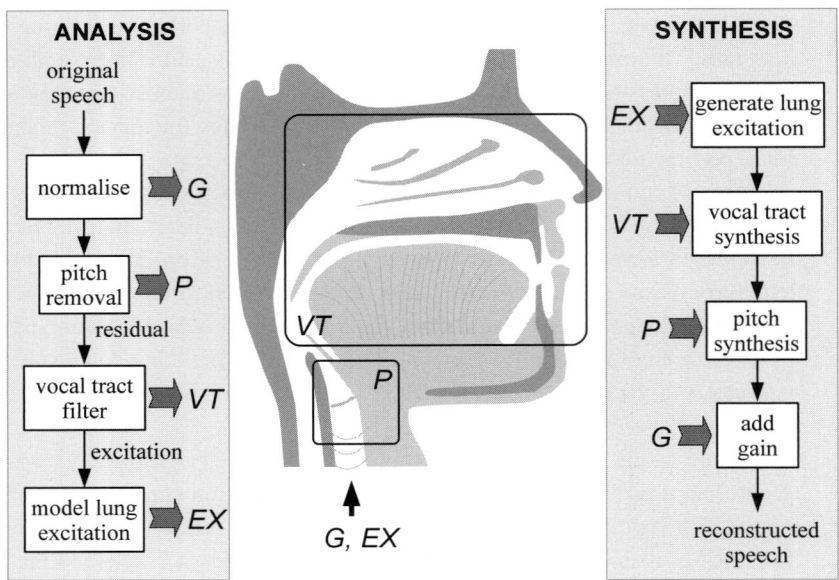

Figure 6.7 Parameterisation of the speech signal into components based loosely upon the human speech production system to convey gain (G), pitch (P), vocal tract filter (VT) and lung excitation (EX) information.

limited by a maximum speed. Muscles cannot move infinitely fast and thus, although human speech varies considerably throughout an utterance, it actually remains pseudo-stationary if we just consider periods of around 30 ms in length (this aspect of speech was discussed previously in Section 2.5.1).

In actual fact, the action of the glottis to generate pitch spikes is often shorter than 30 ms, so through some clever processing (see Section 6.3.2) pitch needs to be removed from the speech signal first – leaving a much lower-energy signal called the *residual*.

Pseudo-stationarity implies that around 240 samples at 8 kHz sample rate (corresponding to a duration of 30 ms) have similar statistics. If pitch has been removed then these samples can be parameterised by a smaller set of values: typically eight or ten linear prediction coefficients, plus an estimate of lung excitation (which could be modelled by random noise). Linear prediction coefficients are generator polynomials for a digital filter that, when stimulated with some input signal, recreates the characteristics of the original samples. The signal that they produce may not appear to be identical when viewed in the time domain, but their frequency response will match the original. Linear predictive coding (LPC) has been used successfully, by itself, in speech coding: the very low bitrate US Federal Standard 1015 2.4 kbits/s algorithm, developed in 1975, is based on LPC. It does, however, have very low perceived quality, and is therefore limited to use in military communications. LPC is most often refined with some other techniques discussed in this section when it is used in modern coding algorithms. We will start by discussing the LPC filter itself, look at how to extract LPC parameters from the original speech signal residual and then go on to consider issues relating to LPC stability and quantisation.

6.2.1.1 The LPC filter

Assuming we have some LPC coefficients that describe the vocal characteristics of a 30 ms or so vector of speech residual, these can be used in two ways. Firstly in a synthesis filter to 'add in' the vocal characteristics to a sample vector. Secondly in an analysis filter to remove the characteristics.

For a Pth-order linear prediction filter represented by P coefficients $a[0], a[1], \ldots, a[P-1]$, the LPC synthesis filter is an all-pole infinite impulse response (IIR) filter, shown for current sample n in Equation (6.1). $x(n)$ is the input audio vector and $y(n)$ is the vector of output audio, which would have the vocal characteristics encoded in a added into it:

$$y(n) = x(n) + \sum_{p=0}^{P-1} a(p) y(n-p). \qquad (6.1)$$

In MATLAB, the linear prediction coefficients are easy to generate using the `lpc()` function, which uses the Durbin–Levinson–Itakura method (which we will explore later, in Section 6.2.2) to solve the recursive equations necessary to generate LPC coefficients.

Assuming we have a MATLAB recording of speech in some vector imaginatively named `speech`, we would normally analyse this one pseudo-stationary window at a time, using overlapping windows. For an 8 kHz sampling rate and pseudo-stationarity of 20–30 ms, we would thus define an analysis window of 160–240 samples (see Section 2.5.1). For illustrative purposes, we will manually segment a speech recording, window it and perform LPC analysis to obtain a tenth-order set of linear prediction coefficients:

```
seg=speech(1:160);
wseg=seg.*hamming(160);
a=lpc(wseg,10);
```

For the sake of demonstration and comparison, here is an example set of LPC coefficients which we can use for testing:

```
a=[1;-1.6187;2.3179;-2.9555;2.8862;-2.5331;
2.2299;-1.3271;0.9886;-0.6126;0.2354];
```

Note that the set begins with a 1.0, and this is standard for LPC coefficient sets within MATLAB. However, when transmitting or storing LPC coefficients of speech, we may delete the leading 1.0 before storage/transmission, always remembering to add it in again during recall/reception.

In z-transform terms, the LPC coefficients are denoted $A(z)$ such that the synthesis filter $H(z)$ is:

$$H(z) = \frac{1}{A(z)} = 1 \bigg/ \left\{ 1 + \sum_{i=1}^{P} a_i z^{-i} \right\}. \qquad (6.2)$$

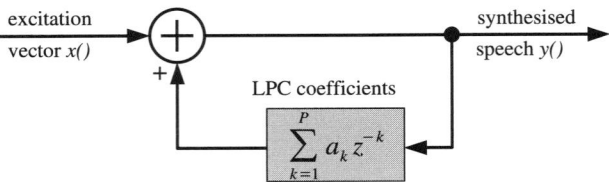

Figure 6.8 Use of LPC coefficients in a synthesis filter.

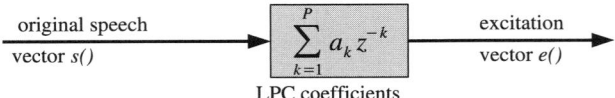

Figure 6.9 Use of LPC coefficients in an analysis filter.

Note the '1+' in the final denominator, which leads to a block diagram structure as shown in Figure 6.8, implementing the direct form of the expression in Equations (6.1) and (6.2), and which mirrors the leading 1.0 of the LPC parameter set in MATLAB.

The LPC analysis filter shown in Figure 6.9 acts to remove the vocal tract information from a signal, but, since this is a digital filter, it has a spectral response which we can analyse or plot. In general, we would do this by substituting z^{-1} in Equation (6.2) with a sweep from $z = 0$ up to $z = e^{-j\omega}$, effectively sweeping from DC up to the Nyquist frequency.

In MATLAB we can cheat by using the freqz() function to do this and plot the magnitude and phase response. A synthesis filter using the LPC coefficients would create the formants and other resonances as seen in the vocal tract, and can be plotted using:

```
freqz(1, a);
```

On the other hand, an analysis filter using the same coefficients would act to remove the resonance from a signal:

```
freqz(a);
```

The freqz() function not only plots a handy magnitude and phase response graph, but can also return these as complex arrays which we can use for subsequent analysis algorithms. As an example, consider the plots of the synthesis and analysis filters constructed from the LPC coefficients represented by the example a vector given in Figure 6.10. The plot on the left clearly shows spectral peaks (which reveal the positions of F1, F2 and F3). The plot on the right shows the response of a filter that would *remove* those formants (i.e. an LPC analysis filter).

6.2 Parameterisation

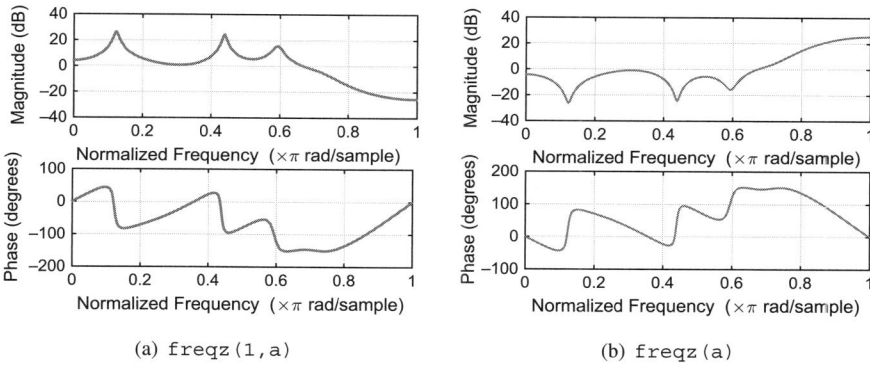

Figure 6.10 Comparison of (a) `freqz(1,a)` and (b) `freqz(a)`.

We can also use `freqz()` to obtain the magnitude response of the synthesis filter at N equally spaced frequencies, and then determine the maximum resonance peak (perhaps corresponding to formant F1 in a speech sample), assuming an 8 kHz sample rate as follows:

```
Fs=8000;    %sample rate
N=100;      %frequency resolution
[H, F] = freqz(1,a,N);
%Plot magnitude with a log scale on the y-axis
semilogy(0:Fs/(N-1):Fs,abs(H));
[y,n]=max(abs(H));
PeakF=(n-1)*Fs/(N-1);
```

The peak frequency of the plot is returned, along with the array index, by the `max()` function, and, because MATLAB indexes arrays from element 1, it is necessary to adjust the returned index, n, when determining the value in Hz.

Both analysis and synthesis forms are used within LPC-based speech compression algorithms – one (or more) at either end. This is indicated in Figure 6.11, which shows a generic LPC-based speech coding algorithm. The same diagram, with some differences in how the excitation is derived or managed, probably represents all CELP class coders (which we will see later in Section 6.4.1) and also the GSM-style pulse-excitation coders (Section 6.3.1).

In the encoder, original speech is normalised to be at some predetermined amplitude range, then the pitch information is extracted to produce a residual. The residual, as mentioned previously, contains vocal tract information which is modelled by LPC. Applying the LPC analysis filter to the residual will result in the vocal tract information being (mostly) removed, leaving a lung excitation signal which is modelled in some way and then transmitted to the decoder.

The decoding process is essentially the reverse of the encoding process, and results in reconstructed speech – the fidelity of which depends on the ability of the algorithm

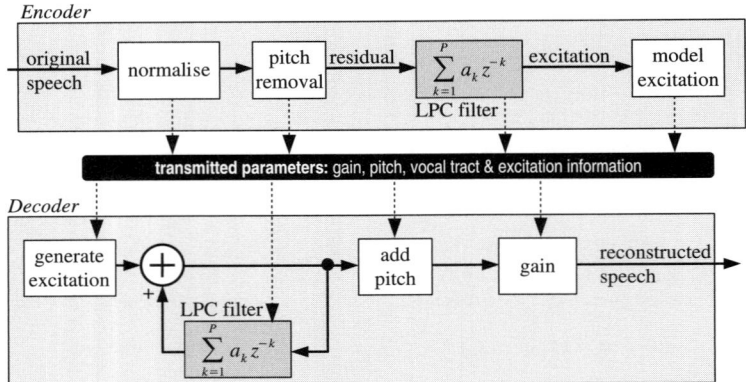

Figure 6.11 The role of LPC analysis and synthesis filters in many speech coding algorithms to convey parameterised vocal tract information from encoder to decoder.

to model the signals being parameterised and on the degree of quantisation applied to the parameters passed from encoder to decoder. Note that in the diagram of Figure 6.11 the extraction process (by which the various parameters are determined) has not been shown.

6.2.1.2 MATLAB implementation

A simple LPC encoder is presented in Listing 6.1. The function `lpc_code()` analyses a speech recording of given sample rate and parameterises it into linear prediction coefficients, long-term predictor values and a lung excitation. There are P LPCs per analysis frame, one LTP lag and one multiplier per four analysis frames and one frame-sized residual vector (named `lung`) to capture the left-over information after removal of the vocal tract and pitch parts. The sample rate is specified as Fs Hz, and a 50% overlap is used for analysis.

Whilst the code is not particularly efficient, it does encode input speech with very good quality. In particular, it uses odd- and even-frame filters (see lines 54 to 58), preserving filter history, to prevent filter edge effects.

Listing 6.1 lpc_code.m

```
function [lpcs,ltps,lung]=lpc_code(speech,Fs,Ws,P)
%Typical values Fs=8000,Ws=floor(Fs*0.02),P=10
Ol=Ws/2;   %50 per cent overlap
L=length(speech);
Nf=fix(L/Ol)-1;
Fn=Fs/2;   %Nyquist
speech=speech/max(speech); %Normalise
SFS=4;   %pitch superframe is 4 normal frames
sfstart=1;
```

6.2 Parameterisation

```
10  SRo=zeros(1,P);
11  SRe=zeros(1,P);
12  Resid=zeros(L,1);
13  PLresid=zeros(L,1);
14  PLhist=zeros(Ws,1);
15  %---define the parameters used
16  lpcs=[];
17  ltps=[];
18  lung=[];
19  for ff=1:Nf
20      sf=1+floor(ff/SFS);
21      n1 = (ff-1)*Ol+1; % start of this frame
22      n2 = (ff+1)*Ol;   % end of this frame
23      wwf=speech(n1:n2).*hamming(Ws);
24      %determine LTP params for each superframe
25      if(ff==1)
26          %get pitch for 1st superframe
27          [B, M] = ltp(speech);
28          ltp_a(1)=1; ltp_a(M)=B;
29          ltps=[B, M];
30      elseif(rem(ff,SFS)==0)
31          [B, M] = ltp(speech(sfstart:n2));
32          ltps=[ltps;[B,M]];
33          ltp_a=ltp_a-ltp_a;
34          ltp_a(1)=1; ltp_a(M)=B;
35          sfstart=n2; %end filter, next superframe
36          % if pitch > than superframe, /2
37          if(M>SFS*Ws/2)
38              M=round(M/2);
39          end
40      end
41      %calc spectral response for each frame
42      a=lpc(wwf,P);
43      lpcs=[lpcs;a(2:P+1)]; %ignore leading 1
44      %Remove the spectral response from speech
45      %(filter memory for odd and even frames)
46      if(rem(ff,2)==0)
47          [Rwwf, SRo]=filter(a,1,wwf,SRo);
48      else
49          [Rwwf, SRe]=filter(a,1,wwf,SRe);
50      end
51      %---construct a residual array
52      Resid(n1:n2)=Resid(n1:n2)+Rwwf;
```

```
53        %filter this to remove pitch
54        if(M > length(PLhist)+1) %pad with zeros
55   [PLwwf, PLhist]=filter(ltp_a(1:M),1,Rwwf,[PLhist;zeros(
        M-length(PLhist)-1,1)]);
56        else
57            [PLwwf, PLhist]=filter(ltp_a(1:M),1,Rwwf,PLhist
                (1:M-1));    %cut short
58        end
59        %construct residual array
60        PLresid(n1:n2)=PLresid(n1:n2)+PLwwf;
61
62        lung=[lung; PLwwf'];
63   end
```

The reverse process, resynthesis from the encoded parameters, is implemented in lpc_decode() presented in Listing 6.2. This uses equivalent parameters and a similar process (including odd- and even-frame filters) to resynthesise speech.

Listing 6.2 lpc_decode.m

```
1  function rsp=lpc_decode(lpcs,ltps,lung,Fs,Ws)
2  Ol=Ws/2;   %50 per cent overlap
3  Nf=size(lpcs,1);
4  P=size(lpcs,2);
5  Fn=Fs/2;   %Nyquist
6  L=Nf*Ol+Ol;
7  SFS=4;   %pitch superframe is 4 normal frames
8  SFo=zeros(1,P);
9  SFe=zeros(1,P);
10 rsp=zeros(L,1);
11 Phist=zeros(Ws,1);
12 ltp_a=zeros(1,SFS*Ws);
13
14 for ff=1:Nf
15     sf=1+floor(ff/SFS);
16     n1 = (ff-1)*Ol+1; % start of this frame
17     n2 = (ff+1)*Ol;   % end of this frame
18     %---reconstruct pitch
19     B=ltps(sf,1);
20     M=ltps(sf,2);
21     ltp_a=ltp_a-ltp_a;
22     ltp_a(1)=1; ltp_a(M)=B;
23     %---residual
```

```
24      PLwwf=lung(ff,:)';
25      %add pitch to the pitchless residual
26      if(M > length(Phist)+1)   %pad with zeros
27          [Pwwf, Phist]=filter(1,ltp_a(1:M),PLwwf,[Phist;
                zeros(M-length(Phist)-1,1)]);
28      else
29          [Pwwf, Phist]=filter(1,ltp_a(1:M),PLwwf,Phist
                (1:M-1));    %cut short
30      end
31      %--VT response, LPCs
32      a=[1,lpcs(ff,:)]';
33      %add in the spectral response
34      if(rem(ff,2)==0)
35          [Swwf, SFo]=filter(1,a,Pwwf,SFo);
36      else
37          [Swwf, SFe]=filter(1,a,Pwwf,SFe);
38      end
39      %---overlap-add the output speech
40      rsp(n1:n2)=rsp(n1:n2)+Swwf;
41  end
```

6.2.1.3 LPC stability issues

Typically the LPC parameters $a(\ )$ are represented in MATLAB as floating point values (in fact, all of the parameters in the listings above are stored in floating point). If these were used in a speech coder to represent speech parameters then they would need to be handled as smaller fixed point values for storage or transmission between encoder and decoder (since floating point values occupy too many bits).

Unfortunately this quantisation process can occasionally produce a set of parameters that results in an unstable synthesis filter in the decoder. This means that the LPC filter output magnitude rises sharply then 'blips' towards infinity before recovering. When listening to speech this effect would be recognisable as a loud pop or squeak. The same effect can sometimes be heard on live news broadcasts from remote locations by satellite phone – pops and blips in the audio disrupting a reporter's voice are the result of errors in the parameterised audio bitstream 'kicking' the LPC synthesis filter into momentary instability.

Of course, in the presence of bit errors in the parameters such things are expected; however, even in the absence of errors, directly quantising LPC parameters often results in instability. We will look briefly at the effect of quantising LPC parameters directly in Section 6.2.6, but note that LPC parameters are never quantised directly – in practice they are always converted into an alternative form. One such form is the reflection coefficients described in Section 6.2.2.

Log area ratios (LARs) are the method of choice for the GSM speech coder, comprising a logarithmic transformation of the reflection coefficients, and other more esoteric mathematical transformations exist. None of these can compete with the quantisation characteristics of line spectral pairs (LSPs), which are worthy of their own discussion in Section 6.2.5.

6.2.1.4 Pre-emphasis of the speech signal

An important practical point to note when performing LPC analysis is that the LPC coefficients that are found are supposed to match the analysed speech signal as closely as possible. However, it turns out that the LPC equations tend to satisfy the lower frequencies while matching the higher frequencies more poorly. Thus it is common to emphasise the higher frequencies prior to LPC analysis.

In fact, when speech is radiated from a human mouth, from an area of high pressure, through a constriction, into a low-pressure area, a spectral roll-off occurs to reduce the amplitude of the higher frequencies. Thus speech recorded from *outside* the mouth will differ from speech recorded from *inside* the mouth (and there do exist tiny microphones that can record speech from within the mouth). Since the LPC filter employed in speech analysis is supposed to model the vocal tract response, it is preferable to allow it to analyse the signal produced by the vocal tract without the influence of lip radiation. We can therefore counteract the effects of lip radiation by performing pre-emphasis before LPC analysis.

Given a speech signal $s(n)$, the pre-emphasis would normally be performed with a single-tap filter having transfer function $(1 - \alpha z^{-1})$ and an emphasis coefficient, α, of nearly 1. A typical value used in research systems is $\alpha = 15/16 = 0.9375$. Thus each pre-emphasised speech sample $s'(n)$ comes from the current and previous input speech samples acted on by the following FIR filter:

$$s'(n) = s(n) - 0.9375 \times s(n-1). \tag{6.3}$$

By performing pre-emphasis of the speech signal in this way prior to LPC analysis, we can better approach the signal that leaves the vocal tract, and can overcome one of the issues with LPC analysis where the coefficients match the higher-frequency components poorly.

Of course, any speech system that outputs something based upon pre-emphasised speech will sound a little strange – we do not normally hear speech from *within* a person's mouth. So, even though the processing can be conducted on pre-emphasised speech, the output must be de-emphasised, to replace the attenuated low frequencies which we had removed.

The de-emphasis filter matches the emphasis filter, reversing the emphasis applied to the speech signal $s'(n)$ to recreate more natural sounding speech $r(n)$. This IIR filter is as follows:

$$r(n) = s'(n) + 0.9375 \times r(n-1). \tag{6.4}$$

In MATLAB we can create a pre-emphasis filter very easily, and apply this using the `filter()` function. If the original speech signal is called s, the pre-emphasised output

6.2 Parameterisation

is to be `es` and the de-emphasised version of this is `ds` then we can easily convert between them:

```
% Create emphasis/de-emphasis filter coeffs
h=[1, -0.9375];
% Apply the emphasis filter
es=filter(h, 1, s);
% Apply the de-emphasis filter
ds=filter(1, h, es);
```

It is instructive to try listening to this on a piece of test speech and to use the MATLAB `psd()` function to conveniently plot a frequency response of the signals at each stage. The slightly nasal sounding `es` is a sound that many speech researchers are familiar with. If, when developing a speech algorithm, you hear something like this, then you will realise that you have lost some of the lower frequencies somewhere, or perhaps forgotten the de-emphasis filter.

6.2.2 Reflection coefficients

Since the LPC coefficients by themselves cannot be reliably quantised without causing instability, significant research effort has gone into deriving more stable transformations of the parameters. The first major form are called *reflection coefficients* because they represent a model of the synthesis filter that, in physical terms, is a set of joined tubes of equal length but different diameters.

In fact the same model will be used for line spectral pairs (Section 6.2.5) but under slightly different conditions. The reflection coefficients quantify the energy reflected back from each join in the system. They are sometimes called partial correlation coefficients or PARCOR after their method of derivation.

Conversion between PARCOR and LPC is trivial, and in fact LPC coefficients are typically derived from input speech by way of a PARCOR analysis (although there are other methods). This method, and the rationale behind it, will be presented here, and can be followed in either [63] or [64]. First, we make the assumption that, given a frame of pseudo-stationary speech residual, the next sample at time instant n can be represented by a linear combination of the past P samples. This linear combination is given by Equation (6.5):

$$x'[n] = a_1 x[n-1] + a_2 x[n-2] + a_3 x[n-3] + \cdots + a_P x[n-P]. \qquad (6.5)$$

The error between the predicted sample and the actual next sample quantifies the ability of the system to predict accurately, and as such we need to minimise this:

$$e[n] = x[n] - x'[n]. \qquad (6.6)$$

The optimum would be to minimise the mean-squared error over all n samples:

$$E = \sum_n e^2[n] = \sum_n \left\{ x[n] - \sum_{k=1}^{P} a_k x[n-k] \right\}^2. \quad (6.7)$$

In order to determine the set of LPC coefficients resulting in the minimum mean-squared error, E, in Equation (6.7), we must differentiate the expression and then equate to zero:

$$\frac{\delta E}{\delta a_j} = -2 \sum_n x[n-j] \left\{ x[n] - \sum_{k=1}^{P} a_k x[n-k] \right\} = 0. \quad (6.8)$$

This results in a set of linear equations containing P unknowns, a, that can be derived from the known speech samples $x[n]$ to $x[n-P]$:

$$\sum_{k=1}^{P} a_k \sum_n x[n-j]x[n-k] = \sum_n x[n]x[n-j], \quad (6.9)$$

where $j = 1, \ldots, P$.

There is of course a choice of methods to solve such a set of equations. Most common are the covariance and autocorrelation methods. The former splits the speech into rectangular windowed frames and minimises the error over each frame of N samples. The latter assumes that the signal is stationary with finite energy, with an infinite summation range (which is acceptable if we window the speech before analysis). Covariance tends to be more accurate for small speech frames, but, with the sharp cutoffs of a rectangular window, can lead to instability. For this reason, most speech coding algorithms opt for the autocorrelation approach (and hence derive partial correlation coefficients, known as PARCORs, along the way).

For the infinite summation, we first note that the following relationships hold:

$$\sum_{n=-\infty}^{\infty} x[n-j]x[n-k] \equiv \sum_{n=-\infty}^{\infty} x[n-j+1]x[n-k+1] \quad (6.10)$$

and

$$\sum_{n=-\infty}^{\infty} x[n-j+1]x[n-k+1] \equiv \sum_{n=-\infty}^{\infty} x[n]x[n+j-k]. \quad (6.11)$$

Using these equivalence relationships, we re-formulate Equation (6.9) as follows:

$$\sum_{k=1}^{P} a_k \sum_{n=-\infty}^{\infty} x[n]x[n+j-k] = \sum_{n=-\infty}^{\infty} x[n]x[n-j]. \quad (6.12)$$

Note the similarity between this and the standard autocorrelation function in Equation (6.13), where $R(k)$ denotes the kth autocorrelation:

$$R(k) = \sum_{n=-\infty}^{\infty} x[n]x[n+k]. \quad (6.13)$$

The set of P linear equations can thus be represented as the following matrix

$$\begin{bmatrix} R(0) & R(1) & R(2) & \ldots & R(P-1) \\ R(1) & R(0) & R(1) & \ldots & R(P-2) \\ R(2) & R(1) & R(0) & \ldots & R(P-3) \\ \vdots & \vdots & \vdots & \ddots & \vdots \\ R(P-1) & R(P-2) & R(P-3) & \ldots & R(0) \end{bmatrix} \begin{bmatrix} a_1 \\ a_2 \\ a_3 \\ \vdots \\ a_P \end{bmatrix} = \begin{bmatrix} R(1) \\ R(2) \\ R(3) \\ \vdots \\ R(P) \end{bmatrix}. \tag{6.14}$$

In practice a window, usually Hamming (Section 2.4.2), is applied to the input speech prior to calculating the autocorrelation functions, and the entire autocorrelation results are usually normalised by dividing by $R(0)$ first. These normalised coefficients are denoted $r(i)$.

Standard techniques exist for the matrix solution, including brute force matrix inversion, the famous Durbin–Levinson–Itakura method and the Le Roux method, which is slightly less efficient but is a compact and easily followed recursive formula [65]:

$$k_{n+1} = \frac{e_{n+1}^n}{e_0^n} \quad \text{for} \quad n = 0, \ldots, P, \tag{6.15}$$

$$e_0^{n+1} = e_0^n - k_{n+1} e_{n+1}^n = e_0^n (1 - k_{n+1}^2), \tag{6.16}$$

$$e_i^{n+1} = e_i^n - k_{n+1} e_{n+1-i}^n \quad \text{for} \quad i = n, \ldots, P, \tag{6.17}$$

where the initial conditions for the recursion are set to $e_i^0 = R(i)$ for each i in the set of P equations. The values of k that are obtained from the Le Roux method are the *reflection coefficients*.

6.2.3 Converting between reflection coefficients and LPCs

Conversion from the reflection coefficients (which in some cases are quantised and transmitted from speech encoder to decoder) to the LPC parameters, which are required for the LPC synthesis filter to recreate encoded speech, is not difficult. The relationship is shown in Equation (6.18) for all P coefficients, where the notation a_j^i indicates the jth LPC coefficient at time instant i:

$$a_j^i = a_j^{(i-1)} + k_i a_{(i-j)}^{(i-1)} \quad \text{with} \quad 1 \leq j \leq i-1. \tag{6.18}$$

In order to perform the reverse conversion from LPC parameters into reflection coefficients, we start with the initial value:

$$k_i = a_j^i$$

and then follow with:

$$a_j^{(i-1)} = \frac{a_j^i - a_i^i a_{(i-j)}^i}{1 - k_i^2} \quad \text{with} \quad 1 \leq j \leq i-1, \tag{6.19}$$

where Equation (6.19) is repeated with i decreasing from P to 1 with initial conditions of $a_j^P = a_j$ for all j's between 1 and P.

6.2.4 LPCs and the tube model

When looking at reflection coefficients, we had briefly noted that they were inspired by considering the vocal tract to be a set of tubes of equal length but different diameters. Just as a single tube resonates at a set frequency (think of an organ pipe, or the sound that can be made when blowing air over the top of a glass bottle), multiple tubes together resonate at complex frequencies. The illustration in Figure 6.12 shows three different sets of interconnected tubes, increasing in complexity from top to bottom. In operation, the main tube resonances will clearly lie at the formant frequencies, while other smaller resonances also contribute to the overall spectral shape of the sound, affecting its perceived quality or texture.

In fact the tube model, which predates LPC by a number of years, was first described by John Kelly and Carol Lochbaum [66], and is now most often known as the Kelly–Lochbaum model (or even more simply as the K–L tube model).

In the most popular usage of this model, 22 interconnected tubes are sufficient to model most of the important detail of the human vocal tract, while an extended 44-tube model is used where frequencies up to 20 kHz are considered important. The only variable quantity in the model is the diameter of the tubes: their lengths, interconnection points and sequence are all fixed once the sample rate is set.

Much more recently, researchers have been able to relate this type of model to real vocal tract geometry. For example, researchers such as Brad Story *et al.* [67] scanned volunteers using magnetic resonance imaging (MRI) while the volunteers were speaking various phonemes. By analysing the MRI data, they can determine the vocal tract shape as it varies from the glottis to the lips, transforming these into an area function by slicing into equal-length segments.

Figure 6.13 illustrates the usefulness of this by comparing five sets of area functions extracted from [67]. The vocal tract tube model diameters are plotted as they change in steps from the glottis to the lips for the vowels /i/, /ɪ/, /ʌ/, /o/ (i.e. the vowels in 'me', 'lid', 'bun' and 'hot' respectively) and the consonant /l/. In addition, LPC coefficients

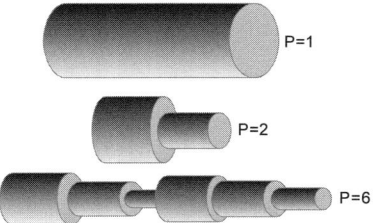

Figure 6.12 First-, second- and sixth-order tube models. Imagine sound entering from one end (the 'glottal' end), resonating through the tubes and exiting at the opposite ('lip' or 'mouth') end.

6.2 Parameterisation

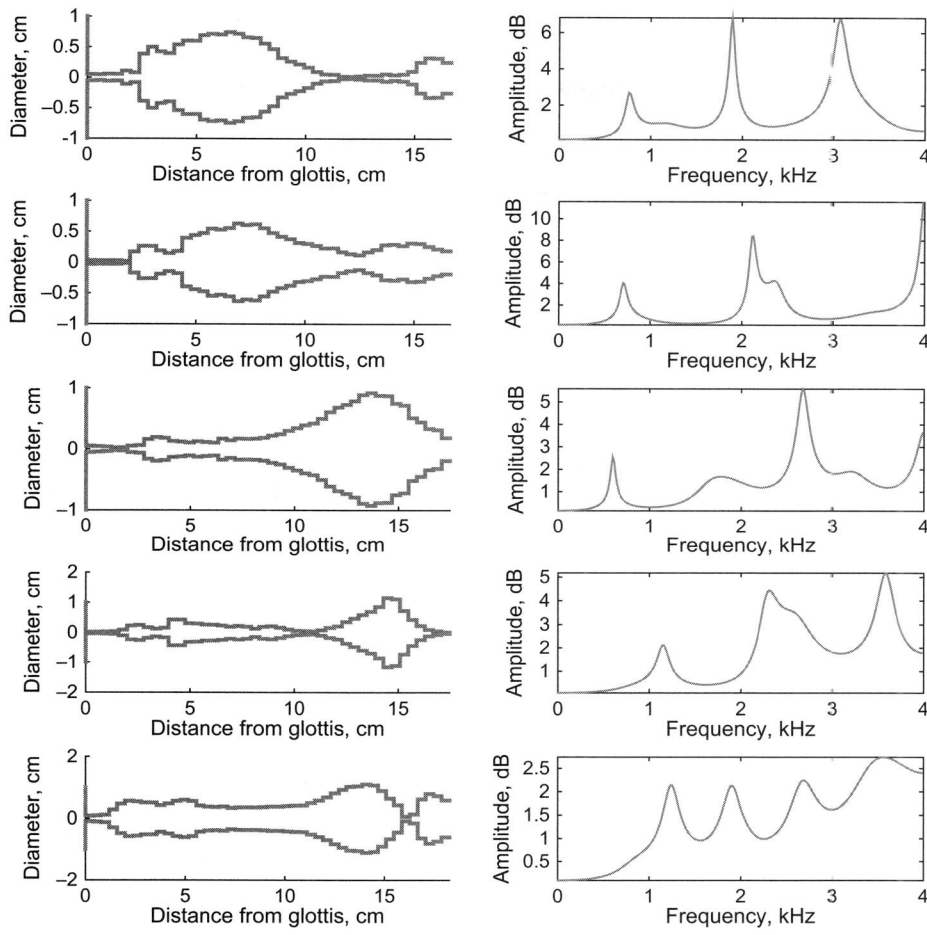

Figure 6.13 Tube cross-sections on the left, and frequency responses on the right for vowels /i/, /ɪ/, /ʌ/, /o/ and consonant /l/ (all in IPA notation).

are derived from the area functions and used to plot an LPC power spectrum from 0 to 4 kHz for each of the sounds. The conversion from area functions to generate the plots shown is described in Infobox 6.3 on page 164.

In this book, we will not consider area functions or real VT shape further, but readers are encouraged to experiment with this representation. When evaluating different areas, it is actually quite difficult to predict how even small changes in area to one part of the tube model can cause large changes in frequency response across the spectrum. What is amazing is the fact that most human brains can learn to control this complex, dynamic and sensitive system in order to create consistent and intelligible speech. It does take a year or more to gain basic control, and a number of years beyond that to master the system – which highlights the difficulties involved, although most of us rarely pause to consider this fact.

Box 6.3 Tube models and VOICEBOX

We will make use of the excellent VOICEBOX package from Mike Brookes of Imperial College Department of Electrical & Electronic Engineering (links to this MATLAB resource, and basic instructions for installation and use, are given on the book website). VOICEBOX contains a large number of extremely useful functions, but we will primarily make use of `lpcaa2rf()` and `lpcrf2ar()` to convert from area functions to reflection coefficients, and then from reflection coefficients to LPCs respectively, as follows, given the vector ph of area functions such as that published for a single phoneme by Story et al. [67].

```
%add free space beyond lips & glottis
sect=0.396825; %in cm
A=[0,ph,100];
rf=lpcaa2rf(A);     %gives reflection coeffs
lpcs=lpcrf2ar(rf); %gives LPC coeffs
ph=ph/pi;        %convert from area to diameter
subplot(2,1,1) %Plot tube shape
oldyu=+1;
oldyl=-1;
for ll=1:L
  startx=(ll-1)*Sect;
  endx=(ll)*Sect;
  startyu=0.5*ph(ll);
  startyl=-0.5*ph(ll);
  line([startx,startx],[oldyu,startyu],'linewidth
     ',2)
  line([startx,endx],[startyu,startyu],'linewidth
     ',2)
  line([startx,startx],[oldyl,startyl],'linewidth
     ',2)
  line([startx,endx],[startyl,startyl],'linewidth
     ',2)
  oldyu=startyu;
  oldyl=startyl;
end
axis tight
%Plot the lower part of the spectral response
subplot(2,1,2)
Fn=4000;
xf=freqz(1,lpcs(1:9),100,Fn*2);
plot([1:100]*Fn/100,log10(xf));
```

6.2.5 Line spectral pairs

Line spectral pairs (LSPs) are a direct mathematical transformation of the set of LPC parameters, and are generated within many speech compression systems, such as the more modern CELP coders (which will be discussed later in Section 6.4.1). LSP usage is popular due to their excellent quantisation characteristics and consequent efficiency of representation. They are also commonly referred to as line spectral frequencies (LSFs) [65].

LSPs collectively describe the two resonance conditions arising from an interconnected tube model of the human vocal tract. This includes mouth shape and nasal cavity, and forms the basis of the underlying physical relevance of the linear prediction representation. The two resonance conditions describe the modelled vocal tract as being either fully open or fully closed at the glottis respectively (compare this with the model of the reflection coefficients in Section 6.2.2). The model in question is constructed from a set of equal length but different diameter tubes, so the two conditions mean the source end is either closed or open respectively. The two conditions give rise to two sets of resonance frequencies, with the number of resonances in each set being determined by the number of joined tubes (which is defined by the order of the analysis system). The resonances of each condition give rise to odd and even line spectral frequencies respectively, and are interleaved into a set of LSPs which have monotonically increasing value.

In reality, however, the human glottis opens and closes rapidly during speech, so it is neither fully closed nor fully open.

Hence actual resonances occur at frequencies located somewhere between the two extremes of each LSP condition. Nevertheless, this relationship between resonance and LSP position lends a significant physical basis to the representation. Figure 6.14 illustrates LSPs overlaid on an LPC spectral plot (made using the lpcsp() MATLAB function given later in Section 6.2.5.3). The ten vertical lines were drawn at the LSP frequencies, and show the odd and even frequencies being interleaved. Both the LSPs and the spectrum were derived from the same set of tenth-order linear prediction parameters which were obtained from a linear predictive analysis of a 20 ms voiced speech frame.

Notable features of Figure 6.14 include the natural interleaving of the odd and even LSP frequencies, and the fact that spectral peaks tend to be bracketed by a narrow pair of lines (explained by the comment previously indicating that the actual resonance frequency lies somewhere between the open and closed model positions, represented by odd and even lines). Local minima in the spectrum tend, by contrast, to be avoided by nearby lines. These and several other properties explain the popularity of LSPs for the analysis, classification and transmission of speech.

6.2.5.1 Derivation of LSPs

Line spectral frequencies are derived from the linear predictive coding (LPC) filter representing vocal tract resonances in analysed speech as we have seen in

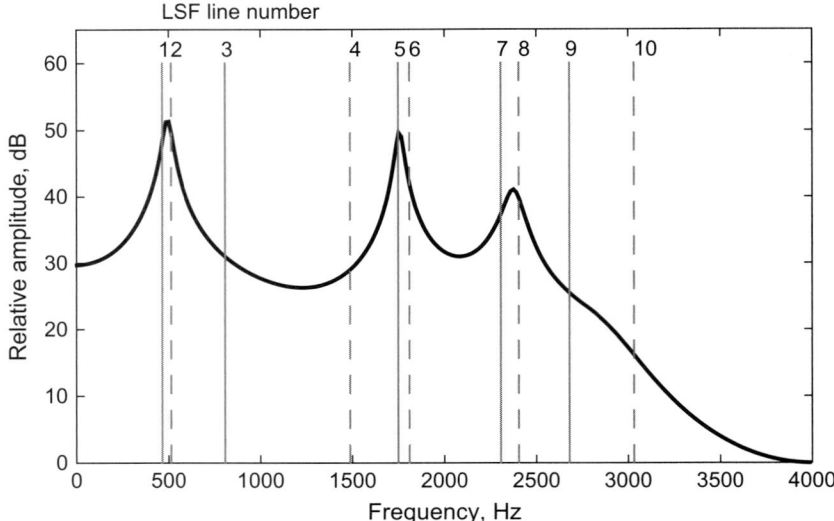

Figure 6.14 Plot of a sample LPC spectrum with the corresponding LSP positions overlaid. Odd lines are drawn solid and even lines are drawn dashed.

Section 6.2.1.1. For Pth-order analysis, the LPC polynomial would be:

$$A_p(z) = 1 + a_1 z^{-1} + a_2 z^{-2} + \cdots + a_P z^{-P}. \quad (6.20)$$

We will define two $(P+1)$th-order polynomials related to $A_p(z)$, named $P(z)$ and $Q(z)$. These are referred to as antisymmetric (or inverse symmetric) and symmetric in turn based on observation of their coefficients. The polynomials represent the interconnected tube model of the human vocal tract and correspond respectively to complete closure at the source end of the interconnected tubes and a complete opening, depending on an extra term which extends the Pth-order polynomial to $(P+1)$th-order. In the original model, the source end is the glottis, and is neither fully open nor fully closed during the period of analysis, and thus the actual resonance conditions encoded in $A_p(z)$ are a linear combination of the two boundaries. In fact this is simply stated:

$$A_p(z) = \frac{P(z) + Q(z)}{2}. \quad (6.21)$$

The two polynomials are created from the LPC polynomial with an extra feedback term being positive to model energy reflection at a completely closed glottis and negative to model energy reflection at a completely open glottis:

$$P(z) = A_p(z) - z^{-(P+1)} A_p(z^{-1}), \quad (6.22)$$

$$Q(z) = A_p(z) + z^{-(P+1)} A_p(z^{-1}). \quad (6.23)$$

The roots of these two polynomials are the set of line spectral frequencies, ω_k. These relate to symmetric and antisymmetric polynomials as will be shown later in Equation (6.25).

6.2 Parameterisation

The complex roots of the polynomials can be shown to lie on the unit circle in the z-plane if the original LPC filter was stable [65], and alternate around the unit circle. Note that *any* equivalent size set of roots that alternate in this way around and on the unit circle will represent a stable LPC filter. This is something with implications which we will exploit for LSP-based speech processing (for example a voice changing application in Chapter 10).

There are many different ways of calculating or estimating roots, including trial and error, Newton–Raphson approximation, evaluating around the unit circle looking for sign changes and so on, with the exact choice of method having implications for accuracy, processing complexity and processing delay. In MATLAB we can simply use the in-built roots() function to perform the calculation.

If we denote the set of complex roots as θ_k, then the line spectral frequencies are determined from solving Equations (6.22) and (6.23):

$$\omega_k = \tan^{-1}\left(\frac{\text{Re}\{\theta_k\}}{\text{Im}\{\theta_k\}}\right), \qquad (6.24)$$

where ω_k are the P line spectral frequencies expressed in radians.

Despite the seemingly complex equations, an example MATLAB script for converting from LPC to LSP is relatively short, as shown in Listing 6.3.

Listing 6.3 lpc_lsp.m

```
1  %This function transforms LPC values to LSPs.
2  %The LPC array begins with 1 (ie a=[1, ....])
3  %The order must be EVEN, length(a) must be odd
4  function lsp=lpc_lsp(a)
5  p=length(a);
6  %derive the coefficients for P'(z) and Q'(z)
7  A(1)=1;
8  B(1)=1;
9  for k=2:p
10    A(k)=(a(k) - a(p+2-k)) + A(k-1);
11    B(k)=(a(k) + a(p+2-k)) - B(k-1);
12  end
13  r1=roots(A);
14  r2=roots(B);
15  for k =1:p-1
16    if (real(r1(k)) < 0)
17      theta1(k)=pi-abs(atan(imag(r1(k))/real(r1(k))));
18    else
19      theta1(k)=abs(atan(imag(r1(k))/real(r1(k))));
20    end
21    if (real(r2(k)) < 0)
22      theta2(k)=pi-abs(atan(imag(r2(k))/real(r2(k))));
```

```
23    else
24       theta2(k)=abs(atan(imag(r2(k))/real(r2(k))));
25    end
26 end
27 p=p-1;
28 for k=1:p/2
29    theta(k)=theta1(k*2);
30    theta(k+(p/2))=theta2(k*2);
31 end
32 lsp=sort(theta);   %Sort into ascending order
```

For testing purposes, an example set of LPC coefficients from tenth-order analysis of speech is given below, with the corresponding LSP converted using the function above (these are the same coefficients as were used previously for the illustration of LPC coefficients in Section 6.2.1.1):

```
a=[1;-1.6187;2.3179;-2.9555;2.8862;-2.5331;
2.2299; -1.3271;0.9886;-0.6126;0.2354];

lsp=[0.3644;0.4023;0.6334;1.1674;1.3725;
1.4205;1.8111;1.8876;2.1032;2.3801];
```

Note that the LPC coefficient array begins with a fixed '1.0' by convention (this was also discussed in Section 6.2.1.1). By contrast, each of the LSP values is different and thus meaningful in itself. As we will examine in Section 6.2.6, they tend to occupy certain frequency ranges for most analysis frames of a particular type.

6.2.5.2 Generation of LPC coefficients from LSPs

Conversion from LSPs to LPCs is a simple process (although the in-depth proof is more difficult – for this refer to the excellent reference book by Saito and Nakata [65]). For this conversion we can, of course, use the ordered LSPs ω_k to recreate the polynomials that they are roots of, namely $P(z)$ and $Q(z)$ [68]:

$$P(z) = (1 - z^{-1}) \prod_{k=2,4,\ldots,P} (1 - 2z^{-1} \cos \omega_k + z^{-2}),$$
$$Q(z) = (1 + z^{-1}) \prod_{k=1,3,\ldots,P-1} (1 - 2z^{-1} \cos \omega_k + z^{-2}), \quad (6.25)$$

and then these can be substituted into Equation (6.21), which expresses $A_p(z)$ as a linear combination of $P(z)$ and $Q(z)$. Since Equation (6.25) involves the cosine of the LSPs, and bearing in mind that some of the more efficient LSP calculation algorithms will yield the roots in the cosine domain [69], it is common also to perform the reverse conversion from LPC to LSP in the cosine domain.

6.2 Parameterisation

If q_k denotes the array of correctly ordered, lowest-first, LSP frequencies ω_k, in the cosine domain:

$$q_k = \cos \omega_k. \tag{6.26}$$

To perform the conversion, we first need to solve the following set of recursive equations as an intermediate step:

$$\begin{aligned}&\text{for } k = 1, \ldots, P \\ & f_k = -2f_{(k-1)}q_{(2k-1)} + 2f_{(k-1)} \\ &\text{for } m = (k-1), \ldots, 1 \\ & f_m = f_m - 2f_{(m-1)}q_{(2k-1)} + f_{(m-2)}.\end{aligned} \tag{6.27}$$

We then apply initial conditions of $f_0 = 1$ and $f_{-1} = 0$, and calculate the coefficients of g similarly as follows:

$$\begin{aligned}&\text{for } k = 1, \ldots, P \\ & g_k = -2g_{(k-1)}q_{(2k)} + 2g_{(k-1)} \\ &\text{for } m = (k-1), \ldots, 1 \\ & g_m = g_m - 2g_{(m-1)}q_{(2k)} + g_{(m-2)}.\end{aligned} \tag{6.28}$$

Once the values of f and g have been determined, they form a second set of equations:

$$f'_k = f_k + f_{(k-1)} \quad \text{for} \quad k = 1, \ldots, P/2, \tag{6.29}$$

$$g'_k = g_k + g_{(k-1)} \quad \text{for} \quad k = 1, \ldots, P/2, \tag{6.30}$$

which are then averaged to form LPC coefficients from the following:

$$a_k = \frac{1}{2}f'_k + \frac{1}{2}g'_k \quad \text{for} \quad k = 1, \ldots, P/2, \tag{6.31}$$

$$a_k = \frac{1}{2}f'_{(k-P/2)} - \frac{1}{2}g'_{(k-P/2)} \quad \text{for} \quad k = P/2+1, \ldots, P. \tag{6.32}$$

Using MATLAB, a function that reads in LSPs (represented in radians), converts them to cosine domain and then replicates the steps above is given in Listing 6.4.

Listing 6.4 lsp_lpc.m

```
%Transforms an array of LSPs to LPCs.
%Note: number of LSPs must be EVEN
%
function a=lsp_lpc(theta)
p=length(theta);
q=cos(theta(1:p));

f1(10)=1;
f1(9)=0;
for i=1:p/2
    f1(10+i)=-2*q(2*i-1)*f1(10+i-1) + 2*f1(10+i-2);
```

```matlab
12     for k=i-1:-1:1
13         f1(10+k)=f1(10+k)-2*q(2*i-1)*f1(10+k-1)+f1(10+k-2);
14     end
15 end
16
17 f2(10)=1;
18 f2(9)=0;
19 for i=1:p/2
20     f2(10+i)=-2*q(2*i)*f2(10+i-1) + 2*f2(10+i-2);
21     for k=i-1:-1:1
22         f2(10+k)=f2(10+k)-2*q(2*i)*f2(10+k-1)+f2(10+k-2);
23     end
24 end
25
26 f1b(1)=f1(11)+1;
27 f2b(1)=f2(11)-1;
28 for i=2:p/2
29     f1b(i) = f1(10+i) + f1(10+i-1);
30     f2b(i) = f2(10+i) - f2(10+i-1);
31 end
32
33 for i=1:p/2
34     a2(i)     = 0.5*( f1b(i) + f2b(i) );
35     a2(i+p/2) = 0.5*(f1b((p/2)-i+1) - f2b((p/2)-i+1));
36 end
37
38 a=[1,a2];
```

The example LSP and LPC coefficient values given in Section 6.2.5.1 can again be used for testing of the function.

6.2.5.3 Visualisation of line spectral pairs

As discussed previously, line spectral pairs are resident in the frequency domain. Their values can be denominated in Hz, more normally in radians, or as the cosine of their radian value. In this book, unless stated otherwise, all LSP values used in equations and MATLAB code will be given in radians.

Whatever the units, each line is located at a particular frequency. Thus a traditional method of visualising the value of lines is to plot them as an overlay on the LPC spectrum. This was illustrated in Figure 6.14. The MATLAB `freqz()` function generates a frequency response from an LPC polynomial, so we shall use that to plot a spectrum, and then overlay the LSPs on top of this, drawn as lines:

```
function lpcsp(a, lsp)
[HH, FF]=freqz(1, a, 100);
semilogy(abs(HH),'m-');
hold on V=axis;
axis([1,length(FF),V(3),V(4)]);
hold on;
lsc=100/pi;
for lp=1:length(lsp)
   line([1+lsp(lp)*lsc,1+lsp(lp)*lsc], [V(3),V(4)]);
end
hold off;
```

An enhanced version of this function called lpcspF() can be found on the website. It produces an identical plot to lpcsp(), but takes an additional argument of sample frequency and embellishes the plot with labelled axes and line numbers. Figure 6.14 used this to plot the LPC and LPS values using the following MATLAB code. It makes use of the example a array presented above (and available on the website as a.m):

```
lpcspF(a, lpc_lsp(a), 8000);
```

6.2.6 Quantisation issues

Since LSPs are most often used in speech coders, where they are quantised prior to transmission, it is useful to explore the quantisation properties of the representation. In order to do this, we can use some representative recorded speech, quantise in turn by different amounts, dequantise to recreate speech, and in each case compare the original and dequantised speech.

In fact, this has been done using a large pre-recorded speech database called TIMIT [70]. Several thousand utterances were analysed, then LPCs and then LSPs were derived and quantised by different amounts. The quantisation scheme used in this case was uniform quantisation, where each parameter is represented by an equal number of bits, ranging in turn, from 4 to 16 bits per parameter.

Tenth-order analysis was used, and a segmental signal-to-noise ratio (SEGSNR, discussed in Section 3.4.2) determined between the original and dequantised speech. Both LSP and LPC quantisation were tested.

Table 6.1 lists the results, where the more positive the SEGSNR value, the more closely the dequantised speech matches the original speech. These results clearly indicate that LPCs are far more susceptible to quantisation effects than are LSPs: down around 5 bits per parameter, the recreated speech resulting from LPC quantisation exhibits sharp spikes of noise and oscillation, totally unlike the original speech. Hence

Table 6.1 SEGSNR resulting from different degrees of uniform quantisation of LPC and LSP parameters.

Bits/parameter	LPC	LSP
4	–	−6.26
5	−535	−2.14
6	−303	1.24
7	−6.04	8.28
8	−10.8	15.9
10	19.7	20.5
12	22.2	22.2
16	22.4	22.4

the huge difference between original and recreated speech evidenced by the SEGSNR value of −535. By contrast the LSP representation, with an SEGSNR of −2.14, indicates quite easily understandable speech. This substantiates the assertion in Section 6.2.5 that LSPs are favoured for their superior quantisation characteristics. Finally, note that both approaches in the table achieve a SEGSNR level of 22.4 when more bits are used for quantisation, plainly indicating the limit of achievable SEGSNR for the analysis process used (i.e. the window function, autocorrelation number of parameters and so on).

LSPs may be available in either the frequency domain or the cosine domain (depending on the method of solving the polynomial roots). Each line's value can be quantised independently (scalar quantisation) on either a uniform or a non-uniform scale [71], which can also be dynamically adapted. Alternatively, lines can be grouped together and vector quantised with either static or adaptive codebooks [72]. Vector quantisation groups sets of lines together and quantises these as a set. Typical sets may be a $(2, 3, 3, 2)$ or a $(3, 3, 4)$ arrangement for a tenth-order system.

Both scalar and vector quantisation can be applied either to the raw LSP values themselves or to differential values, where the difference is either that between a line's current position and its position in the previous frame or that between its current position and its mean position [73]. We can refer to these as the short-term and long-term average differences respectively [74].

An adaptation of long-term average differential quantisation (which uses the distance between the current position and mean position of a line) is to recalculate the nominal position every frame based on an even distribution of nominal positions between the values of the first and last LSPs. This is known as interpolated LSF (or LSF interpolation, LSFI) [75]. A different form of interpolation is that applied by the TETRA standard CELP coder [76], which quantises LSPs which have been interpolated between subframes (of which there are four per standard-sized frame). This approach can provide a degree of immunity to the effects of subframes lost due to burst errors.

An effective quantisation scheme will generally minimise either the signal-to-quantisation noise ratio for typical signals or a more perceptually relevant measure. Such measures could be the commonly used spectral distortion (SD) value (see Section

3.4.2) or similar variants. Some published examples are the LSP distance (LD) [73], LSP spectral weighted distance measure (LPCW) [77], local spectral approximation weight (LSAW) [78] and inverse harmonic mean weight (IHMW) [79].

In all cases, it is necessary to appreciate the dynamics of the signal to be quantised, and optionally to assign different levels of importance to critical spectral regions, either directly or by allocating greater quantisation accuracy to LSPs with a frequency locus within such critical regions. It is possible to match regions of spectral importance to LSP accuracy through the selection of different quantiser resolutions for different lines. For example, lines 9 and 10 in a tenth-order analysis would relate to formant F3, if present. This formant can be considered less important to speech intelligibility than formants F1 and F2. Therefore lines 9 and 10 may be quantised with fewer bits than, for example, lines 5 and 6.

By plotting the LSP line frequency locus for a number of TIMIT speech recordings, as shown in Figure 6.15, we can see that the line localisation in frequency is fairly limited. The figure shows which lines are located predominantly in frequency regions of less importance to intelligibility: these are natural candidates for being quantised with fewer bits than other lines. The plot was obtained through tenth-order LPC analysis on 40 ms frames with 50% overlap for different male and female speakers. These LPC coefficients were then transformed into LSP values, with the relative frequency of their values computed across 40 analysis bins and then plotted in the vertical axis for each of the LSP lines.

Table 6.2 lists the average and median LSP frequency for a flat spectral frame. It also lists the standard deviation between the average frequency of each line and its actual location (averaged over several hundred seconds of speech extracted randomly from the TIMIT database and processed with tenth-order analysis). In the table, the standard deviation of lines 1 and 10 is significantly smaller than that of other lines, such as line 6. In a non-uniform scalar quantisation scheme, this would mean that lines 1 and 10 require fewer quantisation bits than line 6 for equivalent signal-to-quantisation noise ratio.

Similarly to Figure 6.15, a differential analysis has been plotted in Figure 6.16, where the relative frequency of frame-to-frame difference is computed for each line. Again,

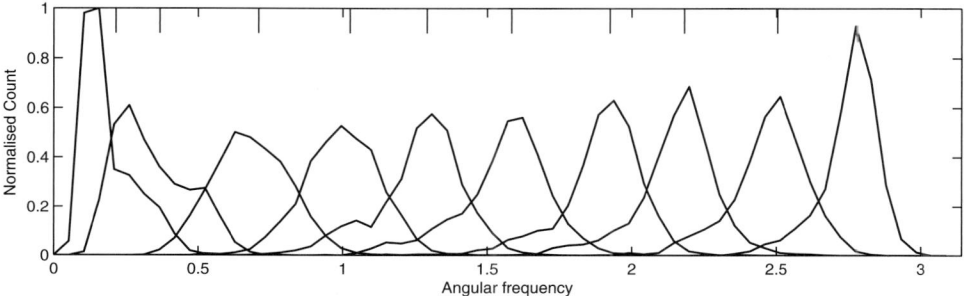

Figure 6.15 A histogram of the relative frequency of each LSP for tenth-order analysis of speech, showing tick marks at the top of the plot corresponding to the mean angular frequency of the ordered lines.

Table 6.2 Average frequency, standard deviation and median frequency for ten line frequencies.

No.	Average (Hz)	σ (Hz)	Median (Hz)
1	385	117	364
2	600	184	727
3	896	241	1091
4	1269	272	1455
5	1618	299	1818
6	1962	306	2182
7	2370	284	2545
8	2732	268	2909
9	3120	240	3272
10	3492	156	3636

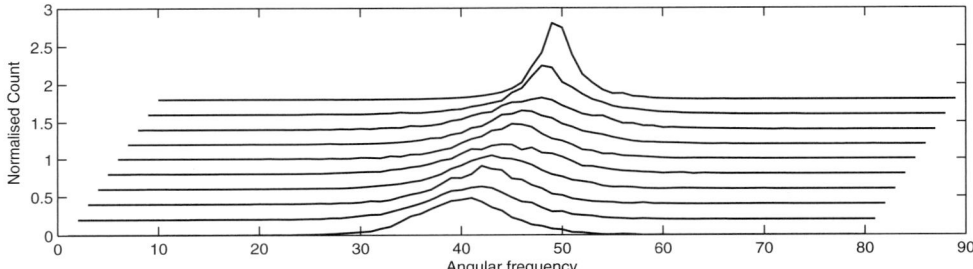

Figure 6.16 A histogram of the relative frequency difference between LSP line locations for consecutive analysis frames, plotted using 80 analysis bins.

it can be seen that the frame-to-frame variation for individual lines is localised. Both Figures 6.15 and 6.16 corroborate the evidence of Table 6.2 in showing that lines 1 and 10 exhibit less variation in their absolute value, but also less frame-to-frame variation, than do the centre lines.

Experimental results suggest that dynamically choosing one or other of the two quantiser bases (frame-to-frame differential or distance-from mean) to quantise and transmit can improve quantiser performance [73, 80]. In practice this implies that the quantised parameters are calculated using both methods for each analysis frame, and the method which results in the closest match is then chosen. A single bit from the quantisation bit pool must then be reserved to flag which method is in use [73].

In a standard CELP coder employing scalar quantisation, around three or four bits are used to quantise each LSP [72, 81], or around 30 or 40 bits for a tenth-order representation. A number of techniques have been published in order to allow representation with fewer bits at equivalent levels of quality. A survey of published results from [68, 75] and other previously mentioned references can assess the spectral distortion associated with a number of different coding methods at between 3 and 3.6 bits per LSP. This is

6.2 Parameterisation

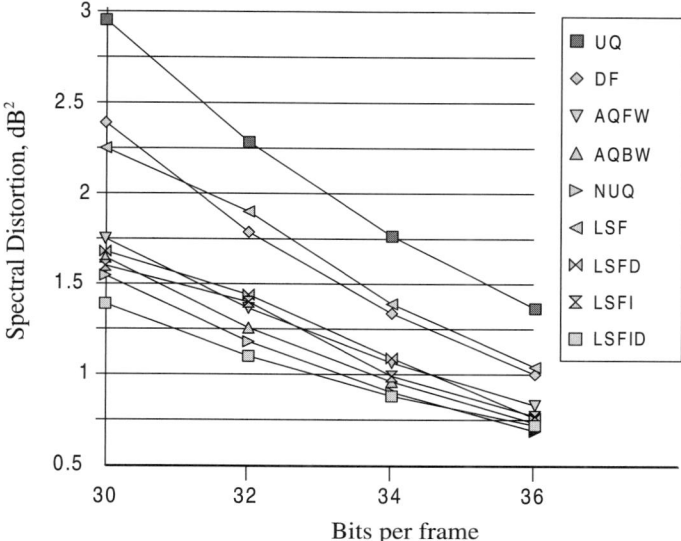

Figure 6.17 A plot of average spectral distortion due to nine different published methods of LSP quantisation for allocations of between 30 and 36 quantisation bits per frame.

shown in Figure 6.17, which plots the spectral distortion due to quantising a tenth-order LSP representation using the following methods:

- uniform scalar quantisation (UQ and LSF);
- differential scalar quantisation (DQ and LSFD/DF);
- adaptive forward sequential quantisation (AQFW);
- adaptive backward sequential quantisation (AQBW);
- non-uniform scalar quantisation (NUQ);
- dynamically interpolated quantisation (LSFI);
- dynamically interpolated differential quantisation (LSFID).

From the plot it can be seen that all methods improve at a similar rate as bit allocation increases, and that differential methods are better than non-differential methods. Furthermore, non-uniform quantisation improves over uniform quantisation. It should be noted that the results shown derive from some slightly different conditions and so the reader should be wary about making direct comparisons (as opposed to relative comparisons).

Vector quantisation (VQ) methods can be used to improve over scalar quantisation [72]. In VQ, a candidate vector is generated comprising the set of parameters to be quantised. The vector is compared, in turn, with each entry in a codebook of equal sized vectors. The distance between the candidate vector and each codebook entry is calculated, usually based on Euclidean distance, or sometimes on a perceptually relevant distance measure (which we will explore in Section 6.5). The index of the codebook entry nearest the candidate vector is then taken as the transmitted parameter.

Split VQ allows the vector of all LSPs for a particular frame to be split into subvectors, and compared with subcodebooks. A considerable amount of research has been performed to determine optimal vector and codebook splits. A typical split for a tenth-order system is into three subvectors of three lower, three middle and four upper LSPs. In fact this has been shown to be optimal [77] for a three-subvector system, whilst a 4–6 split is optimal for a two-subvector system.

The reader should also be aware that this has been an extremely hot research area in recent years, and that some quite impressive results have been achieved which exceed even the performance of the methods presented here.

6.3 Pitch models

The source–filter model is perhaps the ultimate in speech parameterisation, with different processing blocks dedicated to replicating the effects of the human vocal system: LPC/LSP for the vocal tract, random noise (and similar) for the lung excitation and a pitch filter or similar to recreate the effect of the glottis.

Measurements of the human pitch-production system, especially those using microwave and X-ray sensors, reveal the action of the glottis, which is not a smooth action: it does not generate a pure sinewave tone. In actual fact, the pitch waveform is made up of a sequence of very spiky pulses. This is shown in Figure 6.18, where one pulse has been identified from a sequence of several plotted as if isolated from a speech utterance.

There has been quite a lot of research on determining pitch shapes: how these relate to overall vocal quality, speech intelligibility and so on. There is substantial evidence that the fidelity of the pitch pulse shape is important to overall perceived quality and other evidence to indicate that the specific pulse shapes, which vary considerably from person to person, are one of the differentiating factors for speaker recognition (where an automatic system identifies someone through their voice, see Section 9.1).

When coding or compressing speech in a parametric fashion, there are several items of information that are important for pitch, and these are handled differently by the various speech compression algorithms. These are listed below:

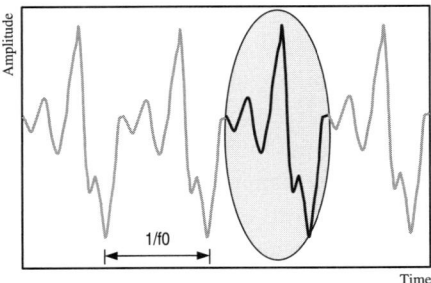

Figure 6.18 A pitch waveform showing several individual pitch pulses. One is circled, and the spacing between pulses, shown, determines the perceived pitch frequency.

6.3 Pitch models

- the actual shape of the pulse;
- relative heights/locations of negative- and positive-going spikes;
- largest spike amplitude;
- the spacing between pulses.

The highest-quality compression algorithms would consider all aspects. Some code only the bottom three items, CELP coders tend to code the bottom two and regular pulse excitation systems code only the bottom one. It goes without saying that more bits are required to code more information, thus many algorithms give priority to only the most important aspects for intelligibility, namely the lower entries on the list.

6.3.1 Regular pulse excitation

Regular pulse excitation (RPE) is a parametric coder that represents the pitch component of speech. It is most famously implemented in ETSI standard 06.10, and currently is the primary mobile speech communications method for over a third of the world's population, by any measure an impressive user base. This is due to its use in the GSM standard, developed in the 1980s as a pan-European digital voice standard. It was endorsed by the European Union, and quickly found adoption across Europe and then beyond.

GSM codes frames of 160 13-bit speech samples (at a sampling rate of 8 kHz) into 260 compressed bits. A decoder takes these and regenerates 160-sample output speech frames. There are many sources of information on GSM, not least the open standard documents, so there is no need to consider full details here. However, we will examine the pitch coding system for GSM 06.10, the traditional or 'full rate' standard.

In GSM, the original speech is analysed to determine vocal tract parameters (LPC coefficients) which are then used to filter the same vector of 160 speech samples to remove the vocal tract information, leaving a residual. The eight LPC coefficients will be transformed into LARs (log area ratios) for transmission.

The residual is then split into four subframes. Each subframe is analysed separately to determine pitch parameters. The analysis is made on the current subframe concatenated with the three previous reconstituted subframes. The reconstituted subframes are those that have been generated from the previous pitch values – those that have been quantised for transmission. Thus they are effectively the subframes as generated by a decoder.

These four subframes (the current one and the three reconstituted ones) form a complete frame which is subjected to long-term prediction (LTP), which is actually quite simple and will be discussed in the next section. When this contribution is removed from each subframe a set of pitch-like spikes remain – assuming of course the presence of pitch in the original speech. An RPE analysis engine compares the subvector of spikes with four candidates, one of which is chosen, along with a location (grid position) to represent the pitch spikes in that subframe.

This pulse train is actually coded by ADPCM before transmission. This entire coding process is known as RPE-LTP, and is shown diagrammatically in Figure 6.19. If there

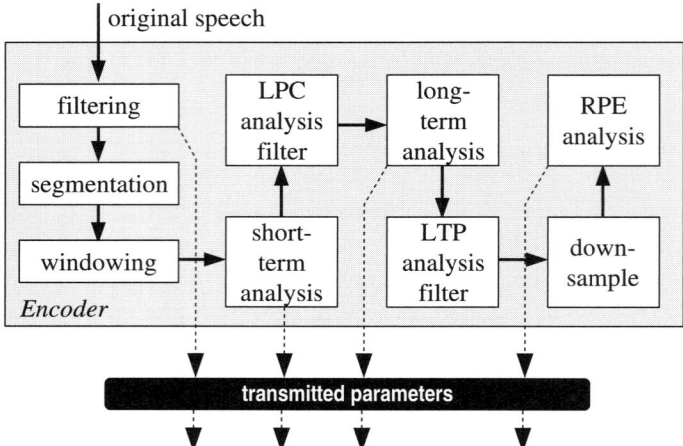

Figure 6.19 GSM RPE-LTP encoder block diagram, showing transmitted parameters.

were no pitch in the original speech (it has been judged to be unvoiced speech), then the residual is represented as random noise instead.

Up to 13 pitch pulses are coded per 40-sample subframe, achieved through downsampling at a ratio of 1:3 from several sequence start positions 1, 2 or 3. As can be imagined, a set of regular pulses is not particularly similar to the pitch waveform shown in Figure 6.18, but is perfectly sufficient, allied with the LTP and LPC analysis, to maintain speech intelligibility.

6.3.2 LTP pitch extraction

We can define a long-term prediction pitch filter by a number of taps and lags. We will illustrate for a simple one-tap pitch filter, which is the most common variant. Given a vector c of audio samples, we can add in a pitch component identified by an amplitude β and lag M as shown in Equation (6.33) to generate a spiky signal x which includes pitch:

$$x(n) = c(n) + \beta x(n - M). \tag{6.33}$$

β scales the amplitude of the pitch component and the lag M corresponds to the primary pitch period. For more complex pitch representations, more taps (with different lags) are used. It is also possible for multiple taps to be fixed around a single lag value to better approximate the shape of the pitch pulse. A three-tap filter with single lag is shown in Equation (6.34):

$$x(n) = c(n) + \beta_1 x(n - M - 1) + \beta_2 x(n - M) + \beta_3 x(n - M + 1). \tag{6.34}$$

Often, fractional pitch lags are used to improve pitch quality further. In all cases, the LTP filter, being IIR, calculates using past output values. M can range in length from less than a subframe to more than a frame, so at times the LTP filter evidently acts on a long sequence of audio (hence the name 'long-term'). This may be problematic in implementation since the pitch can change faster than the other vocal characteristics: remember the four subframes in GSM, each with possibly different pitch values.

Both of the filters described here assume that the lag and tap amplitudes are already known. The derivation of the actual pitch values is named pitch extraction.

6.3.2.1 Pitch extraction

There is actually a very wide variety of pitch extraction methods in the published literature (some are given later in Section 7.2.1), although we will here describe one of the simpler, and more common, methods used in speech coders. This method relies upon minimising the mean-squared error between an LPC residual (containing pitch) and the reconstructed pitch signal resulting from the analysis.

If E is the mean-squared error, e is the residual and e' is the reconstructed pitch signal after analysis, then:

$$E(M, \beta) = \sum_{n=0}^{N-1} \{e(n) - e'(n)\}^2, \qquad (6.35)$$

and, assuming a single tap pitch filter as in Equation (6.33), then:

$$E(M, \beta) = \sum_{n=0}^{N-1} \{e(n) - \beta e(n - M)\}^2, \qquad (6.36)$$

where N is the analysis window size (usually one or more subframes). In order to find the set of β and M that minimises the mean-squared error (i.e. best reproduces the original pitch signal) we need to differentiate the expression and set to zero:

$$\frac{\delta E}{\delta \beta} = \sum_{n=0}^{N-1} \{2\beta e^2(n - M) - 2e(n)e(n - M)\} = 0, \qquad (6.37)$$

so

$$\beta_{\text{optimum}} = \frac{\sum_{n=0}^{N-1} e(n)e(n - M)}{\sum_{n=0}^{N-1} e^2(n - M)}. \qquad (6.38)$$

We can now substitute the optimum β from Equation (6.38) into (6.36) to give the optimum M from:

$$E_{\text{optimum}}(M) = \sum_{n=0}^{N-1} e^2(n) - E'_{\text{optimum}}(M). \qquad (6.39)$$

Because only the second part of the equation varies with respect to M, it must be maximised to minimise the error. Thus the following must be determined with respect to each permissible value of M, and the value at which a maximum occurs must be stored:

$$E'_{\text{optimum}}(M) = \frac{\left[\sum_{n=0}^{N-1} e(n)e(n - M)\right]^2}{\sum_{n=0}^{N-1} e^2(n - M)}. \qquad (6.40)$$

Effectively, this means that the best pitch delay (and the one that is finally chosen) is the one that, averaged over a whole analysis frame, allows the best prediction of that subframe. Once found, the pitch scaling factor β is then chosen similarly, as the optimal scaling factor averaged over the analysis frame:

$$\beta = \frac{\sum_{n=0}^{N-1} e(n)e(n-M)}{\sum_{n=0}^{N-1} e^2(n-M)}. \tag{6.41}$$

This can be illustrated by the MATLAB code in Listing 6.5 which performs essentially the same operations in a straightforward (but fairly inefficient) fashion. The function ltp(sp) returns the pitch tap, M, and multiplier, B, for the given segment of input speech, sp.

Listing 6.5 ltp.m

```
function [B,M]=ltp(sp)
   n=length(sp);
   %upper & lower pitch limits (Fs~8kHz-16kHz)
   pmin=50;    pmax=200;
   sp2=sp.^2;  %pre-calculate
   for M=pmin:pmax
      e_del=sp(1:n-M);
      e=sp(M+1:n);
      e2=sp2(M+1:n);
      E(1+M-pmin)=sum((e_del.*e).^2)/sum(e2);
   end
   %---find M, the optimum pitch period
   [null, M]=max(E);
   M=M+pmin;
   %---find B, the pitch gain
   e_del=sp(1:n-M);
   e=sp(M+1:n);
   e2=sp2(M+1:n);
   B=sum(e_del.*e)/sum(e2);
```

Normally we would analyse speech using this method on a frame-by-frame basis. For example, given a recording of speech, speech at sample rate Fs and window size Ws, we can compute frame-by-frame pitch parameters as follows:

```
Ws=480;
L=length(speech);
Nf=floor(L/Ws);
Bs=[]; %collect pitch multipliers
Ms=[]; %collect pitch taps
```

6.3 Pitch models 181

```
for ff=1:Nf
   seg=speech(1+Ws*(ff-1):Ws*ff).*hamming(Ws);
   [B,M]=ltp(seg);
   Bs=[Bs, B];
   Ms=[Ms, M];
end
```

Plotting the resulting arrays, Bs and Ms, yields some information about the underlying speech, as illustrated for an example speech recording in Figure 6.20. The predominant pitch component is determined from the pitch lag as M/F_s since this is computed in samples.

It can also be interesting to plot the error E that is calculated just before the maximum is found. Often, the plot will show a periodic structure (and illustrates the issue of pitch doubling or halving, if the pitch search range is reduced to exclude the largest peak then very often the second largest peak will be at twice the pitch period, as discussed further in Section 6.3.3).

For computational purposes, note the identical denominator in Equations (6.40) and (6.41), which would only have to be determined once in an efficient implementation. Although the MATLAB code attempts a slight saving in computation by pre-calculating an array of squared input samples, it would actually be more efficient if we performed the correlation by shifting the numerator $e(n)$ only, rather than both the numerator

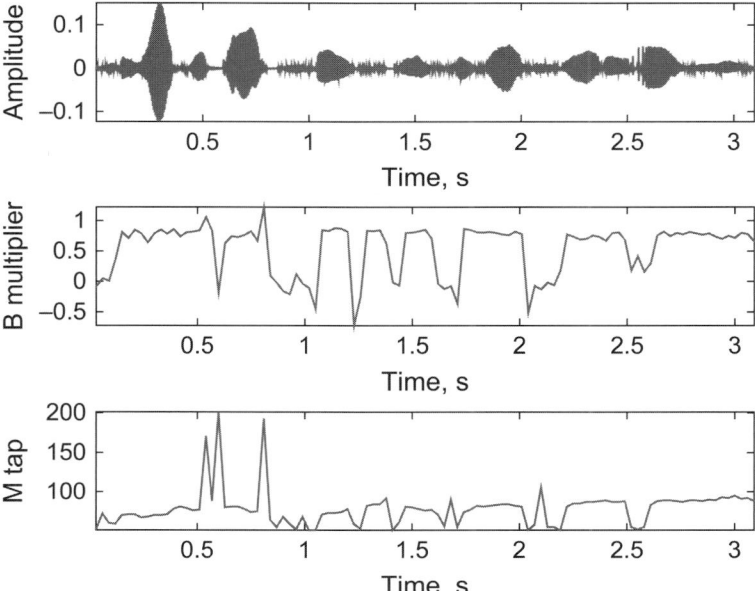

Figure 6.20 An example speech waveform (top) analysed to determine the one-tap pitch multiplier (middle) and pitch lap/lag (bottom).

and the denominator. In fact this method is used in most real-time speech coders. If extreme efficiency is required, the MATLAB function could be modified in such a way – hopefully yielding identical results.

6.3.3 Pitch issues

The pitch extraction method of Section 6.3.2.1, in common with many other methods, often produces an answer equal to half or twice the actual pitch period. This is called *pitch halving* and *pitch doubling* and is the scourge of many engineers working with pitch detection algorithms.

To some extent, setting hard limits on pitch period can provide an answer (i.e. saying, for example, that pitch cannot be less than 50 Hz or more than 300 Hz), but such a range still has ample scope for doubling or halving. Just think of the comparison between a squeaky 5-year-old child's voice and that of a deep bass voice such as those belonging to Paul Robeson or Luciano Pavarotti. Some algorithms disallow sudden shifts in pitch as being unlikely in a real scenario. This is true of speech, but such decisions tend to result in speech systems unable to handle music, singing or DTMF (dual tone multiple frequency) and facsimile signalling.

G.728, for example, limits the pitch slew rate between subframes to ± 6 samples unless the relative strength of the new pitch component is at least 2.5 times greater than that of the previous frame [82]. Any limits used are likely to require empirical testing with a range of subject matter.

6.4 Analysis-by-synthesis

The idea behind analysis-by-synthesis at the encoder is to analyse a frame (or more) of speech, and extract parameters from this. These parameters are then used to create a frame of reconstructed speech. The frames of original and reconstructed speech are then compared to see how closely they match. Some part of the parameter extraction process is then varied to create a slightly different set of parameters, which are in turn compared with the original speech.

Perhaps several hundred iterations are made across a search space, and the best set of parameters (based on how close the match is between original and reconstructed speech) is then transmitted to the receiver. Something to consider is that the parameters may need to be quantised before being transmitted to the decoder. In this case the quantised–dequantised parameters are the ones used by the encoder to check how good the matching is.

Before we look at the most famous of the analysis-by-synthesis coding structures, it is important to remember that 'degree of matching', calculated as a difference between vectors, may not relate at all to how a human perceives the degree of difference. As a very trivial example, imagine a continuous sinewave original signal. Next imagine a version which is delayed by a few degrees. The degree of matching in a mean-squared sense will be quite low – in fact it may be less than the matching between the original sinewave and random noise. However, the perceived difference between the sinewaves is probably zero, but huge when one signal is random noise.

Therefore, most practical analysis-by-synthesis algorithms use a perceptual matching criterion: either a perceptual weighting filter or something like a spectral distortion measure (see Section 3.4.2), rather than a mean-squared match.

6.4.1 Basic CELP

CELP is the logical culmination of an evolutionary process in speech compression algorithms: it can provide excellent quality speech at low bitrates and is a common choice for speech products. It utilises a source filter model of speech, parameterised as we have seen with gain, vocal tract, pitch and lung excitation information.

CELP stands for either *code excited linear prediction* or *codebook excited linear prediction* depending on whom you ask. What is incontrovertible though, is that the technique collectively describes a large variety of similarly structured algorithms which can achieve good performance in the real-life coding of speech. We will begin looking at the basic structure of a CELP coder, and subsequently look briefly at algebraic, adaptive and split variants. Be aware that this has been an area of intense research activity for two decades: many 'mutant forms' of CELP coder, which incorporate some unusual techniques, have emerged in the research literature.

We will start with the basic CELP encoder, designed to decompose a speech signal into various parameters. A block diagram of such a system is shown in Figure 6.21. This shows the speech signal being filtered (including normalisation, yielding gain information), segmented and windowed. It is then analysed for pitch components (represented by LTP parameters) and vocal tract resonances (represented by LPC coefficients). Readers may note the similarity to the RPE system in Figure 6.19, and indeed these two coders do share many characteristics.

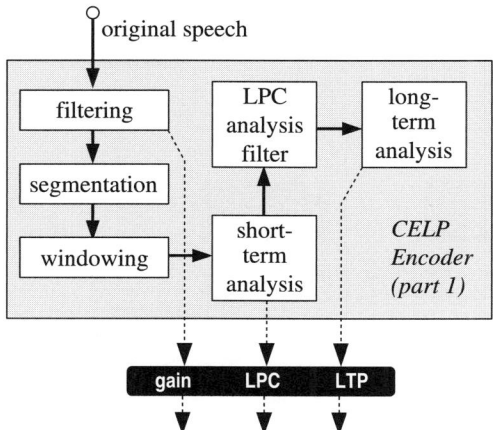

Figure 6.21 A block diagram of part of a CELP encoder, showing original speech being decomposed into gain, LPC and LTP parameters.

Where the CELP and RPE systems differ most greatly is in the handling of the original lung excitation signal. RPE treats this either as white Gaussian noise or as a pulse-train. The CELP coder takes a different approach: it utilises a large codebook of *candidate vectors* at both encoder and decoder, and essentially runs through an iterative process to attempt to identify which of the candidate excitation vectors best represents the actual lung excitation.

At least that is the theory – in practice none of the parameters exactly characterise the required information perfectly. This means that both the LPC and LTP representations, neither being perfect, will 'leak' information. In the RPE encoder, vocal tract information not caught by the LPC analysis is unlikely to be picked up by the LTP and RPE analysis, so that information will be lost to the encoder and consequently not transmitted to the decoder. This contributes to loss of quality in speech processed by such a system.

In CELP, the codebook of candidate excitation vectors can often pick up some of the information which was not caught by the LTP and LPC analysis. So in practice the codebook does not just model lung excitation. Since this mechanism greatly improves the quality of the system over RPE, we will consider in a little more detail exactly how it works.

Following a particular analysis frame of speech through the CELP encoder, first the basic gain, pitch and vocal tract parameters are determined (as shown in Figure 6.21), and then these parameters are used to recreate pseudo-speech as in Figure 6.22, as the output from the LPC analysis filter. The first candidate vector in the codebook, named codeword 0, is used as the lung excitation. Amplification, LPC and LTP synthesis filters add gain, pitch and vocal tract information to the lung excitation in order to derive a frame of pseudo-speech.

This pseudo-speech is compared with the original frame of speech. In fact the comparison simply finds a difference vector between the two, perceptually weights this

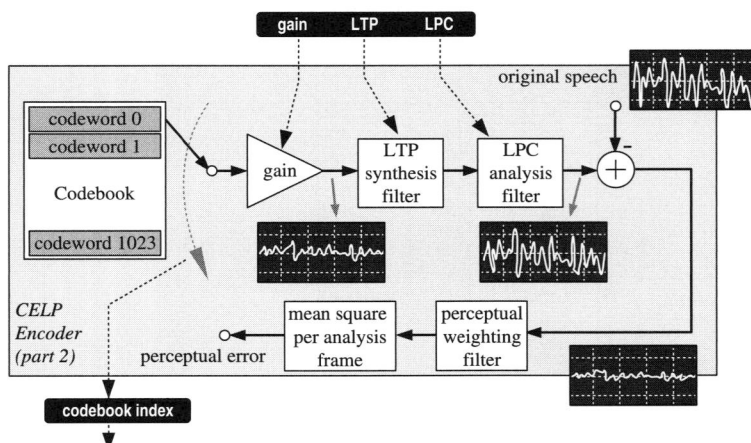

Figure 6.22 A block diagram of the remainder of the CELP encoder. Gain, LPC and LTP parameters were obtained in the first part shown in Figure 6.21, whilst the section now shown is devoted to determining the optimum codebook index that best matches the analysed speech.

6.4 Analysis-by-synthesis

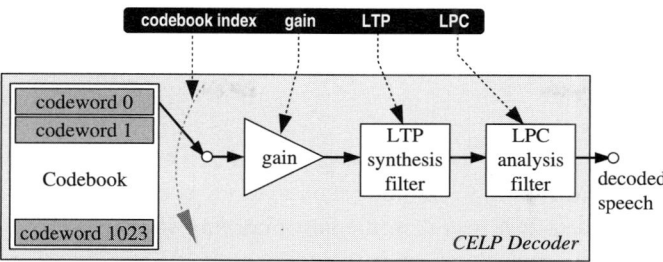

Figure 6.23 A block diagram of the remainder of the CELP decoder utilising codebook index, gain, LPC and LTP parameters to recreate a frame of speech.

(something we will return to later in Section 6.5), and calculates the mean square for that frame: a single perceptual error value for the current input speech frame. The process is now repeated for codeword 1, and again results in a single perceptual error value. Each element in the codebook is now tried in turn.

For the typical codebook shown, the result will be 1024 perceptual error values. Each one is a measure of the difference between the pseudo-speech recreated with that codebook index and the original speech. Thus the codebook index which resulted in the smallest perceptual error value is the one which can be used to best represent the original speech, and this index (0 to 1023) is transmitted from encoder to decoder.

At the decoder, shown in Figure 6.23, the transmitted parameters are used to recreate a frame of decoded speech. The codeword selected in the decoder is that identified by the transmitted codebook index. This codeword is identical to that at the same position in the encoder – and is thus guaranteed to be the best of the candidate vectors. We would therefore expect the frame of decoded speech to be similar to the original analysed speech. It is evident that the decoder is considerably simpler than the encoder – at least 1024 times in the example shown (since it does not need to repeat once for each codebook entry, but just for the indicated codebook entry). This is offset somewhat by the trend to apply post-filtering to the decoder output (not shown here) in order to improve audio quality.

Since the codebook is the differentiating factor in the CELP method, we should explore that in a little more detail. First, however, we need to remind ourselves of the need for quantisation: the raw LPC, LTP and gain parameters need to be quantised in some way. As we have seen in Section 6.2.6, LPC parameters are rarely used as-is. Within CELP coders they are generally transformed to line spectral pairs prior to being output from the encoder.

In fact all of the parameters, with the probable exception of the codebook index, will be quantised, and in the process transformed in some way. Remembering that the encoder incorporates the decoding process in its codebook search loop, it is important to note that the actual parameters used in this part of the encoder are *already quantised* and then dequantised. The main reason is that if the encoder uses unquantised parameters it may well find a different candidate excitation vector from the one it would choose if operating on quantised–dequantised parameters. Since the actual speech output from the decoder has access only to the quantised–dequantised parameters, the encoder must use the same values to ensure the best possible speech is generated.

6.4.1.1 CELP codebooks

As mentioned previously, each codebook is populated by a number of codewords. These are used as candidate vectors within the encoder, where each candidate is examined in turn, and the candidate that results in the best matching speech frame is chosen. For a typical system that analyses speech in 20 ms frames and has a sample rate of 8 kHz, the candidate vectors need to consist of $8000 \times 0.02 = 160$ samples. Quite often, pitch is analysed and represented in subframes that may be 5 ms long – four subframes per frame – and thus the LTP parameters change four times as often as the LPC parameters, but otherwise the processing structure remains unchanged.

In the original CELP technique [83, 84], the candidate vectors in each codebook were generated from a random number generation algorithm – 'seeded' identically at encoder and decoder so that they would contain exactly the same vectors: at the most basic level, a set of 1024 1×160 random numbers. More modern variants will introduce some structure into the codebook [76] – or allow the addition of two vectors to be used as an excitation candidate. Such techniques are known as split codebooks.

Many useful enhancements to basic CELP rely on the fact that the CELP encoder (Figure 6.22) actually contains a decoder as part of its structure. That means that, for the same codebook index, the encoder pseudo-speech is identical to the decoder output speech. This is exploited in allowing the encoder to predict the state of the decoder, something necessary for any adaptive system. One such system allows the codebook to adapt: ensuring that both encoder and decoder codebooks adapt equally (keep in step) is tricky, but necessary to ensure performance.

Although speech quality improvement has always been the main driving factor behind the advance of the CELP technique, computational complexity reduction has been another significant factor. A third factor has been minimisation of processing latency. Three well-known enhancements to CELP are now discussed that address the quality, computational complexity and processing latency issues.

6.4.2 Algebraic CELP

As mentioned previously, the computational complexity of the CELP coder is large: this is due to the need to synthesise pseudo-speech for every vector within the codebook. This already-complicated process must be repeated hundreds or thousands of times for each analysis frame. Whilst many methods attempt to reduce the amount of computation in the CELP coder by reducing the number of loops (see Section 6.4.3), algebraic CELP (often known simply as ACELP) attempts to reduce the amount of computation required within each loop.

Referring to Figure 6.22, the action required for each excitation vector is to amplify it, perform LTP synthesis (to put in the pitch component) and perform LPC synthesis (to add in the vocal tract components). Each of these is a linear time-invariant (LTI) process, which means that mathematically it does not matter in which order they are performed. However, in terms of processing complexity, it is often better to perform the amplification last (so that the numbers being processed by the filters do not become too large to be easily represented), and we will demonstrate that, in ACELP at least, it is better to perform the LPC filter first.

The filtering process, as described in Section 2.3, consists of a series of multiplications. Filter coefficients are multiplied by input values, and summed to produce output values. In the CELP encoder, this is done twice for each codeword – first in the LPC filter and second in the LTP filter. These may have ten and two coefficients respectively, so clearly the LPC filter is the most computationally complex one. If it were possible to reduce the complexity of the LPC filter, there would be a correspondingly large reduction in overall complexity.

In ACELP, the codewords are carefully constructed so that they are sparse (meaning most elements are zero, with a few individual non-zero elements scattered around), with each element in the vector being constrained to $+1$, 0 or -1. These codewords are then fed directly into the LPC synthesis filter. Let us remind ourselves what the filter looks like mathematically:

$$y(n) = a(1) \times x(n) + a(2) \times x(n-1) + a(3) \times x(n-2) + \cdots \\ + a(m+1) \times x(n-m).$$

In this case, $x()$ contains the input data, $y()$ contains the output data, and the filter coefficients are $a()$. If we rewrite the filter on the basis of some illustrative input vector $x = [+1, 0, 0, -1, 0, 0, -1]$, meaning $x(1) = +1$, $x(2) = 0$, $x(7) = -1$ and so on, then the filter equation reduces to:

$$y(n) = a(1) + 0 + 0 - a(4) + 0 + 0 - a(7).$$

This equation involves no multiplications, simply a summation of all non-zero elements (and, with it being sparse, there are few numbers that need to be summed). ACELP generally operates with 80% of the elements in the codebook set to zero. This is significantly simpler and faster to calculate than the full filter equation that would be required with a codebook vector of random elements.

A further advantage of ACELP is that it is entirely possible to calculate the position (and value) of each non-zero element in the candidate vector on-the-fly. Such an action requires some minor additional processing to generate the vectors, but does mean that the static codebook storage is no longer required. For memory-constrained embedded systems, this may well be a significant consideration.

6.4.3 Split codebook schemes

Unlike ACELP (above, Section 6.4.2) which reduces the computational complexity of the LPC filtering process, split codebook schemes attempt to reduce the 'search space'. By default the search space is the entire codebook since each element in the codebook is tested in turn, and there may be very many such elements. If there were some way to reduce the number of candidate excitation vectors tested, then complexity can be reduced.

Two main methods exist. The first one is to order the codebook in such a way that the codewords are arranged by known approximate characteristic. Through one of several different predictive processes, it is possible to determine the rough spectral or time-domain characteristic of the vector required for a particular speech analysis frame. Only codewords fitting the required description need be tested as candidates. Occasionally such a system may make a suboptimal choice, but, given a sufficiently good predictive algorithm, such events may be rare.

The second method is to use two or more different codebooks with orthogonal properties. Each orthogonal property is found separately, and the two codebooks can be searched independently. As an example of a dual-codebook system, codebook 1 – perhaps carrying a selection of spectral distributions – is searched with codebook 2 set to some arbitrary value. Once the best match from codebook 1 is found, this is maintained whilst codebook 2 – perhaps carrying temporal envelope information – is searched. Ideally the two are orthogonal (meaning that, whichever codeword from one codebook is used, the best match from the other codebook will not change for a particular frame of analysed speech).

At one extreme, a split system could be a mixed excitation linear prediction (MELP) coder – although this name is typically reserved for coders with split excitation that do not utilise a codebook (and hence do not perform analysis-by-synthesis) [85]. VSELP (vector sum excited linear prediction [86]) is a common split system with ordered codebooks that are searched independently, with the best codeword from each added together to provide the excitation vector. In this case, each codebook output is subject to an individual scaling factor. Less important for this discussion, but a fact which should be mentioned for consistency, is that in the VSELP coder one of those codebooks is used to replace the LTP pitch filter. This codebook is pitch-adaptive, and is obviously searched, and optimised, before the other two codebooks of candidate vectors are searched.

6.4.4 Forward–backward CELP

When using a speech coder such as CELP in a real application, it will naturally take some time for a particular sound of speech to travel from the input microphone through the system and arrive at the output loudspeaker. Apart from the transmission of speech through wiring (or fibre optics, or wireless), processing takes some finite time.

The first delay comes about due to collection of audio samples into a buffer before they are processed: for a speech system working on a 30 ms analysis frame at a sample rate of 8 kHz, each frame contains $8000 \times 0.03 = 240$ samples. Processing can typically not begin until all of those 240 samples are available, and naturally the final sample in the buffer was collected 30 ms after the first sample. Even with no other processing, such a system will delay audio by 30 ms. The output buffering arrangements might well affect the system in a similar way. These latencies between input and output ignore the operation of any coder or decoder and all propagation delays.

Next we look at the operation of the CELP encoder of Figure 6.22. Looping around 1024 times, an entire decoding operation must be performed, and then perceptual weighting and mean-squared calculation. None of this can begin until the sample input buffer has been filled, and then the following process is likely to require a significant amount of processing time. This does, of course, depend on the clock and calculation speed of the underlying hardware. To put this into perspective, the original inventors of CELP found that their Cray-1 supercomputer required 125 seconds of processing time to process just a single second of speech [84]. Of course, computers are far more powerful today than in the 1980s; however, such processing is still very far from instantaneous.

As CELP began to be adopted in real systems, figures of 200–300 ms latency were observed. Unfortunately, at these latencies, human conversations become rather strained: people begin to speak over one another, and feel uncomfortable with the long pauses in conversation. A useful figure of merit is that most people will not notice a latency of 100–150 ms, but beyond this it starts to become intrusive to conversation.

Clearly, a reduced latency CELP was required: both for the input/output buffering and for the processing times. The solution was found, and standardised as ITU G.728, boasting submillisecond processing latency. The primary technique used to achieve the latency reduction was the forward–backward structure.

Despite the unusual name, forward–backward CELP refers to the order in which processing is performed. This is perhaps best illustrated in a timing diagram. First we consider the basic CELP timing structure as shown in Figure 6.24. It can clearly be seen that speech has to be gathered (buffered), then analysed for gain, LPC and LTP parameters before these are used in the lengthy codebook search loop. Once this is complete, the previously determined parameters and codebook index can be output for transmission. The decoder receives these and uses them to synthesise speech, which it then outputs. Quite clearly, the latency from input to output is dominated by the codebook search loop. By contrast, the forward–backward CELP structure, illustrated in Figure 6.25, manages to perform the codebook search in parallel with the speech buffering. This is an interesting concept, only possible because the encoder uses the

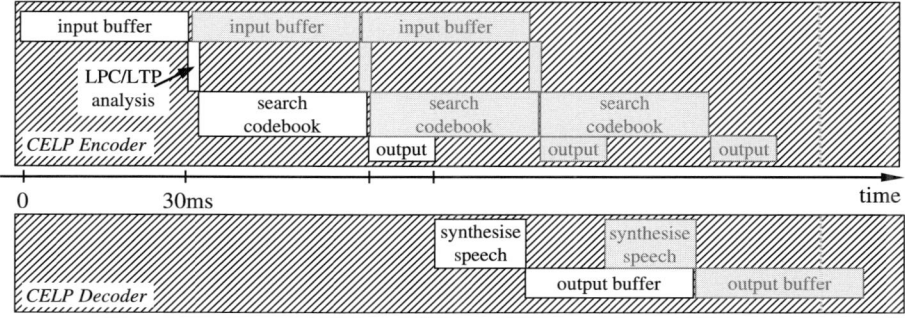

Figure 6.24 Timing diagram for basic CELP encoder and decoder processing, illustrating processing latency.

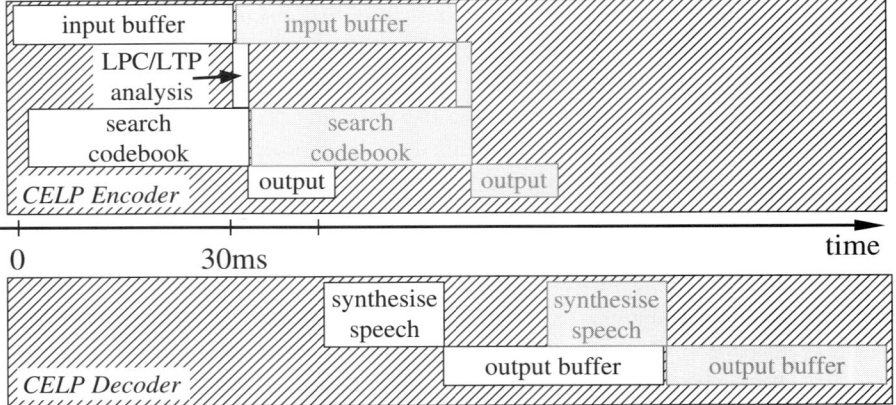

Figure 6.25 Timing diagram for forward–backward CELP encoder and decoder processing.

gain, LPC and LTP parameters found in the *previous* speech frame for the synthesis part of the encoder. The diagram clearly illustrates the significant reduction in processing latency that this provides.

Note that the actual amount of processing is unchanged – it is just organised differently so that the output speech can be made available earlier. In practice, if latency were

---

Box 6.4 Speech coding standards based around CELP and RPE

There is an entire spectrum of speech coding and compression systems, and then another entire spectrum designed to handle music and general audio. The tendency is for the higher-rate algorithms to have higher perceived quality – but there are also significant differentiating factors in the computational complexity and in the underlying techniques used: the class of CELP-based systems contains many weird and wonderful performance-enhancing ideas in addition to the basic codebook excited linear prediction structure.

Name	Description
G.723.1	The basic ITU CELP standard at 5.3 or 6.3 kbits/s
G.728	Low-delay CELP, 16 kbits/s sampled at 8 kHz
G.729	Algebraic CELP 8 kbits/s standard
TETRA	CELP 4.8 kbits/s
FS1015	LPC-based tenth-order system at 2.4 kbits/s
FS1016	Basic CELP at 4.8 or 7.2 kbits/s
GSM 06.10	Regular pulse excited, 13.6 kbits/s
MELP	Mixed excitation CELP 1.6 to 2.4 kbits/s
QCELP	Basic CELP at 8 and 13 kbits/s

Many variants and extensions of these original standards exist.

a particular issue, the frame size would be reduced to 20 ms and a system of overlapping frames used (with half of each analysis frame already having been collected earlier, the buffering delay is reduced).

6.5 Perceptual weighting

The perceptual error weighting filter (PEWF) is a common enhancement used within speech coders. In this context it has a particular meaning which may not be quite the same as its interpretation elsewhere. Based on its context of speech coding, it makes use of linear prediction parameters, which themselves encode vocal tract information. It uses these parameters to 'strengthen' resonances, and thus to increase formant power in encoded speech.

The idea is that, since the formant regions are more relevant to human perception, the weighting process improves the perception of these. Whilst this argument is true of speech, for general music systems, perceptual weighting more often involves either the use of a perceptual model or simply the application of a digital version of the A-weighting filter. Here we will present a typical PEWF as found within a CELP speech coder. The LPC synthesis filter is termed $H(z)$, and two bandwidth expansion factors are used, ζ_1 and ζ_2, with the relationship $\zeta_1 < \zeta_2 \leq 1$. Then the weighting filter $W(z)$ is defined as:

$$W(z) = \frac{1 - H(z/\zeta_1)}{1 - H(z/\zeta_2)}. \tag{6.42}$$

Remembering that the LPC synthesis filter is defined as:

$$H(z) = \sum_{k=1}^{P} a_k z^{-k}, \tag{6.43}$$

then the frequency scaled version will be:

$$H(z/\zeta) = \sum_{k=1}^{P} \zeta^k a_k z^{-k}. \tag{6.44}$$

Taking Equation (6.44) in difference form, and substituting into Equation (6.42), the PEWF is quite simply realised in discrete terms as:

$$y[n] = x[n] + \sum_{k=1}^{P} a_k \{\zeta_2^k y[n-k] - \zeta_1^k x[n-k]\}. \tag{6.45}$$

This can be applied at the output of a CELP speech decoder to slightly enhance the intelligibility of voiced speech, and can also be used within the CELP encoder to accentuate the importance of any formant regions within the codebook search loop (i.e. to weight the mean-squared matching process toward any formants that may be present). Typically the values of ζ_1 and ζ_2 are very close to unity. In the past the author has used $\zeta_1 = 0.95$ and $\zeta_2 = 1.0$ for several systems.

6.6 Summary

This chapter has presented a survey of the main techniques in speech communications – primarily related to the quantisation of the speech signal and the trade-offs that this involves in terms of quality and computational complexity. Both time-domain waveform quantisation and parameterisation were discussed, before a presentation of pitch modelling and extraction techniques. Finally, the important and pragmatic audio engineering subject of analysis-by-synthesis was explored, culminating in the CELP codec family.

Although we have already been introduced to most of the main methods of speech and audio analysis in our discussion of speech communications, several important topics remain to be covered in the next chapter as we consider advanced audio analysis.

Bibliography

- *Principles of Computer Speech*
  I. H. Witten (Academic Press, 1982)

- *Linear Prediction of Speech*
  J. Markel and A. Gray (Springer-Verlag, 1976)

- *Speech Processing: A Dynamic and Optimization-Oriented Appraoch*
  L. Deng and D. O'Shaughnessy (CRC Press, 2003)

- *Speech and Audio Signal Processing: Processing and Perception of Speech and Music*
  B. Gold and N. Morgan (Wiley, 1999)

- *Fundamentals of Speech Signal Processing*
  S. Saito and K. Nakata (Academic Press, 1985)

- *Introduction to Digital Speech Processing*
  L. R. Rabiner and R. W. Schafer (Now Publishers Inc, 2007)

- *Computer Speech Processing*
  F. Fallside and W. Woods (Prentice-Hall, 1985)

- *Spoken Language Processing: A Guide to Theory, Algorithm and System Development*
  X. Huang (Prentice-Hall, 2001)

Questions

Q6.1 Describe the characteristics of a 'toll quality' speech system.

Q6.2 What is the specific improvement that adaptive delta-modulation brings to standard delta-modulation?

Q6.3 Explain the purpose of the perceptual error weighting filter within the CELP analysis-by-synthesis loop.

Q6.4 Working in MATLAB, create a 1 s long array of random noise, ensuring its mean is 0, maximum is $+0.5$ and minimum -0.5. Use the `filter()` command to implement an LPC synthesis filter which applies the spectral shape of the example LPC coefficients given on page 151 to the random data. Now window the array and extract a new set of LPC coefficients, but this time extract only six coefficients instead of ten. Use the `freqz()` function to visualise both the original and the smaller set of LPCs. How do the spectral shapes differ?

Q6.5 Convert the smaller set of LPCs from Question 6.4 into LSPs using `lpc_lsp()` and then plot both the LSP and the LPC data using `lpcsp()` to visualise the resonances. Compare these with the `freqz()` plot.

Q6.6 In a typical speech coder, if LPCs are computed over 20 ms analysis frames, approximately how long would the LTP pitch computation be made over?

Q6.7 If I wanted to use just 160 bits of data (arranged as ten separate 16-bit integers) to represent the spectral shape of a pseudo-stationary window of unvoiced speech, would it be better to use an LPC or an LSP representation? Explain why.

Q6.8 Identify a physiological significance of the pairs of odd and even line spectral frequencies, when related to the human vocal tract?

Q6.9 With reference to the diagram of the human head in Figure 6.7, which parameter out of G, P, VT and EX is likely to be most different between two phonemes, one of which is spoken and the other of which is whispered? Explain why.

Q6.10 Compute the size of data file that would result when storing a 10 minute speech recording in (a) 16-bit 44.1 kHz stereo, (b) 128 kbits/s MP3 audio, (c) G.728 format and (d) G.723.1 format, referring to Infobox 6.4 on page 190 for the latter two formats.

Q6.11 In the common tube model of the human vocal tract, which element is constant: (a) the diameter of each tube segment, (b) the volume of each tube segment or (c) the length of each tube segment? Can you identify which feature of the model leads to the fixing of that characteristic?

Q6.12 Explain what is the main difference between algebraic CELP and standard CELP, and identify one reason why this difference is important.

Q6.13 When extracting LTP filter parameters from a segment of speech, which parameter is likely to halve when the pitch doubles, (a) the tap or (b) the multiplier?

Q6.14 Comment on how well a regular pulse excitation (RPE) coder models the shape of the physical underlying pitch pulses when it encodes pitch information.

Q6.15 With reference to the LPC and LSP plot of Figure 6.14, what is the significance of line spectral frequencies becoming close, such as lines $\{1, 2\}$ and $\{5, 6\}$, and how is this evident in the sounds that the parameters represent?

7 Audio analysis

Analysis techniques are those used to examine, understand and interpret the content of recorded sound signals. Sometimes these lead to visualisation methods, whilst at other times they may be used in specifying some form of further processing or measurement of the audio. In this chapter we shall primarily discuss general audio analysis (rather than speech analysis which uses knowledge of the semantics, production mechanism and hearing mechanism implicit to speech).

There is a general set of analysis techniques which are common to all audio signals, and indeed to many forms of data, particularly the traditional methods used for signal processing. We have already met and used the basic technique of decomposing sound into multiple sinusoidal components with the fast Fourier transform (FFT), and have considered forming a polynomial equation to replicate audio waveform characteristics through linear prediction (LPC), but there are many other useful techniques we have not yet considered.

Most analysis techniques operate on analysis windows, or frames, of input audio. Most also require that the analysis window is a representative stationary selection of the signal (stationary in that the signal statistics and frequency distribution do not change appreciably during the time duration of the window – otherwise results may be inaccurate). We discussed the stationarity issue in Section 2.5.1, and should note that the choice of analysis window size, as well as the choice of analysis methods used, depends strongly upon the identity of the signal being analysed. Speech, noise and music all have different characteristics, and, while many of the same methods can be used in their analysis, knowledge of their characteristics leads to different analysis periods and different parameter ranges of the analysis result.

Undoubtedly, those needing to perform an analysis will require some experimentation to determine the best methods to be used, the correct parameters to be interpreted and optimal analysis timings.

We will now introduce several other methods of analysing sound that form part of the audio engineer's standard toolkit, and which can be applied in many situations. We will also touch upon the analysis of some other more specialised signals such as music and animal noises before we discuss the use of tracking sound statistics as a method of analysis.

7.1 Analysis toolkit

Each of the following methods operates over an analysis frame of size N. In some cases, the measures are computed in the time domain, with the consequence that the frame does not need to be windowed prior to analysis and there is no need for the analysis frames to be overlapped. Frequency-domain methods tend to require windowing and overlapping.

7.1.1 Zero-crossing rate

There is nothing particularly advanced or complicated about the zero-crossing rate (ZCR). It is a 'poor man's' pitch determination algorithm, working well in the absence of noise, and is designed to be very simple in terms of computation. It works well for a noise-free and simple waveform like a sinewave, the idea being to count how many times the waveform crosses the zero-axis in a certain time: the number of crossings per second will equal twice the frequency. If we define sign{ } to be a function returning $+1$ or 0 depending upon whether the signal is greater than zero or not, then the ZCR for the ith analysis frame, of length N, can be determined as:

$$\text{ZCR}_i = \frac{1}{N} \sum_{n=0}^{N-1} |\text{sign}\{x_i(n)\} - \text{sign}\{x_i(n-1)\}| \,. \tag{7.1}$$

The first term of $1/N$ converts the crossing *count* to a crossing *rate*: the proportion of the samples which cross the zero-axis. Most of the time, in a system with fixed frame size, where ZCR values are compared from frame to frame, the division is unnecessary (because we are interested in the comparison between fixed-sized frames, which can be done by comparing crossing count, rather than computing the crossing rate), allowing a small saving in computation. ZCR is illustrated in Figure 7.1(a), where the zero crossings of a sinewave are counted over a certain analysis time. In this case the

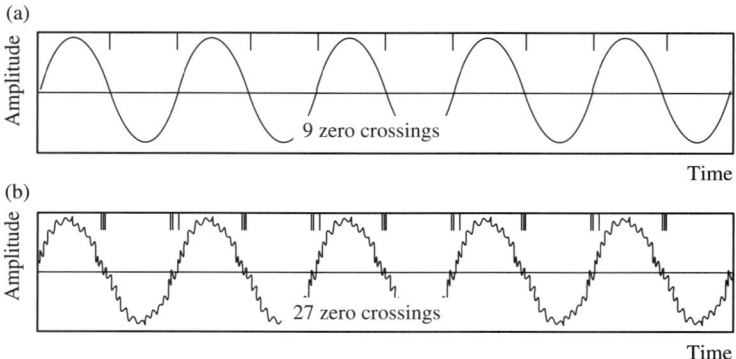

Figure 7.1 Zero-crossing rate calculation illustrated for a pure sinewave (a) and one corrupted with noise (b). Zero crossings are indicated by tick marks at the top of each plot. The noise-corrupted sinewave below exhibits many more spurious zero crossings due to additive noise taking the signal backwards and forwards over the trigger amplitude several times each period.

fundamental frequency of the sinewave causes nine crossings across the plot. However, in the presence of additive noise, the 'wobble' in the signal as it crosses the zero-axis causes several false counts. In Figure 7.1(b) this leads to an erroneous estimate of signal fundamental frequency – in fact, if we used crossing rate to estimate frequency, the result would be three times too high.

In MATLAB, determining the ZCR is relatively easy, although not particularly elegant:

```
function [zcr]=zcr(segment)
zc=0;
for m=1:length(segment)-1
   if segment(m)*segment(m+1) > 0
      zc=zc+0;
   else
      zc=zc+1;
   end
end
zcr=zc/length(segment);
```

To illustrate how this works, the MATLAB zcr() function above was applied to a recording of speech sampled at 16 kHz. The speech was segmented into non-overlapping analysis windows of size 128 samples, and the ZCR was determined for each window. Both the original waveform and the frame-by-frame ZCR score are plotted in Figure 7.2. There is clearly quite a good correspondence between the ZCR measure and the shape of the speech waveform – specifically, higher-frequency regions of the recorded speech, caused by phonemes with higher-frequency components such as the /ch/ sound, demonstrate a higher ZCR score.

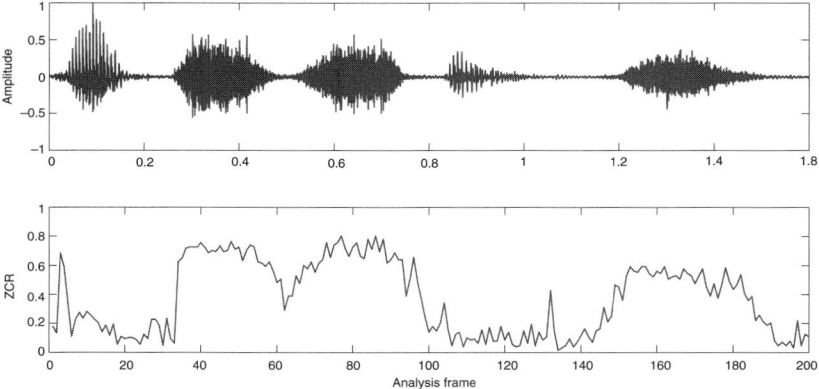

Figure 7.2 The upper graph shows a recording of a male speaker saying 'it's speech', which contains several stressed high-frequency sibilants and the lower graph is the frame-by-frame ZCR corresponding to this. The excursions of the ZCR plot closely match the high-frequency speech components, namely a breathy /i/ at the start, both /s/ sounds and the final /ch/.

Audio analysis

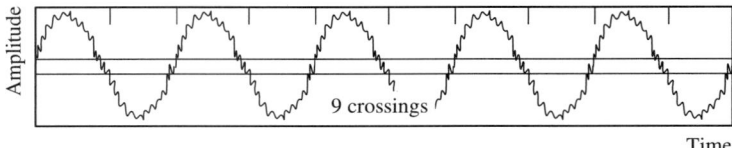

Figure 7.3 Threshold-crossing rate illustrated for a noisy sinewave, showing that the spurious zero crossings of Figure 7.1(b) are no longer present.

However, as noted above, the ZCR measure is easily confused by noise. Therefore a threshold region centred on the zero-axis is commonly applied, and full crossings of the threshold are counted instead of zero crossings. In essence, this introduces a region of hysteresis whereby a single count is made only when the signal drops below the maximum threshold and emerges below the minimum threshold, or vice versa. This is usually called threshold-crossing rate (TCR), and is illustrated in Figure 7.3.

In practice, the advantage of TCR for noise reduction can often be achieved simply by low-pass filtering the speech and then using a ZCR. This knocks out the high-frequency noise or 'bounce' on the signal. Since ZCR is used as a rough approximation of the fundamental pitch of an audio signal, bounds for filtering can be established through knowing the extent of the expected maximum. In speech, it has to be stressed that the filtered ZCR (or TCR) measure provides an approximate indication of the content of the speech signal, with unvoiced speech tending to result in high ZCR values, and voiced speech tending to result in low ZCR values. Noise also tends to produce high ZCR values, and thus it is difficult to use ZCR for analysis of noisy speech, significantly limiting its use in practical applications.

In most ways, TCR results are very similar to ZCR. In the absence of noise, a TCR plot for a speech recording would resemble that for ZCR as in Figure 7.2.

7.1.2 Frame power

This is a measure of the signal energy over an analysis frame, and is calculated as the sum of the squared magnitude of the sample values in that frame. For speech frame i, with N elements, denoted by x_i, the frame power measure is determined from:

$$E_i = \frac{1}{N} \sum_{n=0}^{N-1} |x_i(n)|^2 . \tag{7.2}$$

As with the case of ZCR in Section 7.1.1, the division by N will often be unnecessary in practice, if the measure is used to compare between frames.

In MATLAB, frame power has a very simple formula:

```
[fpow]=fpow(segment)
  fpow=sum(segment.^2)/length(segment);
```

Frame power provides a compact representation of the volume of amplitude variations in speech. As we saw in Section 3.3, unvoiced speech is spoken with less power than voiced speech, and for this reason frame power can be used as a rough indicator of voiced/unvoiced speech.

The simplicity of the MATLAB function above hides the fact that this calculation requires a multiplication to be performed for each and every sample within the analysis window. In implementation terms this can be relatively 'expensive', prompting simplification efforts which led directly to the AMDF below. In fact the similarity of the two measures is illustrated in Figure 7.4, which plots the frame power and AMDF measures together for an example recording of the first seven letters of the alphabet. The speech was recorded with a 16 kHz sample rate, with analysis performed on non-overlapping 128-sample frames. Each of the plots is scaled to a maximum of 1.0 for comparison purposes.

7.1.3 Average magnitude difference function

The average magnitude difference function is designed to provide much of the information of the frame power measure, but without multiplications:

$$\text{AMDF}_i = \frac{1}{N} \sum_{n=0}^{N-1} |x_i(n)|. \tag{7.3}$$

In MATLAB, it is again very simple:

```
function [amdf]=amdf(segment)
  amdf=sum(abs(segment))/length(segment);
```

An illustration of AMDF obtained for a sequence of recorded speech is shown in Figure 7.4. This plots the waveform for a 16 kHz sample rate recitation of the letters A to G. Below, both AMDF and frame power are plotted for 128 sample analysis frames. Note the correspondence of the two measures: both have a high output level when speech power is expected to be high (such as the /a/ sound in the letter A as well as the /c/ and /ee/ sounds of the letter C) and both output low measures when speech power is expected to be low (for example, the letter F and gaps between words).

Although the correspondence between frame power and AMDF appears to be quite close in this plot, it should be noted that the AMDF output is higher than frame power during the gaps between words. This is an indicator that the AMDF may be less immune to confusion by noise than the frame power measure.

7.1.4 Spectral measures

In Chapter 2 we looked at how to use an FFT to determine a frequency spectrum. In MATLAB we created and then plotted the spectrum of a random vector (see

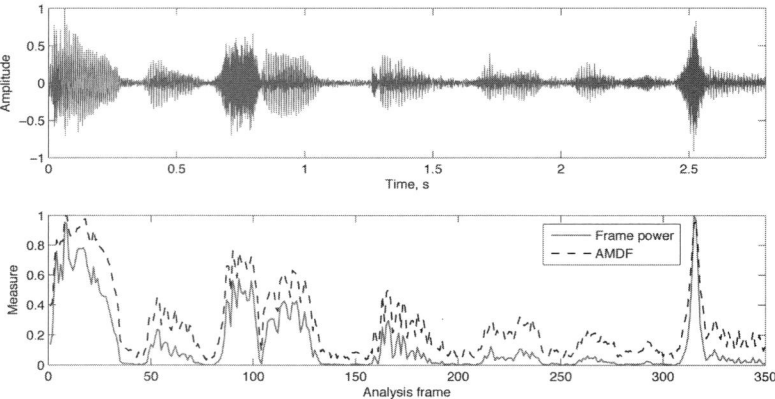

Figure 7.4 Average magnitude difference function (AMDF) and frame power plots (lower graph) for a recitation of the alphabet from A to G (plotted as a waveform on the upper graph). The duration of the letter C is enough for there to be almost two amplitude 'bumps' for the /c/ and the /ee/ sounds separated by a short gap, spanning the time period from approximately 0.7 to 1.2 seconds.

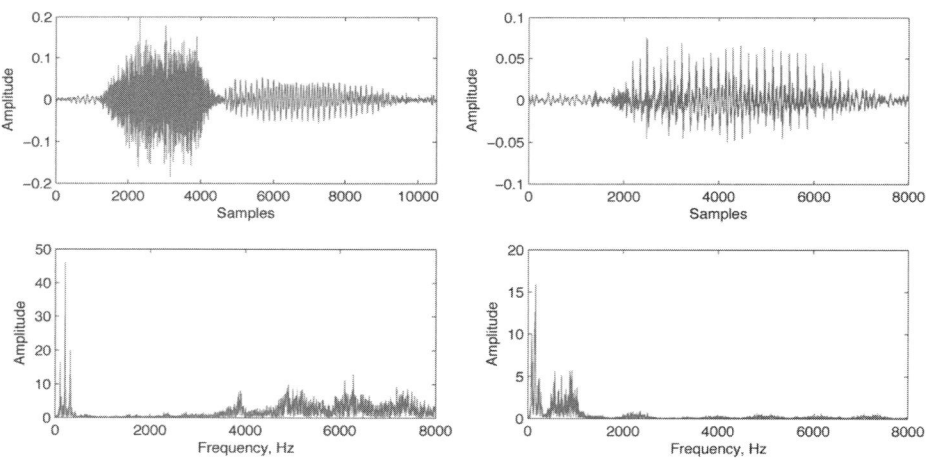

Figure 7.5 Plot of two 16 kHz sampled speech utterances of spoken letters C (left) and R (right), with time-domain waveform plots at the top and frequency-domain spectra plotted below them.

Section 2.3). If, instead of plotting the spectrum directly, we were to analyse the spectral components, we could use these to build up a spectral measure.

To illustrate this, we will plot the spectra of two different regions of a speech recording, examine these and analyse further. Both the time- and the frequency-domain plots are shown in Figure 7.5, for the spoken letters C and R. From the time-domain waveform plots in the figure, it can be seen that the C is probably louder than the R (it has higher amplitude) and also has a slightly longer duration. The frequency-domain plots show some spikes at low frequencies – most likely formant frequencies – but also show that the C has more high-frequency components than the R. Most of the signal power in the R seems to be below about 1 kHz, but much of the signal power in the C seems to be above this.

In MATLAB we can devise a measure to compare the low-frequency spectral power with the high-frequency spectral power. To do this, first we need to derive the spectra (as plotted in the lower half of Figure 7.5) and name them fft_c and fft_r to denote the spectra of spoken letter C and spoken letter R, respectively:

```
Nc=length(speech_letter_c);
Nr=length(speech_letter_r);
fft_c=fft(speech_letter_c);
fft_c=abs(fft_c(1:Nc/2));
fft_r=fft(speech_letter_r);
fft_r=abs(fft_r(1:Nr/2));
```

At this point, we could plot the spectra if required. Remember, we are plotting only the positive frequency components: the FFT output would be as long as the original speech array, so we take the lower half of this only for simplicity, and we take the absolute value because the FFT output is complex. When trying this example, MATLAB may complain that you need to ensure that only integers are used for the 'colon operator' indexing. This would happen if the length of the original speech arrays was not even (in fact, considering the further subdivision we are about to perform, we should really ensure that the speech array sizes were a multiple of four). In the real world, instead of Nr/2 or Nr/4 we really should use floor(Nr/2) and floor(Nr/4)[1] assuming we wanted robust code – but at this point we will put up with such possible complications to keep the code simple. However, by all means use the floor() function if MATLAB complains.

Next we can simply sum up the frequency elements within the required ranges (in this case, the lower half frequencies and the upper half frequencies respectively):

```
c_lowf=sum(fft_c(1:Nc/4))/(Nc/4);
c_highf=sum(fft_c(1+Nc/4:Nc/2))/(Nc/4);
r_lowf=sum(fft_r(1:Nr/4))/(Nr/4);
r_highf=sum(fft_r(1+Nr/4:Nr/2))/(Nr/4);
```

For the example spectra plotted in Figure 7.5, the results are telling. Splitting the spectrum in half, the mean absolute lower half frequency component for R is 0.74, and that for C is 0.87. For the mean absolute higher half frequency components, R scores 0.13 whereas C scores 2.84. However, it is the ratios of these that are particularly meaningful. The letter C has a high-frequency to low-frequency ratio of 3.3, but the letter R scores

---

[1] floor() simply rounds down the given value to the next lowest integer.

only 0.18. These figures indicate that much of the energy in the spoken letter C is higher frequency, whilst much of the energy in the spoken letter R is lower frequency. Indeed we can visualise this by looking at the spectral plots, but the important point is that we have just used the analysis to create a measure that could be performed automatically by a computer to detect a spoken C or a spoken R:

```
c_ratio=c_highf/c_lowf;
r_ratio=r_highf/r_lowf;
```

Although this is a relatively trivial example, it is possible, for one speaker, to identify spoken letters by segmenting them (isolating an analysis frame that contains a single letter, and even subdividing this), performing an FFT, then examining the ratio of the summed frequency components across different regions of the frequency spectrum. Unfortunately this technique cannot normally be generalised to work reliably for the speech of many different people or in the presence of different background noises.

7.1.5 Cepstral analysis

The cepstrum was introduced in Section 2.6.2.2 where an example of the technique was presented as a useful method of visualising speech signals. As with many other visualisation methods, the useful information that the human eye can notice in a plot can also be extracted and analysed by computer.

The usefulness of the cepstrum derives from the fact that it is the inverse FFT of the logarithm of the FFT. In general terms, this means that the frequency components have been ordered logarithmically. In mathematics, one of the principles of the logarithm is that, if something is the multiplicative combination of two items, then, in the logarithmic domain, these items are combined additively. Put another way, if a signal under analysis, $y(t)$, can be said to be equal to $h(t)$ multiplied by $x(t)$, then:

$$y(t) = h(t) \times x(t),$$
$$\log[y(t)] = \log[h(t)] + \log[x(t)]. \qquad (7.4)$$

Relating back to speech signals, $x(t)$ may well be a pitch component, while $h(t)$ is a vocal tract component. In the time domain these are related multiplicatively, but in the cepstrum domain they are related additively. In a cepstral plot then the pitch component, for instance, would be visible in its own right, separated from the vocal tract component.

Exactly this aspect has has been illustrated in Figure 7.6, plotted using the method of Section 2.6.2.2. The most likely position of the fundamental pitch period component, at index position 64, has been selected.

Cepstral analysis is also used for many other purposes than pitch detection. One of the more useful, and traditional, uses is in the extraction of spectral envelope information for speech analysis. In general, the spectral envelope is the smoothed shape of the frequency

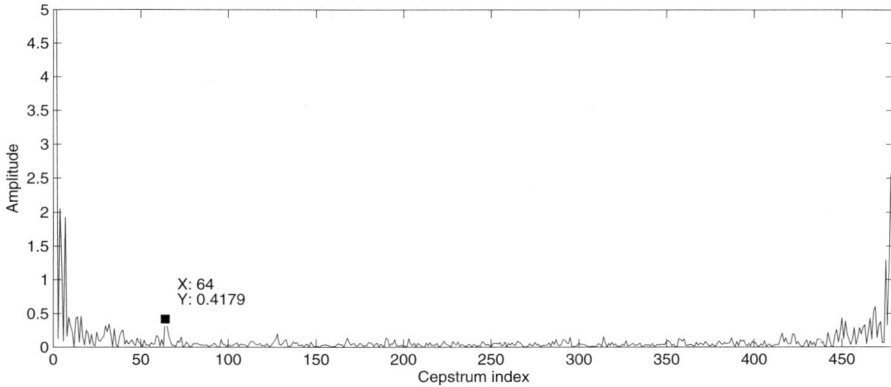

Figure 7.6 Cepstral plot of a segment of voiced speech, amplitude against cepstral index for a 480-sample analysis window. The likely pitch component has been selected, at index position 64.

plot, where the smoothing process really means ignoring the higher-frequency components. In the cepstral domain, this can be performed by discarding all cepstral coefficients related to frequencies higher than the envelope frequency.

If this is performed, and then the FFT plotted, the smoothing process can be quite obvious. MATLAB code which performs this process is provided below, and illustrated, acting on a 480-sample window of voiced speech, named `segment`, in Figure 7.7.

```
len=length(segment);
%Take the cepstrum
ps=log(abs(fft(segment)));
cep=ifft(ps);
%Perform the filtering
cut=30;
cep2=zeros(1,len);
cep2(1:cut-1)=cep(1:cut-1)*2;
cep2(1)=cep(1);
cep2(cut)=cep(cut);
%Convert to frequency domain
env=real(fft(cep2));
act=real(fft(cep));
%Plot the result
pl1=20*log10(env(1:len/2));
pl2=20*log10(act(1:len/2));
span=[1:fs/len:fs/2];
plot(span,pl1,'k-.',span,pl2,'b');
xlabel('Frequency, Hz');
ylabel('Amplitude, dB');
```

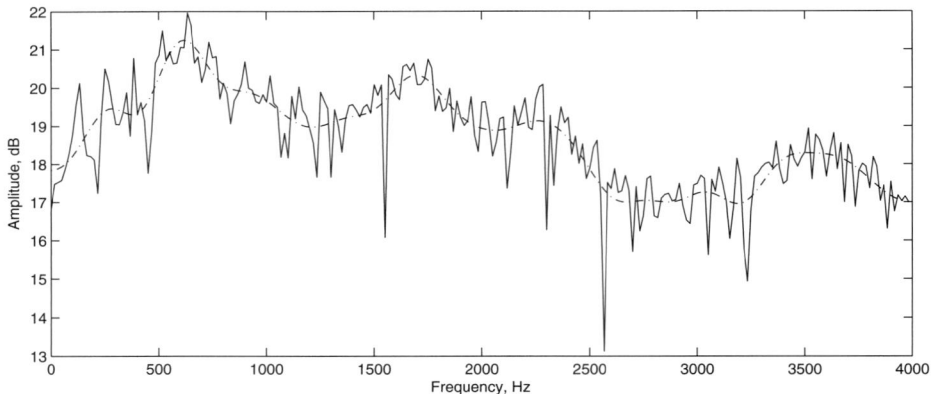

Figure 7.7 Frequency plot of a segment of voiced speech (solid line) overlaid with the frequency envelope obtained from the first few cepstral coefficients (dashed line).

Notice that the MATLAB code, as well as discarding the unwanted cepstral coefficients in the `cep2` array [87], accentuates all of the components between DC and the one immediately below the cutoff frequency by being doubled, to better scale to the required envelope.

Apart from performing the cepstrum manually, MATLAB contains much faster functions to calculate the cepstrum and its inverse in the signal processing toolbox, namely `cceps` and `iceps`. These could have been used synonymously with the operations in the MATLAB code above.

7.1.6 LSP-based measures

Analysis can also be performed using LSP-based measures since usable and interpretable information can be extracted from raw LSPs. These analysis techniques are normally used for speech, although, to demonstrate their wider application, examples will also be given illustrating the analysis of a musical instrument and of an animal sound.

For the following sections, it is assumed that LSP analysis, as described in Section 6.2.5, has been performed, and that a set of line spectral pairs is available for an analysis window. We will first consider instantaneous LSP measures which are means of interpreting a single set of lines from one analysis frame, before we consider the longer-term trends of these measures.

7.1.6.1 Instantaneous LSP analysis

Although there are many possible methods of analysis, in this section we will consider just three features which can be used for individual analysis frames of speech. Later, in Section 7.1.6.2, we will then extend the analysis to the evolution of the measures through time. In the following equations, subscript i refers to one of the p LSPs ω_i which represent speech in analysis frame n.

7.1 Analysis toolkit

The overall shift in LSP location from one frame to the next, referred to as the gross shift, *Shift*, indicates predominant spectral movements between frames:

$$Shift[n] = \left\{\sum_{i=1}^{p} \omega_i[n]\right\} - \left\{\sum_{i=1}^{p} \omega_i[n+1]\right\}. \tag{7.5}$$

In MATLAB this would be:

```
function [shift] = lsp_shift(w1,w2)
    shift=sum(w1) - sum(w2);
```

The mean frequency of all LSPs representing an analysis frame is also a useful measure of frequency bias, *Bias*:

$$Bias[n] = \frac{1}{p}\sum_{i=1}^{p} \omega_i[n], \tag{7.6}$$

or in MATLAB:

```
ffunction [bias] = lsp_bias(w)
    bias=sum(w)/length(w);
```

It is also possible to specify a nominal LSP positioning, and calculate a deviation *Dev* between this reference LSP positioning $\bar{\omega}_i$ and the LSPs under analysis, raised to an arbitrary power β. This is measured as follows:

$$Dev[n] = \sum_{i=1}^{p} (\omega_i[n] - \bar{\omega}_i)^{\beta}. \tag{7.7}$$

A suitable MATLAB function for calculating the deviation is shown below:

```
function [dev] = lsp_dev(w,bar_w,b)
    dev=sum( (w-bar_w).^b );
```

where $\bar{\omega}_i$ would often be a set of nominally positioned LSPs equally spaced across the frequency domain:

$$\bar{\omega}_i = i\pi/(p+1) \quad \text{for} \quad i = 1,\ldots,p. \tag{7.8}$$

In MATLAB, this is again easy. Assuming that $p = 10$ we would simply create the reference LSPs using:

```
bar_w=[1:10]*pi/11
```

206   **Audio analysis**

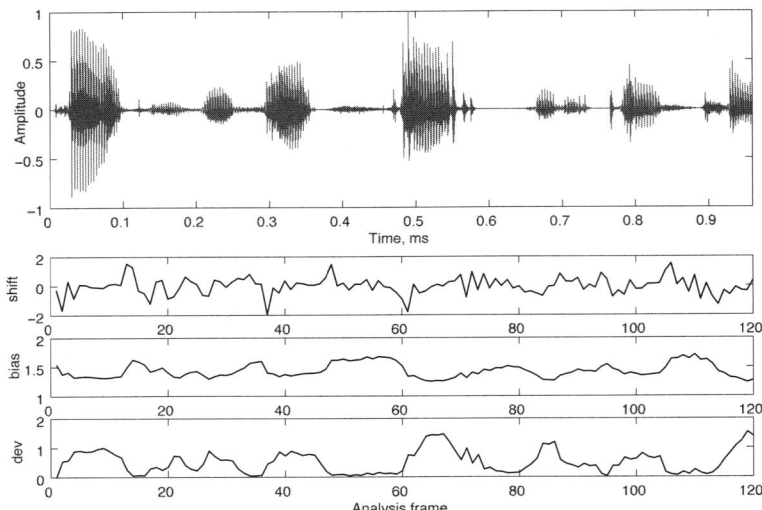

Figure 7.8 Deviation, bias and shift LSP analysis features collected for a 16 kHz sampled speech utterance, with waveform plotted on top, extracted from the TIMIT database.

With the p lines distributed evenly across the spectrum, if this were transformed into LPC coefficients and their power spectrum calculated, it would be flat. *Dev* thus determines how close each frame is to this distribution, such that, with $\beta = 2$, it becomes the Euclidean distance between the actual and comparison distributions. Odd values such as $\beta = 1$ or $\beta = 3$ attribute a sign to the deviation from $\bar{\omega}_i$, so that a positive measure denotes high-frequency spectral bias and a negative measure specifies a low-frequency spectral bias.

Each of these measures provides useful information regarding the underlying speech signal. They are illustrated when applied to a speech recording from the TIMIT database [70] in Figure 7.8 (the deviation plot is given for $\beta = 2$).

Shift indicates predominant LSP frequency distribution movements between consecutive frames. Considering an example of two adjacent frames containing unvoiced and voiced speech, the LSP distributions in the two frames will be low-frequency and high-frequency biased respectively. We saw a similar effect when comparing the spoken C and R spectra in Section 7.1.4. A large difference between the two frames gives a large measure value. The shift measure, as shown in Figure 7.8, peaks at obvious speech waveform changes and may thus be advantageous for speech segmentation.

Bias indicates frequency trends within the current frame – that is, whether the spectrum of the current frame is high-frequency or low-frequency biased. It is similar to the deviation measure which determines how close the LSP distribution of the current frame is to a predetermined comparison distribution. In Figure 7.8 this registers high values for fricatives, indicating the predominance of their high-frequency components.

Where the speech is contaminated by noise of a particularly well-defined shape, if the comparison distribution, $\bar{\omega}$, of Equation (7.7) is set to represent a spectrum of this shape, then the *Dev* measure may be reasonably insensitive to noise when averaged. In other words, analysing the noise itself will produce a zero mean output.

We can also use LSP data to estimate the position of spectral peaks within an analysis frame. Peaks are located approximately halfway between pairs of closely spaced lines, with the peak power related to the closeness of the lines. A ranking of the first few most closely spaced line pairs in terms of narrowness will generally correspond to the ordering of the corresponding peaks by power. It must be noted, however, that in some cases, especially where three lines are close together, the correspondence is far less predictable. For unvoiced frames, other speech resonances are similarly reflected in the LSP distribution, although the visual correspondence when plotted is far less dramatic than in the case of strongly voiced speech.

7.1.6.2   Time-evolved LSP analysis

Each of the instantaneous measures can be tracked to build up a statistical model of the signal under analysis. Similarly, spectral peaks, detected by a measure of line-closeness, can be tracked and post-processed to extract likely formant positions. Analysis of LSP feature changes over time may be combined with consideration of traditional speech features to track dynamic speech characteristics.

Applications of speech feature tracking can include the classification of speech analysis frames into voiced, unvoiced or non-speech periods. Evidence suggests that instantaneous speech features derived from LSP analysis are similar, under low-noise conditions, to zero-crossing rate [88].

A simple example of speech classification uses a binary LSP decision measure of upward or downward deviation. This LSP vote measure counts the number of lines in a frame above their median value, and is a computationally simple alternative to Equation (7.7). It has been used by the author to perform phonetic classification of speech [88]. In this application the LSP vote measure was used to categorise speech frames into six classes: voiceless fricatives and affricatives, fricatives, plosives, nasals and glides, and silence.

At this point we should note that one of the most popular applications of continuous LSP analysis is in the speech recognition field. In general, older speech recognition systems used techniques such as template matching, statistical modelling or multi-feature vector quantisation. More modern systems, by contrast, are generally tied to 'big data' approaches that we will discuss in Chapter 8. It is possible to use LSP-derived data with most speech recognition techniques (big data-based or otherwise), although the statistical matching of a hidden Markov model (HMM) acting on MFCC data (see Section 5.5) is currently still much more common than systems that use LSP-derived features [89].

Finally, continuous LSP analysis has been used to collect statistics on the line frequency distribution over time in a CELP coder and relate this to the language being conveyed. It has been shown that LSP statistics differ markedly across several languages [90], leading to the potential for LSP-based features to be used in language detection algorithms.

7.2 Speech analysis and classification

The analysis of speech is an important requirement of many different applications and the classification of speech into various categories is a necessary part of many techniques. A full list would be lengthy, but the following subset of basic techniques indicates the sheer range of applications of speech analysis and classification:

- detecting the presence of speech;
- detecting voiced or unvoiced speech;
- finding boundaries between phonemes or words;
- classifying speech by phoneme type;
- language detection;
- speaker recognition;
- speech recognition.

Classification is an important, and growing, area of speech research which relates to the machine 'understanding' of speech (where understanding can range from knowing whether speech is present right through to understanding the meaning or emotion conveyed by the spoken word).

In order to begin such classification of speech, it is usually necessary to first perform some form of measurement on the speech signal itself. For example, detecting voiced or unvoiced speech might require the determination of speech power and pitch, perhaps through examination of LSP data. Many methods can potentially be used for analysis of speech, and extensive empirical testing is almost always required to determine the best subset of measures to be used for a particular application, whatever that may be.

By and large, new classification tasks will make use of many of the same basic speech measures as described in our analysis toolkit (Section 7.1). Measures can be used in different ways, to different ends, and most often several measures must be combined together.

In addition to the several methods in our toolkit, there are some methods which have been very much reserved for speech analysis, particularly pitch detection-related methods, since pitch is so important to speech communications.

In the following subsections, we will examine techniques for the analysis and extraction of pitch from speech.

7.2.1 Pitch analysis

Section 6.3.2.1 discussed the use of long-term prediction to determine the pitch of a speech waveform, a method commonly used in speech coders. There are, however, many alternatives. The most accurate require some form of human involvement, or a measurement device such as an accelerometer or electroglottograph (EGG) on the throat or glottis producing the pitch. From most recorded speech waveforms alone, it is currently impossible to identify a definitive answer to the question 'what is the pitch?', but nevertheless some algorithms appear to get close to what we would consider to be a reference standard. Tests of these pitch analysis algorithms tend to rely on consensus answers from human and machine measurement (or a simultaneous EGG reading). As an example, many methods have been surveyed, and compared, in an excellent paper by Rabiner *et al.* [91].

Some of the more mainstream techniques reported by Rabiner and others operating purely on recorded sound include the following:

- time-domain zero-crossing analysis (perhaps thresholded – see Section 7.1.1);
- time-domain autocorrelation (the method used in Section 6.3.2.1);
- frequency-domain cepstral analysis (see Section 2.6.2.2);
- average magnitude difference function-based methods (see Section 7.1.3);
- simplified inverse filtering technique (SIFT);
- LPC- and LSP-based methods (such as those in Section 7.1.6);
- time–frequency-domain analysis as explained in Section 7.2.2.

We will not reproduce the extensive work of Rabiner here, but will introduce a modern alternative, the use of time–frequency distribution analysis. Whilst being more complex than most, if not all, of the other methods mentioned, indications are that this method of analysis is promising in terms of achieving better accuracy than the traditional techniques.

7.2.2 Joint time–frequency distribution

Joint time–frequency distribution (TFD) analysis originally emerged in the radar signal processing field, but has started to be adopted for speech processing in recent years [92]. Its good performance in tracing frequency transitions as time progresses has been noted by speech researchers. Given the importance of pitch in speech systems, it is thus little surprise that TFD analysis has been attempted for pitch determination.

We will discuss four joint time–frequency distributions, as described in [93], for use in pitch determination. These are the spectrogram time–frequency distribution (STFD), Wigner–Ville distribution (WVD), pseudo-Wigner–Ville distribution (PWVD) and reassigned smoothed pseudo-Wigner–Ville distribution (RSPWVD). They all attempt to identify the characteristics of frequency as it changes with time from slightly different points of view. Each one of these TFD algorithms has its own strengths and weaknesses relating to the sensitivity to detect pitch features and the implementation complexity (and these have been well established in the research literature).

The STFD computes the spectrogram distribution of a discrete-time signal x. It corresponds to the squared modulus of the short-time Fourier transform (see Section 2.6). As such it is quite easily described as an integral across analysis window w in Equation (7.9):

$$S_x(t, v) = \left| \int x(u)w(u - t)e^{-j2\pi vu} du \right|^2. \tag{7.9}$$

The instantaneous frequency energy at time t is then given by:

$$E_x(t) = \frac{1}{2} \int S_x(t, v) dv. \tag{7.10}$$

The Wigner–Ville distribution (WVD) is a bilinear distribution that yields a different analysis perspective to the STFD. At any particular time, the signal contributions from the past are multiplied by the signal contributions from the future in order to compute and correlate between these left (past) and right (future) parts of the signal. It has the ability to give a clearer picture of the instantaneous frequency and group delay than the spectrogram does:

$$\mathrm{WV}_x(t, v) = \int x\left(t + \frac{\tau}{2}\right) x\left(t - \frac{\tau}{2}\right) e^{-j2\pi v\tau} d\tau. \tag{7.11}$$

Similarly, the instantaneous frequency energy at time t is then given by:

$$E_x(t) = \frac{1}{2} \int |\mathrm{WV}_x(t, v)|\, dv. \tag{7.12}$$

Although the WVD is perfectly localised on linear signals due to the forward–backward nature of the analysis, if there are several frequency components existing simultaneously, noise will be found in the WVD due to the phenomenon of cross-term interference between those components. It also provides equal weighting to both past and future components.

The pseudo-Wigner–Ville distribution (PWVD) is an advance on the WVD since it emphasises the signal properties near the time of interest compared to far away times. A window function $h(\tau)$ that peaks around $\tau = 0$ is used to weight the Wigner–Ville distribution toward an emphasis of the signal around time t. The PWVD is defined as:

$$\mathrm{PWV}_x(t, v) = \int h(\tau) x\left(t + \frac{\tau}{2}\right) x\left(t - \frac{\tau}{2}\right) e^{-j2\pi v\tau} d\tau. \tag{7.13}$$

Again, the instantaneous frequency energy calculation at time t is relatively simple:

$$E_x(t) = \frac{1}{2} \int |\mathrm{PWV}_x(t, v)|\, dv. \tag{7.14}$$

Finally, a fourth method exists as a refinement to PWVD. This is the reassigned smoothed PWVD (refer to [93] for its application). It uses a centre-of-gravity reassignment method with both time and frequency domain smoothing.

These methods have each been tested within a GSM structure for the determination of pitch [93]. The time–frequency distribution algorithms, with a little post-processing, were used as a replacement for the original pitch analysis used for the RPE structure. This overall system was then tested to determine what changes (if any) resulted in the intelligibility of speech conveyed. In each case the same speech and other conditions were used for the tests, and the Chinese diagnostic rhyme test (CDRT) – a Chinese language equivalent of the diagnostic rhyme test (DRT) of Section 3.4.3 – was used to assess intelligibility. Results indicated a significant improvement in intelligibility for the speech that differed in terms of sibilation, and smaller less significant improvements in several other speech classes.

Overall it seems that the results are promising; TFD might well become another candidate for pitch determination in speech communications systems. However, at present other methods, particularly the autocorrelation-based systems, are the most popular.

7.3 Some examples of audio analysis

We have already discussed a number of methods of analysis in this chapter, as well as previously. The methods at our disposal include both simple and complex techniques, including the following:

- Simple waveform plots (as easy as `plot(speech)`) for a quick visualisation of a recording.
- Zero and threshold crossing rate (ZCR and TCR) in Section 7.1.1, useful for determining the fundamental period of a waveform, usually computed over sequential frames.
- Frame power and AMDF (average magnitude difference function) in Sections 7.1.2 and 7.1.3 respectively, useful for frame-by-frame analysis of speech to detect boundaries between phonemes and words.
- Various spectral methods including taking a discrete or fast Fourier transform over an entire recording (Section 2.3 and Section 7.1.4) or over smaller analysis windows (Infobox 2.5 on page 34). Also used to form the traditional spectrogram plot of Section 2.6 (and Figure 2.9) which is particularly good at visualising different types of speech or sound (to the trained eye).
- Cepstral analysis in Section 7.1.5, useful for separating out different periodicities that might be mixed into one recorded signal (including pitch and formants).
- Line spectral pair (LSP)-based measures in Section 7.1.6, which are widely used in speech processing, especially when combined with speech compression algorithms.
- Time–frequency-domain analysis, which we first met in Infobox 2.5 on page 34, but expanded more formally in Section 7.2.2. This is useful for analysing non-stationary signals or signals that rapidly change between different types or regions.

Most methods are computed on a frame-by-frame basis (usually with overlap) and can be visualised in MATLAB in a wide variety of ways. A measure that gives a single scalar value each frame (such as ZCR, power, AMDF) can be shown as a simple line plot, as in Figures 7.2 and 7.4 – both figures also plotted the waveform, which is effectively a plot of the sound amplitude for each sample (i.e. plotting one scalar value per sample instead of one per frame).

The spectrum of a sound computed over an entire recording can hide information (refer back to Infobox 2.5 on page 34), so it is usually computed on a frame-by-frame basis in most applications, rather than once for the entire recording. This means that each analysis frame of N samples, instead of being represented by a single scalar, is now represented by a vector of information, of length perhaps $N/2$ or N. The situation is similar for LSPs, which would typically yield 10 or 12 values for each frame, as was illustrated in Figure 6.14 (on page 166) – but now imagine this repeated for each analysis frame through a recording. We would, in fact, need to construct a plot that shows each LSP as it changes over time (a brief glance at Figure 7.10 on page 217 reveals exactly that, with the upper part of the figure plotting LSP frequencies as they change from frame to frame).

Spectral information is probably the most interesting – and useful – from an audio analysis perspective. Thus the remainder of this section will follow some example spectral analyses of a speech recording. This section is by no means a complete survey of visualisation methods, but should be considered a means of introducing some diversity into the examination of auditory information.

We will begin by loading a speech array – in this case a recording from the TIMIT database (that has already been converted to Wave file format):

```
[speech,fs]=audioread('SA1_converted.wav');
```

We will use the MATLAB `spectrogram()` function to convert the one-dimensional speech vector into a spectral representation, performing an FFT on successive sample frames that are Hamming windowed. In this case, we will use 1024 sample windows (the second argument), advancing by 128 samples each step (the third argument) so that the analysis frames are overlapped by $(1024 - 128)/128 = 87.5\%$. The fourth argument specifies that a 1024-point FFT is used:

```
spec=spectrogram(speech,1024,1024-128,1024);
```

To be clear, MATLAB reports the sizes of these arrays as follows:

```
>> size(speech)
ans =
       49544           1
>> size(spec)
ans =
         513         380
>>
```

The input speech array is thus 49 544 samples in length, and has been divided into $\lfloor (49\,544 - 1024 + 128)/128 \rfloor = 380$ frames. The 1024-point FFT for each frame has been converted to a vector of 513 real values, with the first value being the DC (0 Hz) energy.

We have met the `spectrogram()` function before, and in general it is used to give a fast and easy visualisation of the spectral content of the input vector. With no output arguments (i.e. removing the `spec=`) causes it to plot directly rather than return the spectral array. Since the spectrogram is actually an image, we can view this easily using the image viewing tools in MATLAB, such as `imagesc()`, although many alternative methods exist. Let us create a few interesting plots:

```
subplot(4,1,1)
plot(speech)
subplot(4,1,2)
imagesc(flipud(10*log10(abs(spec))))
subplot(4,1,3)
contour(10*log10(abs(spec)),6)
subplot(4,1,4)
mesh(10*log10(abs(spec)))
axis tight % make plot completely fill the X-Y axes
view(40,80) % change viewpoint of 3D plot
```

The output from this selection of MATLAB code is shown in Figure 7.9. We have not added any titles or axes (which would have needed the commands `title("a title")`, `xlabel("x label text")` and `ylabel("y label text")` respectively.

Note that all except the top plot take the logarithm of the absolute signal, so that they are plotting intensity levels in decibels, which is normal for auditory visualisation. The contour plot (the third one from the top in Figure 7.9) is particularly interesting because it is possible to specify the number of contours, not just the six that are shown. This allows more or less detail to be revealed in the plot, and is useful for revealing regions of similar intensity.

When plotting in MATLAB, especially with three-dimensional plots, it is sometimes necessary to clear the plot window with command `clf` between plots.

Obviously this section has not been a comprehensive survey of visualisation methods, but has given a few useful ideas which can be expanded upon. Some of these plots will also be useful in subsequent chapters. The reader is encouraged to experiment with the plotting capabilities of MATLAB because it is sometimes only when we look at information in a different or unusual way that relationships and features in the data – that were previously hidden – can become apparent.

7.4 Statistics and classification

Statistics is simply the collation of information from current and past events, plus some rudimentary mathematical analysis. In practice, this is very commonly applied to individual speech and audio analysis data. It begins by comparing the analysis score from a current window with the mean score (this is a first-order analysis), then proceeds to compute the standard deviation of the analysis score (this is second-order analysis), and, in many cases, the analysis continues to third order or beyond. Statistics are actually extremely important for a number of research domains relating to the natural world, not least for speech and audio analysis (as will become particularly clear in Chapters 8 and 9).

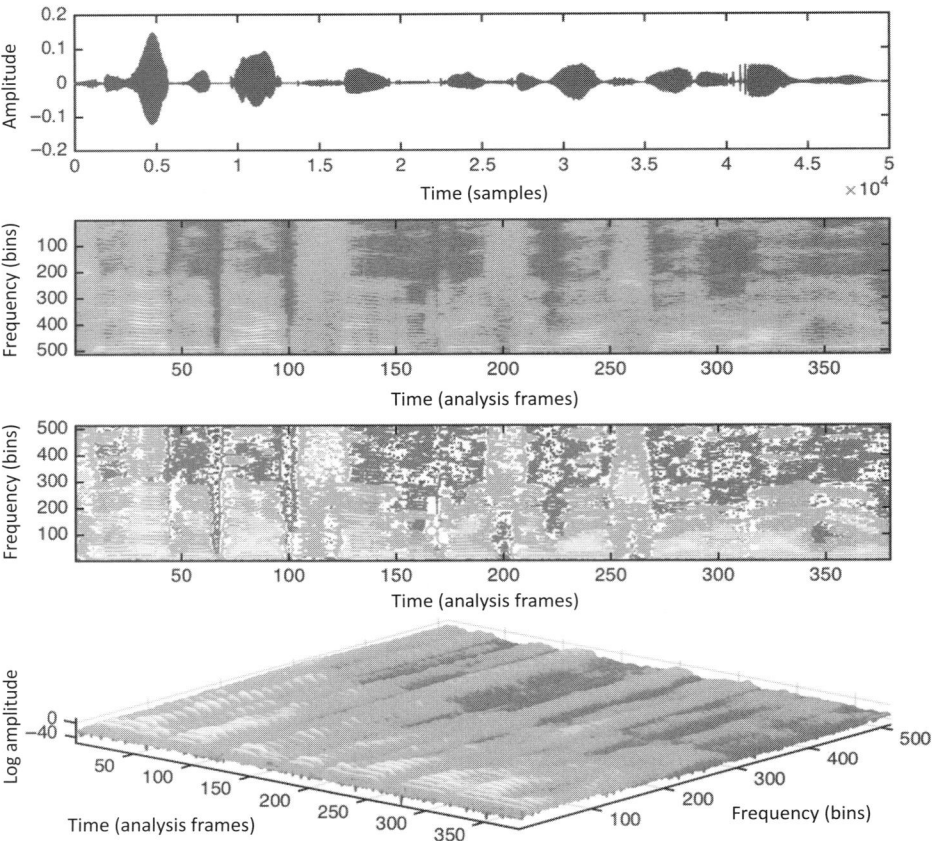

Figure 7.9 From top, plots showing speech waveform, spectrogram, contour spectral plot and three-dimensional mesh spectral plot.

Consider some of the analytical tools (measures or scores) discussed so far in this chapter, and how they relate to something like voice activity detection (see Section 9.2). We have seen in the current section that the scores we obtain from analysing speech cannot provide a definitive indication of the presence or absence of speech when used in isolation. In fact, they need to be combined with other scores ('fused'), as well as tracked over a longer time period (perhaps over hundreds of analysis windows), and this process of tracking needs to be statistical.

Remember that no system which correctly classifies unconstrained speech can be correct 100% of the time. However, scores approaching this are feasible (and big data/machine learning techniques in Chapter 8 are generally able to achieve the best performance).

One of the most important issues to consider is what happens when a classification or detection system goes wrong – and this might be just as important in some applications

7.4 Statistics and classification

as when the system works. In general, the developer needs to bear in mind four conditions regarding the accuracy of a binary classification or detection system (positive–negative match to some criteria):

- True-Positive classification accuracy
  the proportion of positive matches classified correctly;
- True-Negative classification accuracy
  the proportion of negative matches classified correctly;
- False-Positive classification accuracy
  the proportion of negative matches incorrectly classified as positive;
- False-Negative classification accuracy
  the proportion of positive matches incorrectly classified as negative.

Accuracy can usually be expected to improve through having a larger sample size for analysis (or training, in the case of machine learning methods); however, the method of analysis is important. For example, although we can say that speech occupies certain frequency ranges, at certain volumes, there are many types of sound that also occupy these time/frequency locations. Dog growls, chairs scraping across floors and doors slamming could appear like speech if spectral measures alone are used. By and large, we would therefore need to use more than one basic measure as mentioned – perhaps spectral distribution and amplitude distribution (AMDF measure). Unfortunately, those measures would be confused by music – having similar frequency ranges to speech, and similar amplitude variation too. There are thus other aspects that we need to look for.

Specifically for speech, we often need to turn to higher-order statistics. These relate to the usage, generation and content of the speech signal itself, including the following:

First is the *occupancy rate* of the channel. Most speakers do not utter long continuous monologues. For telephone systems, there should generally be pauses in one person's speech – gaps which become occupied by the other party. We know they must be there, because humans need to breathe (although some recipients of a monologue will swear that the person they have been listening to didn't stop to breathe, even for a moment). Normal to-and-fro flows in dialogue can be detected when they occur, and used to indicate the presence of speech. For this, however, we may require analysis of several minutes' worth of speech before a picture emerges of occupancy rate.

The *syllabic rate* is the speed at which syllables are formed and uttered. To some extent, this is a function of language and speaker – for example, many native Indian speakers might have a far higher syllabic rate than most native Maori speakers, and, irrespective of origin, most people do not exhibit a high syllabic rate when woken in the early hours of the morning. However, the vocal production mechanisms are muscle-controlled, and are only capable of a certain range of syllabic rates. Patterns varying at a rate within this range can be detected, and the fact used to classify speech.

Most languages have a certain *ratio of voiced and unvoiced* speech. Degree of voicing can be affected by several conditions, such as sore throat, and speaking environment (think of speaking on a cellphone in a library), but there is still a pattern of voiced and unvoiced speech in many languages. In Chinese, all words are either totally voiced (V), unvoiced–voiced (UV–V) or V–UV–V. In English the pattern is more random, but there *is* a pattern, and, given sufficient raw material for analysis/training, it can be detected and used for classification.

Finally the speech *cadence* is similar to syllabic rate but is effectively the rate at which words are spoken. Obviously this depends upon many factors, not least the length of words being spoken; however, it maintains a surprisingly narrow range of about 2 Hz to 7 Hz for normal speech. This can be determined from the frequency of gaps between words. But this is not as easy as it may appear since the gaps themselves are often extremely difficult to detect automatically, and often practically non-existent. Some word patterns have longer gaps inside words than between them. We will touch upon some of these issues once again when we consider some different types of speech classification later.

7.5 Analysing other signals

Arguably the predominant application of audio analysis has been related to speech communications applications, with a notable (and economically important) extension into music compression, such as MP3. However, there is no reason why the techniques and toolkit introduced in this chapter should not be applied elsewhere. As an example, one of the early algorithms produced by the author was capable of detecting and classifying dog barks, and in 1997 the author was contacted by a team of economists who wished to apply LSP analysis to the stock market.[2] To illustrate the wide use of audio analysis, the following subsections discuss some simple examples of music and animal sound analysis.

7.5.1 Analysis of music

Many researchers have investigated musical instrument recognition and coding. For example, Krishna and Sreenivas [94] evaluated three methods of recognising individual instrument sets. One of these methods was based upon LSPs and was found to be the superior method among those tested. The authors stated several advantages of LSPs, including their localised spectral sensitivities, their ability to characterise both resonance locations and bandwidths (characteristics of the timbre of the instruments), and the important aspect of spectral peak location.

In order to examine some of these claims, it is easy to use MATLAB for analysis. As an example, MATLAB was used to record a violin open A string (tuned to 440 Hz), sampled at 16 kHz.

---

[2] Sadly nothing was subsequently heard from the economists, from which we can conclude that either the technique did not work, or it was so successful that they immediately retired from research before writing up, and did not wish to share their key to unlimited wealth.

Those readers with a violin handy may wish to record a short segment using the methods of Chapter 2. It is important to ensure that the player uses an uninterrupted and smooth bowing action during the recording because we want a stationary characteristic if possible. Given such a recording in a vector named `violin`, we can perform a *P*th-order LPC analysis, convert to LSPs and plot the result.

```
P=16;
Ws=256;    %window size=256 samples
vseg=violin(1:Ws).*hamming(Ws);
a=lpc(vseg, P);
w=lpc_lsp(a);
lpcsp(a, w);
```

There should be a very obvious spectral peak located at 440 Hz, with several harmonics visible. Figure 7.10 shows a plot of this analysis, ranging from 12th to 48th order, revealing greater and greater levels of spectral detail as the order increases. In this case the analysis shows a 4096-sample section. In each case there is a reasonably good correspondence of the LSP positions to spectral peaks – in particular the location of narrow line pairs around many major peaks. As more detail is revealed through

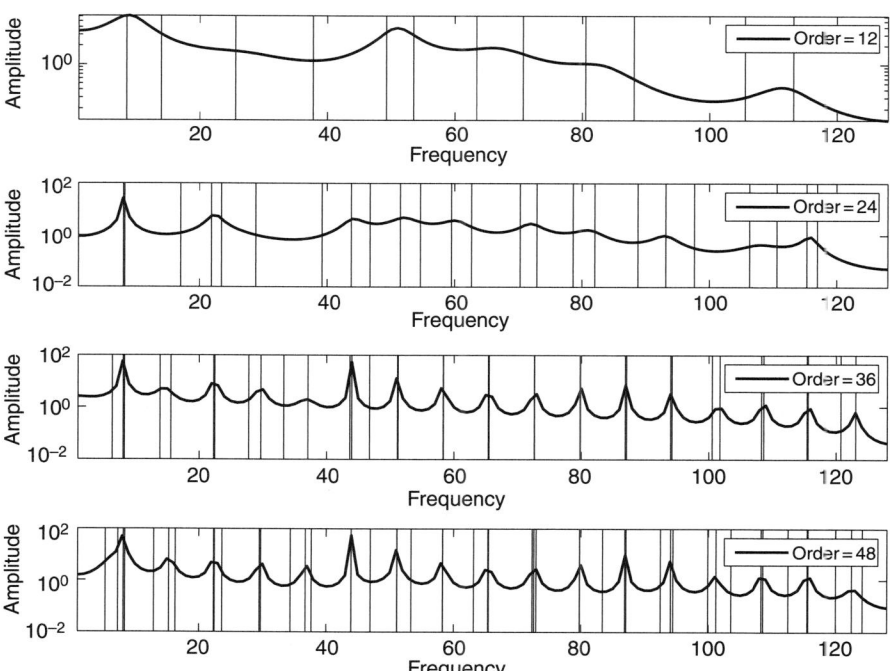

Figure 7.10 LSP locations and LPC power spectrum of a violin open A string plotted for various orders of analysis, against a frequency index from 1 to 128.

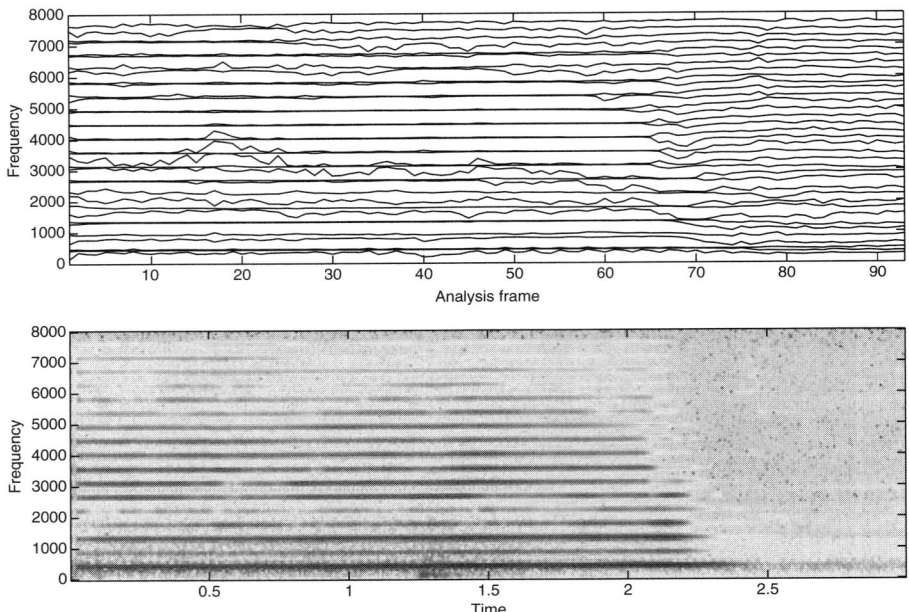

Figure 7.11 LSP tracks for a violin note played on the open A string sampled at 16 kHz (top) compared with a spectrogram of the same recording (bottom).

increased order analysis, the harmonic structure of the played note becomes clearer, although a triple (rather than a pair) at the fundamental of the 48th-order plot probably indicates that the analysis order is greater than the note complexity warrants.

Another plot shows the note as it was played, analysed through a time evolution. This is shown in Figure 7.11, where the upper graph displays the 36th-order LSP values as they evolve over time (note that, although it looks as if they may do, the lines do not cross, since they are always monotonically ordered), for each 512-sample analysis window. Below this, a spectrogram of the same recording is plotted, using a 256-sample window size with 50% overlap. Note the LSP narrowness around the obvious 440 Hz fundamental and harmonics shown in the spectrogram. At the end of the played note, some resonance continues to sound a 440 Hz fundamental but with only weak harmonics (once the abrasion of the rosin-coated horsehair bow on the aluminium-clad string has ceased). During this decay period the upper LSPs gradually lose their tendency to pair up and begin to flatten out. However, the fundamental continues to be marked by an ongoing narrow spacing of the lower two lines. The close visual correspondence between the spectrogram and the LSP evolution plot in Figure 7.11 supports the assertion of Krishna and Sreenivas [94] mentioned previously that LSP values are usable for instrument recognition. Most likely both fundamental and harmonic frequency identification can be determined through automatic analysis of LSP data, especially when combined with other measures such as ZCR or frame power.

7.5 Analysing other signals

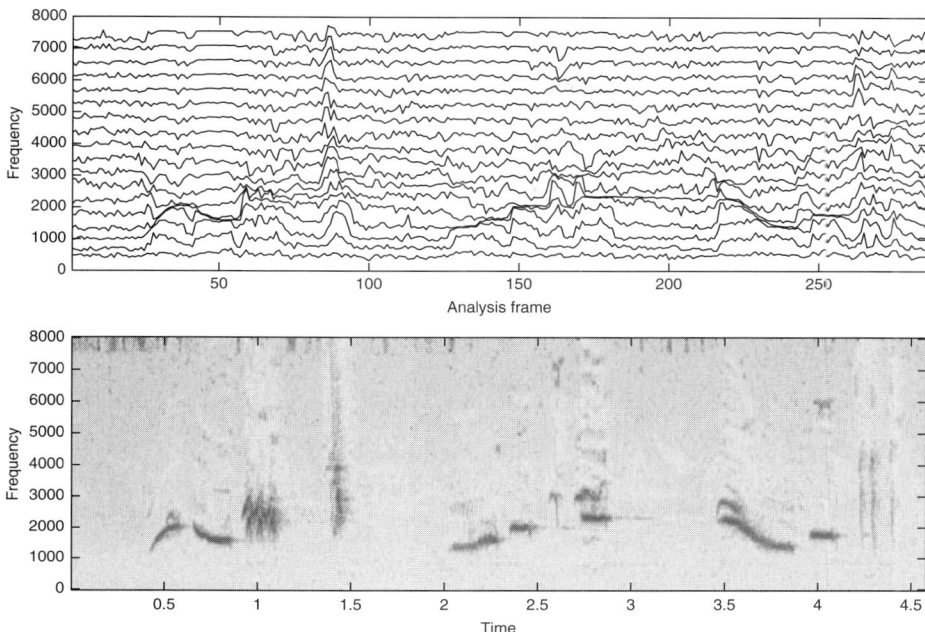

Figure 7.12 LSP tracks for a few seconds of a blackbird's song sampled at 16 kHz (top) compared with a spectrogram of the same recording (bottom).

7.5.2 Analysis of animal noises

A recording of birdsong from the BBC was analysed in a similar fashion to the analysis of the violin note in Section 7.5.1 [95]. After some experimentation, an 18th-order LPC analysis was found to be optimal for LSP track plotting as shown in Figure 7.12.

The analysis covers approximately 4.5 seconds of a blackbird's warble, and shows a clear visual correlation between the birdsong syllables shown in the spectrogram and in the plotted LSP trajectories. The spectrogram had 50% overlapped windows of size 256 samples, and non-overlapped 256-sample windows were used for the LSP tracking. Other authors report that harmonic analysis has been shown to be useful in the classification of birdsong [96], and, since both plots in Figure 7.12 so clearly show harmonic effects (look for LSP pairs shifting together), the potential for such analysis using these methods is clear.

As a final analysis of animal noise, a recording of a very large and angry dog barking was obtained, at some risk to the author, using a 16 kHz sample rate. Again this was analysed by obtaining first LPC coefficients and then a set of LSPs, for a succession of 64-sample analysis frames. The analysis order was set to eight in this instance since the pitch of the dog bark was rather low, and strongly resonant. Figure 7.13 plots the time-domain waveform above the time-evolution plot of the eight LSP tracks. A bar graph overlaid upon this depicts frame power, and both power and LSP frequencies were scaled to be between 0 and 1 for ease of plotting.

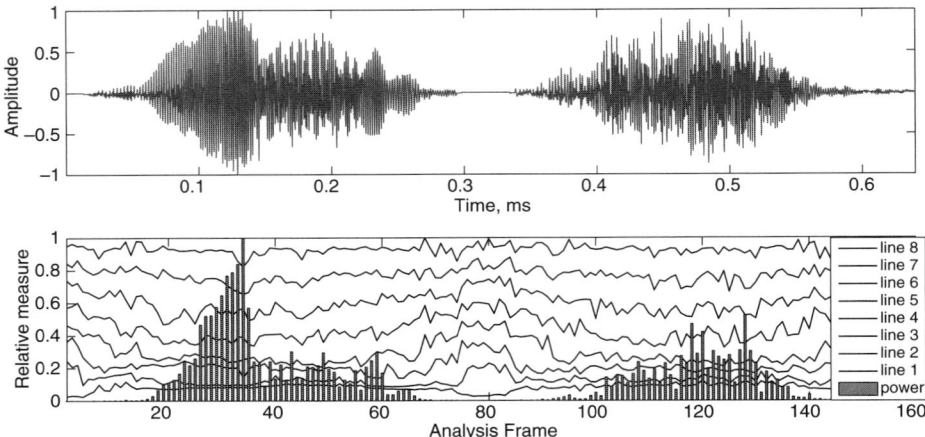

Figure 7.13 Waveform plot (top) and LSP tracks (below) with overlaid frame power plot (bars on lower graph) for a double dog bark, recorded at 16 kHz and an analysis order of eight.

The dog bark waveform of Figure 7.13 resembles human speech – in fact a recording of a human shouting 'woof woof' may look similar, although this will be left as an exercise for the reader. The narrow pairs of LSPs observable during the periods of highest power indicate the resonance of strong formants during the loudest part of the bark. The similarity to human speech would indicate that methods of analysis (and processing) used for human speech could well be applicable to animal noises. In particular, vocally generated animal noises, produced using the same mechanisms as human speech, would be more amenable to speech processing methods than those produced in other ways, such as the abrasive sounds of grasshoppers, dolphin squeaks, pig snorts, snake rattles and so on.

7.6 Summary

This chapter described the 'analysis toolkit' of standard techniques available to the audio engineer, along with the methods of integrating these techniques into an analysis system. The important aspect of pitch extraction for speech analysis was also discussed with reference to several useful techniques.

Apart from speech, examples were given of some of these techniques applied to other non-speech audio signals, namely a musical instrument recording and animal noises.

Bibliography

- *Acoustic Analysis of Speech*
  R. D. Kent (Singular, 2nd edition 2001)

- *Speech and Audio Signal Processing: Processing and Perception of Speech and Music*
  B. Gold and N. Morgan (Wiley, 1999)

- *Advances in Speech, Hearing and Language Processing*
  Ed. W. A. Ainsworth (JAI Press, 1990)

- *Applications of Digital Signal Processing to Audio and Acoustics*
  Ed. M. Kahrs (Springer, 1998)

- *Music: A Mathematical Offering*
  D. Benson (Cambridge University Press, 2006)
  This book is a fascinating study of the close relationship between mathematics and music. Although not describing the practical computational analysis of sound or music, the theoretical background and methods of analysis and science of music and of musical instruments are well described. This book starts with the human ear, and progresses beyond the Fourier transform to music technology.

Questions

Q7.1 Capture two short recordings of sound, the first one being a sustained 'o' sound, spoken at high pitch (i.e. 'ooooooo'). The second recording should hold the same sound spoken at low pitch. Use Audacity or another tool to convert the formats, saving them in mono at 8 kHz sample rate. Now working in MATLAB, load the two files and create side-by-side plots of the waveform and spectrogram. Can you see evidence of the difference in pitch between the two signals?

Q7.2 Using the material from Question 7.1, select a region of length 4096 from each recording within which the spectral characteristics appear to be quite constant (when looking at the spectrogram). Create a loop that extracts 50% overlapping analysis windows from this region in each file. Compute the ZCR and AMDF score for each window, and plot the waveforms and scores. Can you see evidence of the difference in pitch from either of the plot types?

Q7.3 Looking at the definition of the ZCR, can you describe how it would be affected if speech is recorded using a faulty microphone with a 50% DC offset, i.e. the mean of the recording is not 0, but is halfway between zero and the largest peak. Does this imply that such sounds need to be conditioned before applying a ZCR measure?

Q7.4 Identify the difference between (a) ZCR and threshold crossing rate, and (b) AMDF and frame power.

Q7.5 Working in MATLAB, perform a cepstral analysis on the two recordings used in Questions 7.1 and 7.2. From a plot of the cepstra, can you see any evidence of the difference in pitch between the recordings?

Q7.6 When building a fingerprint door lock, designers are told that it is more important to minimise the number of false positives than it is to minimise the false negatives. In terms of a registered user 'Reg' and a group of unregistered people who try to gain access ('robbers'), explain the difference between false positive and false negative outcomes.

Q7.7 The spectrogram is a type of time–frequency distribution display. Name two other methods of time–frequency distribution analysis that can be used for speech.

Q7.8 In the higher-order plots at the bottom of Figure 7.10, the peaks appear to be very regularly spaced. Explain why that might be so (remember that the plot is for sustained violin bowing of a single musical note).

Q7.9 If the violinist was not able to sustain a single note, but instead drifted from one note to the next over a period of a few seconds, explain what you think would be the effect on the plot of Figure 7.10.

Q7.10 In a normal and friendly English telephone conversation between two friends, what range of channel occupancy rate and cadence might be expected? Which one of ZCR, frame power, cepstral analysis, spectrogram or LSP analysis would be the simplest measure for estimating occupancy rate and cadence, assuming a completely noise-free environment?

8 Big data

The emphasis of this book up to now has been on understanding speech, audio and hearing, and using this knowledge to discern rules for handling and processing this type of content. There are many good reasons to take such an approach, not least being that better understanding can lead to better rules and thus better processing. If an engineer is building a speech-based system, it is highly likely that the effectiveness of that system relates to the knowledge of the engineer. Conversely, a lack of understanding on the part of that engineer might lead to eventual problems with the speech system. However, this type of argument holds true only up to a point: it is no longer true if the subtle details of the content (data) become too complex for a human to understand, or when the amount of data that needs to be examined is more extensive than a human can comprehend. To put it another way, given more and more data, of greater and greater complexity, eventually the characteristics of the data exceed the capabilities of human understanding.

It is often said that we live in a data-rich world. This has been driven in part by the enormous decrease in data storage costs over the past few decades (from something like €100,000 per gigabyte in 1980,[1] €10,000 in 1990, €10 in 2000 to €0.1 in 2010), and in part by the rapid proliferation of sensors, sensing devices and networks. Today, every smartphone, almost every computer, most new cars, televisions, medical devices, alarm systems and countless other devices include multiple sensors of different types backed up by the communications technology necessary to disseminate the sensed information.

Sensing data over a wide area can reveal much about the world in general, such as climate change, pollution, human social behaviour and so on. Over a smaller scale it can reveal much about the individual – witness targeted website advertisements, sales notifications that are driven from analysis of shopping patterns, credit ratings driven by past financial behaviour or job opportunities lost through inadvertent online presence. Data relating to the world as a whole, as well as to individuals, is increasingly available, and increasingly being 'mined' for hidden value.

Many observers have commented on the increasing value of data, while businesses are now tending to believe that the data is more valuable than the hardware and infrastructure used to obtain it. This has driven the provision of many free services (e.g. free

---

[1] Sharp-thinking readers might note that this date predates the Euro currency – the figures given here are approximate, and have been converted from advertised sales rates for hard discs using current exchange rates from the original currencies.

Internet access, free search and other services, free email, free mobile applications and games, free video and telephone calls and so on), which in turn are used to gather yet more valuable data.

Moving back to the speech field, there was a growing sense in the research community around the turn of the century that the performance of speech technologies such as automatic speech recognition (ASR) systems had plateaued. In the early days of research, progress had been rapid, but in recent years researchers had had to work harder and harder to unlock smaller and smaller improvements in performance. Although speech researchers believed we had a relatively clear understanding of the speech signal per se, it was increasingly difficult to design good algorithms to exploit that understanding. The easiest aspects of our understanding had already been investigated, turned into workable technology, and published by earlier researchers, so we often had only the most difficult aspects left to choose from. Even those were becoming fewer as researchers published better and better solutions (but which achieved smaller and smaller improvements).

Remember that this research background coincided with an explosive increase in the amount of recorded speech and audio data mentioned above, and hence a natural response of researchers was to begin to make use of that recorded data. Machine learning techniques were thus employed to analyse recordings of speech and audio, and hence infer rules and models relating to their characteristics. The nature of the research changed from relying primarily upon a researcher's understanding of speech and hearing to relying mainly upon the quality of their data and their ability to use computational techniques to learn from that data. Data collection and computational intelligence thus increased in importance, but the result was step change improvements in performance.

It seems that speech and audio research tends to span several steps:

- A new idea or technique is invented in a speech or audio related area.
- Understanding of the technique and its underlying data leads to rules that can turn the idea into an application.
- Improvements in performance, stemming from improvements in the data processing related to that technique. At first, performance increases rapidly, but this improvement soon slows as the most obvious (and easy) ideas from other domains are exploited. Eventually progress is stalled as performance reaches a plateau.
- Big data, allied with machine learning, is applied and progress resumes.
- Better machine learning methods plus more and better data are used to increase performance.
- Eventually even the machine learning methods plateau (although at a much higher level than the original plateau). Researchers then turn their attention to more difficult problems – such as 'robust' methods (ones that work in noise), or with different languages, children's speech, multiple speakers, difficult situations, overlapping speech and so on. Thankfully, the performance of the original method is now so poor for the more difficult problem that researchers have ample scope to try more techniques, and thus publish even more papers.

Perhaps this sounds overly cynical; however, the fact remains that big data techniques have primarily been used to achieve step improvements in performance for existing problems. A prediction for the future is that big data analysis will lead to the new ideas themselves and eventually spawn totally new problem domains that have not yet been identified by human ingenuity.

8.1 The rationale behind big data

One of the pioneers of computing, Charles Babbage, is purported to have written something to the effect that 'errors caused by insufficient data are less significant than those caused by having no data'.[2]

The basic idea behind big data is that the more good data is available, the better able we are to exploit it. While there are many examples that can illustrate the power of big data (and we shall meet some more below), one that the author has personal experience of is the iFlytek 'voice cloud'. iFlytek is a spinoff company from the speech laboratories of the University of Science and Technology of China (USTC),[3] which has grown to become one of the largest speech companies in the world. In mid 2010 iFlytek launched a 'voice cloud' service which now provides ASR capabilities that underpin many services (such as a voice assistant, automated telephone response systems and so on, operated by more than 5000 different service companies worldwide, but mainly in China). The basic principle of the system is that the voice cloud receives a speech recording, recognises the spoken words and then returns text, and does so very quickly (within 100 ms mainly). However, the key point is that it is a machine learning approach which is driven by big data – the ASR models are updated regularly based on the correctness of the system. Around 500 transcribers are busily employed by the company to listen to recordings and then manually decide what was actually being said. If the ASR output that was generated for that recording does not agree with the transcribers, then adjustments are made to the ASR model. The big data, in this case, is taken from the more than 10 million speech recordings that it receives every day.

Gradually, sentence by sentence, day by day, the model correctness increases. Figure 8.1, constructed by interpolating public data, charts the recognition accuracy of the voice cloud over about two years, along with the number of users of the system. Very clearly, the recognition accuracy is being driven by the number of users – although we can see a plateau forming. The scale for number of users is logarithmic, so, if you examine the actual numbers, it is taking more and more users to deliver smaller and smaller incremental improvements in accuracy. However, the amazing fact is that *the system has already exceeded human accuracy* for many scenarios: it can operate over a greater range of dialects and colloquial speech and in more noise than most humans

---

[2] Apparently he actually wrote 'The errors which arise from the absence of facts are far more numerous and more durable than those which result from unsound reasoning respecting true data.'

[3] Now the Chinese National Engineering Laboratory for Speech and Language Information Processing (NELSLIP), USTC, Hefei, Anhui, China.

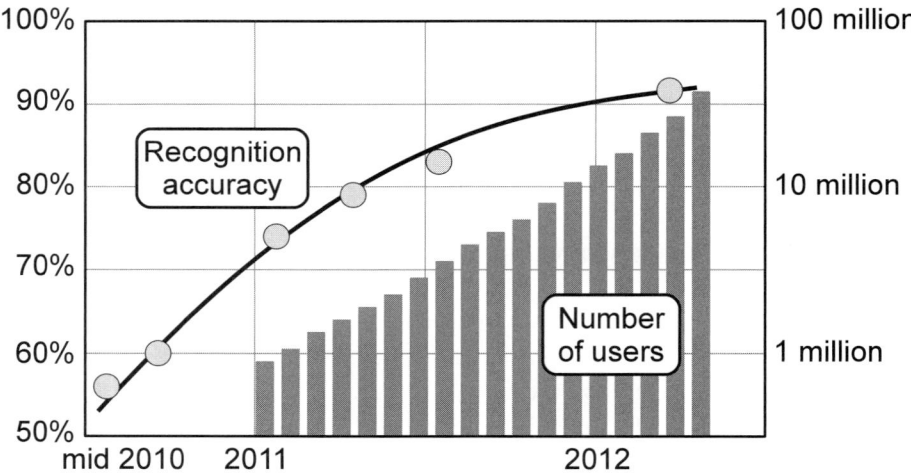

Figure 8.1 Combined plot showing approximate speech recognition accuracy and number of users of the iFlytek voice cloud, plotted using publicly released data.

can, exceeding 90% accuracy in 2014. This is all thanks to the power of big data and machine learning (but not forgetting some very intelligent engineers too).

8.2 Obtaining big data

There are a number of sources for big data, which may include making use of existing information, purchasing data (there are many companies providing such a service, often for surprisingly hefty fees), conducting a campaign to record the required data and so on. Some of the most interesting data is that which already exists and has been captured – inadvertently or otherwise – along with other information.

For human related data, a number of companies operate using a kind of trade-off model. They meet a human need in return for capturing human data. Google is one such company, supplying free search, email, messaging, blogging, video storage and other services in return for the right to analyse the data for their own commercial purposes. Other companies have developed games or mobile applications, which are supplied freely in return for a user's data (or for the right to advertise to a user, although often that right to advertise has been sold to a third party which subsequently collects the data too). Data is valuable, and this means big data is big business.

For the applications we discuss in this chapter, it is not necessary to be as big as Google or iFlytek. In most cases the data collection required is not difficult, and there are many repositories of free or open access data that could be used. However, anything that impacts humans has an ethical dimension (and this definitely includes experimentation as well as data collection), which we will explore further in the next subsection.

8.2.1 Ethical considerations relating to data

The question of ethics is clearly far bigger than we can (or should) cover in this book. However, it is very important, especially when we are dealing with human data, to acknowledge this topic.

This section applies primarily to those who intend to collect new data involving humans or to conduct experiments that involve humans. However, even those who make use of data recorded and provided by others have some responsibility in ensuring that they use that data in ethical ways too. For example, some of the common principles that apply both to human experimentation (such as listening tests) and to data collection (such as recording speech) are as follows:

- Ensure all participants are aware what the procedure entails, have given written permission to undergo it, and are aware what the data will be used for.
- The procedure should be beneficial to the participants, or to a segment of society that they are closely related to.
- The procedure should not be harmful to the participants.
- Participants' data should be stored in such a way that the individual participants cannot be identified from the stored information (i.e. it should be anonymised).
- The information should be used only for the stated purposes (which could, however, be very broad).
- Participants should have the right to request that their data be deleted, as well as the right to know the eventual outcome of the exercise.
- The basic dignity of the participants should always be upheld.

Many organisations, including most universities, have a human ethics committee that must give permission before any human-related procedures are undertaken. Speech and audio experiments and data collection campaigns are not usually problematic since these do not involve invasive procedures, or generally carry any meaningful risk of harm to the participants. However, procedures must be followed, clearance sought and obtained before any activity takes place, and all actions must be carried out with a view to the human ethics involved.

8.3 Classification and modelling

There are a number of components that go hand in hand with a 'big data' approach (apart from the large amounts of data, of course). These include the following:

- **Data storage** (which might involve quantisation or simplification if the storage requirements are too large).
- **Access** to the data, and what form it is stored and organised in.
- The transformation from raw data to representative **features**.
- One or more **machine learning** techniques.
- An **evaluation methodology**, including a method of scoring.

Sometimes these aspects are not as obvious as they might appear. We will briefly consider each in turn before digging more deeply in the following sections.

8.3.1 Data storage and access

It should not come as a surprise to hear that big data can be 'big' in terms of storage requirements. Many public databases are gigabyte-sized, while those used commercially may well be terabyte-sized or larger.

Typically, the raw captured data is reduced in size before storage. This is either by transformation to representative features (see next subsection), or to a reduced size intermediate form. Reducing size includes removing un-useful parts of the data (such as silence regions in speech), removing spoiled or low-quality records, and changing format through quantisation, sample rate adjustment or compression. Quantisation uses a lower resolution to store data with reduced precision, while sample rate reduction limits the highest frequency that can be represented in the data. The former is a relatively low-complexity operation whereas the latter requires anti-alias filtering and smoothing, which are both higher complexity. However, neither is as involved as compression and decompression. Compression also tends to alter the characteristics of the data in ways that might easily affect a machine learning algorithm. Furthermore, all of these transformations are one-way, meaning that it is impossible to recreate the original recording from the transformed file (and therefore it is very important to ensure that any transformation is correct before it is applied). With such caveats, it is no surprise that many data sets include the data in as raw and unprocessed a format as possible, hence increasing the already large data storage requirements.

In addition to storage requirements, the data must be accessible. This includes the indexing and archival methods used as well as the physical connectivity and security. The author can relate an example of obtaining and storing some large amounts of data on computer that may illustrate the difficulties and possible solutions. In this case, the data was recorded speech, on a word-by-word basis. There were several male and female speakers, who would record a fixed sequence of words which were spoken, and then whispered. Several directories ('folders' to Apple users) were created, with the names M01, M02, F01, F02 and so on, for the male and female speakers. For each participant, a single stereo recording was made in 16-bit at 44.1 kHz for all words, spoken and whispered. Within each directory, subdirectories were created with the names 'speech' and 'whisper', and the long recordings were then manually edited, cutting out and saving the various words. These isolated words were saved individually into the appropriate subdirectories, named after their content word, i.e *hello.pcm*, *ghost.pcm*, *hat.pcm* and so on. All of this seems very sensible, perhaps.

Now we skip forward two decades. The author requires the data for analysis and is delighted to find that he still has the files from his previous tests, and especially that they weren't corrupted in transferral across several storage media. The files are stored logically, but there is no indication of the sample rate, bit precision or number of channels (which we could probably guess by listening to a few files – but should have stored them as *.wav* originally), no indication of the recording equipment, location,

8.3 Classification and modelling

actual amplitude or what processing, if any, has been undertaken on the files. Looking at the directory structure, it is obvious which are male and female recordings, but no data has been recorded about the actual speakers, such as age, origin (i.e. accent) and so on. The data should be anonymised so that the original speakers cannot be deduced by anyone outside the experimental team, but such information is useful and should have been recorded. When using the stored data in a MATLAB-based experiment, code was used to identify the contents of a directory and then read in the files one-by-one.

As an example, the following MATLAB code looks inside the named subdirectory (*/Documents/MATLAB/MyFiles/* in this case) to identify the names of all files stored in there, and then loops through each file in turn. The full path name is read out (but the commented line would read each audio file into MATLAB in turn):

```
dir_name='~/Documents/MATLAB/MyFiles/';
sub_d=dir(dir_name);
for fi=2:length(sub_d)
    % 2 because the first 2 files are current
    %and parent dir, in unix
    fname=sub_d(fi).name;
%   [wave,fs]=audioread([dir_name,fname]);
    fprintf('%s\n',[dir_name,fname])
end
```

We had briefly mentioned the issue of accessibility in the discussion above. While it is assumed that the data is stored on some computer-readable medium, the various possible storage media have some quite different access speeds and abilities. Especially when using multi-processor or multi-machine processing (which is highly probable in big data systems, and possible using either Octave or MATLAB with the parallel processing toolbox), it may be necessary to ensure that the data is stored as 'close' as possible (in computer access speed terms) to the machine that will be reading the data, in order to minimise data transfer costs. This might involve splitting a larger database into blocks that are distributed across a number of computers. The cost is primarily in network bandwidth, which is finite, and ignorance of which might easily result in a system that is slowed down significantly by computers waiting for data to be read before it can be processed. It is particularly important in a cloud-based system, which is distributed by nature, or any system that requires access using wireless connectivity.

8.3.2 Features

Raw data, such as recorded audio, is generally of very high dimensionality and needs to be transformed to lower dimensionality and more concentrated features before it can be handled by a machine learning algorithm (such as those we will see in the following subsection). These features are thus a lower-dimensionality representation of the important aspects of the data. There are a number of factors at work here. Firstly, not all of the raw data is useful to us – for example, speech recordings may have silence at the beginning and end, and may include frequency regions which convey hardly any useful information. It would definitely be sensible to remove such regions if possible. Secondly, machine learning algorithms are usually 'trained' by representative data before they can be used for purposes such as classification. This training process involves adjusting various internal parameters, weights, biases and so on, which are often updated incrementally. A useful way of looking at this is to view the machine learning algorithm as distilling intelligence from data. The algorithm has to work harder to distill information from raw data rather than from more concentrated data. In addition, concentrated data contains less 'other' material with which to confuse the algorithm, which might become 'sidetracked' by the wrong information. To put this another way, if it is possible to transform the raw data into a more concentrated and lower-dimensionality representation, then the machine learning algorithm will generally converge better (i.e. it will learn faster and better), which means that less input data is needed to train it, and the end result will be better trained.

Basically, features are important – they need to be able to represent the important parts of the data in as compact a fashion as possible.

Figure 8.2 shows a block diagram of a typical machine learning process. Firstly, a large amount of raw data is transformed to representative features. Usually (but not always) those features have a lower dimensionality than the input data does. The data

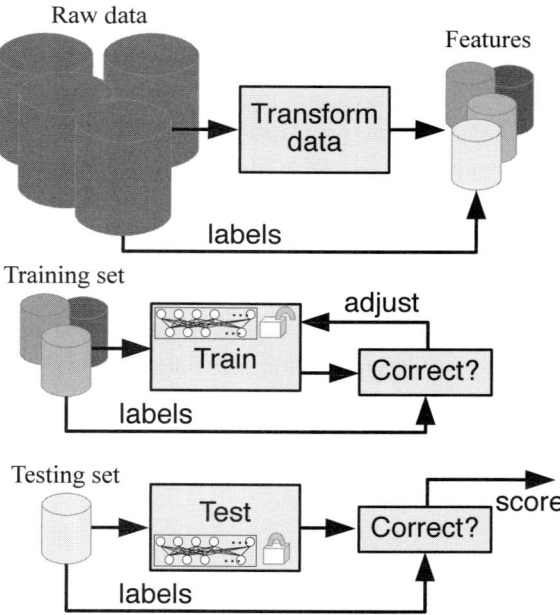

Figure 8.2 Block diagram showing data preparation, training and testing phases in a typical machine learning project.

8.3 Classification and modelling

Table 8.1 Example features and where they might be used.

Feature	Dimensionality	Used where	Reference
MFCCs	Usually 13	ASR, language/speaker ID, sound detection, diarisation	Section 5.5
ΔMFCCs	With MFCCs	– as above –	
ΔΔMFCCs	With the above two	– as above –	
Pitch	Usually 1	To accompany MFCCs	Section 6.3.2
Δpitch	& ΔΔpitch	– as above –	
Energy/power	Usually 1	Commonly used	Section 7.1.2
LPCs	Usually 10 or 12	ASR and similar	Section 6.2.1
LSPs	– as above –	ASR and similar	Section 6.2.5
Filterbank	Up to 40	Sound detection, diarization	Section 5.3.1
Spectrogram	Two-dimensional	Sound detection, VAD	Section 2.6.0.2
Auditory image	Two-dimensional	Sound detection	See [97–99]
i-vector	200~1024	Diarization, language and speaker ID	See [100]

is generally split into a training set and a testing set, used to train the machine learning algorithm and then evaluate performance respectively. We will look at this in more detail in the following subsection. For now we are mainly concerned with the features themselves.

Some common feature types and their usual dimensionalities are shown in Table 8.1. We have already discussed most of these (although not in terms of a big data approach or a machine learning context). Often they are used in combination – for example, pitch and energy often accompany MFCCs because the pitch represents longer-term sound repetition and energy represents absolute amplitude (MFCCs are usually calculated on scaled input frames). Each feature also has a natural duration over which it is normally computed. We have often discussed this when we looked at the individual features, and noted that it also depends on the application. For example, an application that wants to capture phonetic information should be sure to capture features at a rate that is at least as fast as the phonemes change (so one window should be one phoneme length or shorter).

Each of these features is computed over a frame – usually windowed – with the exception of the final three. Spectrograms have three pieces of timing information associated with them: the size of an analysis frame, the step (or overlap) from one frame to the next and the total number of frames, which defines the duration of the information being analysed. The same is true of the auditory image.

The i-vector [100], by contrast, is an interesting feature that aims to represent different sized items of information by a constant sized feature. As an example, this might be used for words (with a single i-vector for one word) or for entire sentences (with one i-vector for an entire sentence). In some applications like speaker ID, language ID or diarization, the i-vector length might be constrained to represent no more than a particular duration of sound (for example, one i-vector per second).

8.3.3 Machine learning

Figure 8.2 shows a block diagram of a typical machine learning process. Firstly, a large amount of raw data is transformed to representative features. Usually (but not always) those features have a lower dimensionality than the input data does. The features for the entire data array are then split into at least two blocks – one block of the transformed data (usually the largest block) is now used to train a machine learning system. During training, features are input to the system which attempts to classify or identify them. For each input feature in turn, the system outputs its 'guess', which is then compared with the actual correct labels for that feature. The difference between correct and predicted output labels is then used to adjust the internal machine learning model parameters. Gradually, over many iterations of labelled training data, if all goes well, the internal model should converge to an accurate model of the real system, so that its guess of output for each feature becomes correct most of the time. Once all the training data has been used up (or the error rate – the proportion of correct output – becomes good enough), the model parameters are locked, and the testing block of data is used to evaluate the performance of the internal model. Almost always, this evaluation should use different data that used by training, otherwise the system is performing a recall task rather than a classification task.

There are many variants of machine learning system in the world today. Some, like the above, are classifiers that attempt to correctly classify new input data into a labelled class, having been trained on a data set of previously labelled data. This is called supervised learning. Other systems use unsupervised learning algorithms which work on unlabelled data, generating their own classes within which they can classify data. It may come as no surprise to hear that there are even more variants of these techniques, including semi-supervised learning, transfer learning and reinforcement learning. Even the labels themselves may vary between unlabelled data, through sequentially numbered classes, to labels that are just as complex as the input data.

As well as there being a profusion of variants to the basic system, there are hundreds of ways in which machine learning techniques fail. It is impossible to quantify all of those reasons, but a brief survey of some of the most common reasons in the speech domain would include the following:

- Insufficient training data – the model does not converge.
- Too much training data – leads to over-training (errors start to increase with more training data).
- Overly simplistic features – they do not contain enough information to train an accurate model, and are thus not able to allow a network to discriminate between classes.
- Overly complex features – these will require far more training data because the model needs to be more complex, and thus lead into the 'insufficient training data' problem.
- Mis-labelled data – many big data systems in particular exhibit dubious labelling quality, which can lead to inaccurate training.

- Unrepresentative training data – most of the techniques we use require that the statistical distribution of the training data is identical to that of the underlying process (or testing set – since that is what we are evaluating), as well as the fact that the training data is a uniformly distributed sample of the underlying process. In practice, this means that there should be a roughly equivalent number of instances of each class within the training set.
- Skewed feature-means or distributions – most techniques require that the input data is normalised or standardised prior to testing or training (get it wrong and ASR accuracy could easily drop by 10% or so).
- Underlying data is not normally distributed – in practice, researchers tend to just assume that the underlying processes are Gaussian distributed, yet that may not be true for digital systems (it is more likely to be true of analogue systems). Often though, a non-Gaussian process can be very well approximated by a sum of Gaussians.
- Incorrect model – if the complexity of the model, or assumptions made in forming the model, do not reflect reality then the system could not possibly be trained successfully.
- A malformed problem – which might include using a one-against-many classifier to model a process in which multiple class labels are permissible (i.e. using a classifier that gives a single best-guess output class for each input, out of many classes to choose from, and yet the training/testing data sometimes has two or more labels for a single feature).

This is just a taster of what can go wrong in machine learning, and doesn't even mention the various parameters that can be tuned. Learning systems tend to have a multitude of parameters such as learning rate, batch size, initial weights, biases, choice of non-linear function, distance measures etc. that can be adjusted to yield some quite large changes in performance. Often, these parameters need to be adjusted on a trial-and-error process, or may be chosen on the basis of the past experience of the researcher (i.e. non-scientifically!).

Consider this question: should the success of the trial-and-error process of adjusting these parameters be evaluated using the training data set or the testing data set? If using the former, the final testing result might be unexpected (i.e. a good set of parameters for the training data might not be the same as a good set of parameters for the testing data). But using the latter means that the final result is optimised for the given testing data, which also might not be representative of reality. Thus, it is normal to reserve a *development data* set – i.e. a third block of input data in addition to the testing and training data. Initial training uses the training data, adjusting (tweaking) of various parameters is performed by evaluating the development data set, then only the final score is assessed using the testing data set.

8.3.4 Evaluation methodology

One of the most interesting aspects of big data is in the method of scoring or evaluating between techniques – essentially deciding which technique is best. It can be problematic – and is exacerbated by the fact that the data is often quite uncontrolled or raw. Even for a simple task such as detecting the presence of speech, this can be problematic. As a trivial example, consider a 10 minute recording of speech in high levels of noise, which contains just 1 minute of speech with the remainder being silence (background noise). System (a) is a fairly good speech detector that manages to correctly detect 80% of the speech, but mistakenly identifies 25% of the non-speech parts as being speech. If we simply considered accuracy over time, this would score a mean accuracy of $(0.8 \times 1 + (1 - 0.25) \times 9)/10 = 0.755$, so it is 75.5% accurate overall. That doesn't sound so bad, given the high levels of noise. Compare this with system (b) which is just a zero, i.e. it always says 'non-speech' no matter what the input is. In this case it will miss the 1 minute of speech, but be correct for the 9 minutes of non-speech, achieving a 90% accuracy. What amazing performance!

To give another problematic example, consider a sound detection system. A sound detector is built that aims to identify the type (or class) of sound that is present every 50 ms, given several possible classes that it has been trained to identify. The system is evaluated on real sounds and it is found that the output is often incorrect, in particular many of the 'doorbell' sounds were mistakenly classified. Investigating further, the researcher discovers that 90% of 'doorbell' sounds have been misclassified as 'telephone' sounds. Listening to the raw testing and training data, she notices that many of the 'telephone' sounds include recordings of old style telephone bells, while much of the 'doorbell' testing material contains the sound of a ringing bell. Thus the system very easily mistakes these two sound classes – yet it may be technically correct in the misclassification since the doorbell sound and the telephone sound are both ringing bells. Hence we see that some incorrect results are more incorrect than others, and we should probably consider that a 'doorbell' misclassified as a 'telephone' (bell) is a higher-quality incorrect result than, for example, a 'doorbell' misclassified as a baby crying. Furthermore, a probabilistic classifier selects an output class and also assigns it a probability – we should probably give a higher score to a correct output with a high probability score, than to a correct output with a low score (and likewise to an uncertain incorrect output over and above a certain incorrect output). Practically speaking, the researcher mentioned above should probably merge the 'telephone' and 'doorbell' classes (hence improving the results), or move most of the sound recordings to a new class called 'bell'.

These examples therefore illustrate that it is very important to think carefully about how any system is evaluated. In Chapter 9, when discussing various applications of this machine learning technology, we will see that many of the application domains have thankfully adopted their own generally accepted classification evaluation scoring methods.

8.4 Summary of techniques

8.4.1 Straight line classifier and the perceptron

A perceptron is a binary classifier that can lean how to separate two classes of information, if they are separable by a straight line or plane (i.e. linearly separable). This

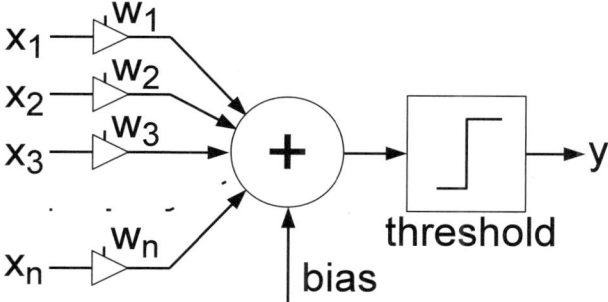

Figure 8.3 Illustration of a single perceptron gathering information from n inputs, using a set of weights and bias to produce a single output, y.

algorithm was invented in the 1950s and has been used in many ways since, not least as the basis of the multi-layer perceptron network (see Section 8.4.2), and can even be found in some of the most modern machine learning such as the convolutional neural network (see Section 8.4.8).

The structure of the perceptron, shown in Figure 8.3, is supposedly similar to that of a human nerve cell: it takes input from a number of sources (which could be from the outputs of other nerve cells, or from sense organs – known as feature vectors in machine learning). It weights each of the inputs before summing the weighted values together, adding a bias term and then using a thresholding function to decide whether to trigger its output. During a learning phase (acting on training data), the weights are gradually and iteratively updated, moving the classification boundary until a correct classification result is found. The weights are then fixed and the system used to classify other data – usually beginning with some testing data for which we compare the known correct output and the perceptron output to test how well it is able to classify.

Mathematically speaking, given an N-dimensional input vector $\mathbf{x}_n = \{x_0, x_1, \ldots, x_N\}$ and a single binary output, y, the internal structure makes use of N weights, $\mathbf{w}_n = \{w_0, w_1, \ldots, w_N\}$ and bias b to compute:

$$y = \varphi \left\{ \sum_{l=1}^{N} w_l x_l + b \right\} = \varphi \left\{ \mathbf{w}^T \mathbf{x} + b \right\}, \tag{8.1}$$

where φ is a thresholding function such as $\varphi a = 1/(1 + e^{-2a})$. Sometimes the bias, b, is given the name w_0 to make the calculations a little more convenient, and we will adopt that convention here. In practice, there are several ways to train a perceptron making use of iterative gradient descent processes, one of which we will demonstrate in MATLAB.

The following MATLAB code creates some simple two-dimensional test data for a binary classification demonstration. Don't worry about understanding the first line containing `repmat()` functions – this is just a compact way of creating a large grid of points in the coordinate range $(0, 0)$ to $(1, 1)$. This can be seen in the top left of

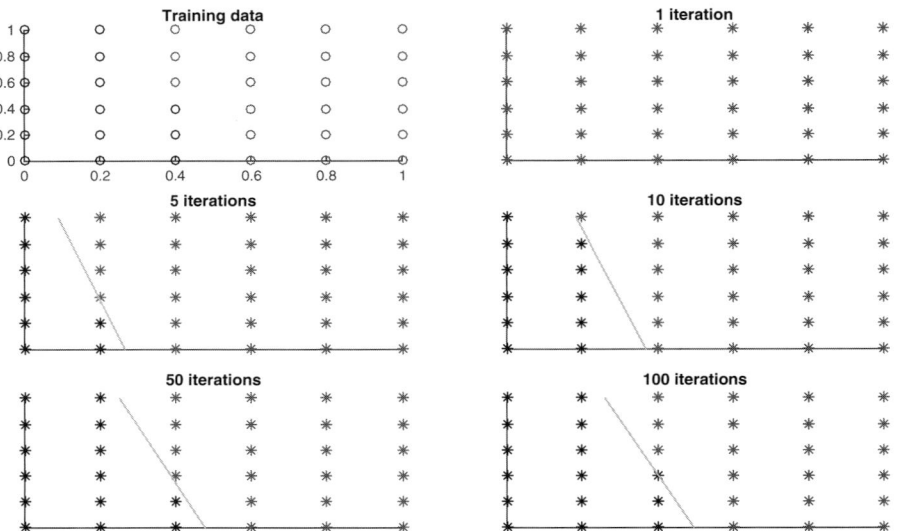

Figure 8.4 Illustration of a perceptron learning to classify two-dimensional data. The training data is plotted top left, and the classifier output data shown after 1, 5, 10, 50 and 100 iterations respectively, with class 1 symbols plotted in black and class 0 symbols plotted in grey.

Figure 8.4, where it is evident that all points are assigned to class 0 apart from several in the lower left corner which have been manually assigned to class 1:

```
%Make a 6x6 square matrix of 2D training data
xin=[[reshape(repmat([0:0.2:1],6,1),[1,6*6])];[repmat
    ([0:0.2:1],1,6)]]';
N = length(xin);
%most are class 0, except a few at lower left
yout=zeros(length(xin));
yout(1:12)=1;
yout(13:15)=1;
```

Next, a two-dimensional perceptron is defined. This contains three randomly initialised weights, w (one for each dimension and one for the bias term), and then iteratively updates the weights based on the distance, delta, between the actual classification output they produce and the correct classification output:

```
%Define a 2D perceptron
b = -0.5;
nu = 0.5;
w = 0.5*rand(3,1); %-1*2.*rand(3,1);
iterations = 100;
```

8.4 Summary of techniques

```
%Loop around the training loop
for i = 1:iterations
    for j = 1:N
y = b*w(1,1)+xin(j,1)*w(2,1)+xin(j,2)*w(3,1);
        z(j) = 1/(1+exp(-2*y));
        delta = yout(j)-z(j);
        w(1,1) = w(1,1)+nu*b*delta;
        w(2,1) = w(2,1)+nu*xin(j,1)*delta;
        w(3,1) = w(3,1)+nu*xin(j,2)*delta;
    end
end
```

The code above works well enough (given sufficient iterations) to train a perceptron to model any linearly separable classification on the input data:

It is much easier to understand this if we can visualise the process. To do so, we can create two subplots, one of which plots the correct classification of the training data:

```
subplot(2,1,1)
hold on
for i=1:N
  if(yout(i)==1)
    plot(xin(i,1),xin(i,2),'ko') %class 1
  else
    plot(xin(i,1),xin(i,2),'ro') %class 0
  end
end
```

Next we create a second subplot with a 1024×1024 resolution in which we can explore how the perceptron classifies every point from 0 to 1:

```
NX=1024; %x resolution
NY=1024; %y resolution
dsc=zeros(NX,NY); % plot surface
%classify over this surface
for x=1:NX
  dx=(x-1)/(NX-1); %range 0:1
  for y=1:NY
    dy=(y-1)/(NY-1); %range 0:1
    z=b*w(1,1)+dx*w(2,1)+dy*w(3,1);
    dsc(x,y)=1/(1+exp(-2*z));
  end
end
```

```
%classify the original xin points
for j = 1:N
  y=b*w(1,1)+xin(j,1)*w(2,1)+xin(j,2)*w(3,1);
  z(j) = 1/(1+exp(-2*y));
end
%plot the xin points and classifications
hold on for i=1:N
  if(z(i)>0.5)
    plot(xin(i,1)*NX,xin(i,2)*NY,'k*') %class 1
  else
    plot(xin(i,1)*NX,xin(i,2)*NY,'r*') %class 0
  end
end
%overlay the decision line on the plot
contour(transpose(dsc > 0.5),1)
```

The output of this code is presented in Figure 8.4, although with six subplots instead of two. The top left subplot contains the original training data while other subplots show the classification results after 1, 5, 10, 50 and 100 iterations. Comparing the various plots, it can be seen that the classification performance gradually improves as the number of iterations increases, culminating in a perfect classification output after 100 iterations.

8.4.2 Multi-layer perceptron

A single perceptron, as we saw in Section 8.4.1 is a machine that can learn by adapting its weights, usually iteratively, to move a classification boundary (a straight line or flat plane) through the multi-dimensional feature space until it correctly separates the classes in the training data set – assuming that they *can* be separated by a line or plane. While the single perceptron has been convincingly surpassed by newer and much better machine learning methods, its use within the multi-layer perceptron (MLP) is still popular in many domains.

An MLP is a stack consisting of two or more layers of individual perceptrons with connections between layers. By stacking the perceptrons together, it is possible to learn more complex decision boundaries that are effectively made up of a number of straight line or plane segments. An example MLP is shown in Figure 8.5 – a three-layer fully interconnected network which is set up to classify five-dimensional input data into six output classes.

In particular, it is made competitive through the power of the back-propagation training algorithm. Back-prop (as it is often abbreviated) allows the internal weights and biases within the machine to be updated iteratively, working on a layer-by-layer basis, from the output back. As with the single perceptron, weights are updated based on a fraction of the difference between what they *actually* produce and what they *should*

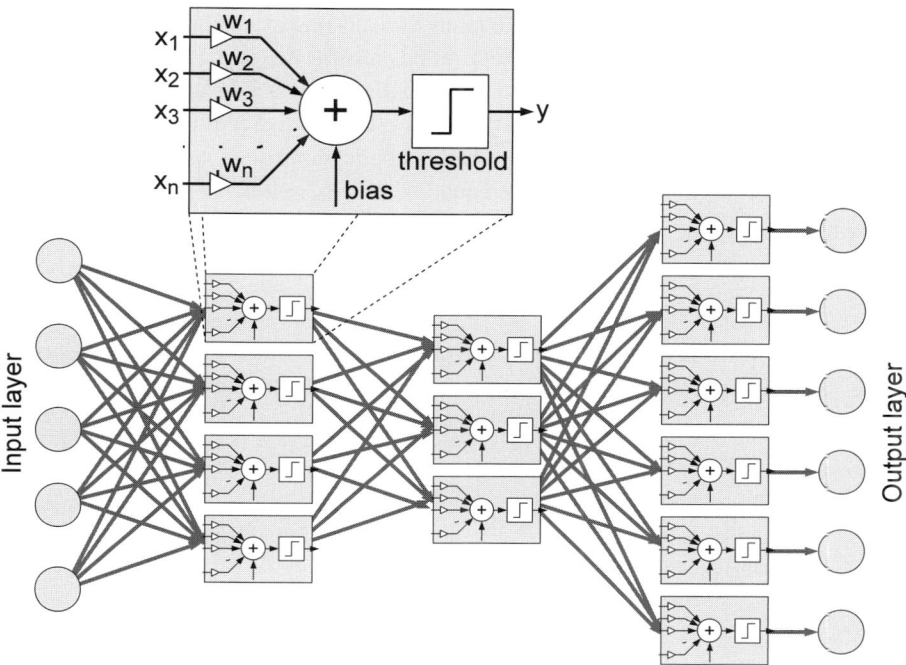

Figure 8.5 Illustration of a multi-layer perceptron network with five-dimensional input data, three internal layers (comprising four, three and six perceptron nodes respectively) and six outputs.

produce for a given training dataset. The back-propagation training method will be carried forward to many of the more modern machine learning systems that we will encounter. While it is important, we will not present the mathematics of the algorithm here since this can be found explained in many books and online tutorials (please refer to the Chapter 8 bibliography on page 264 for several sources of information on the MLP and back-propagation methods).

8.4.3 Support vector machine

An SVM classifier attempts to create a plane or hyperplane (the equivalent of a plane when the input data dimension $D > 3$) that completely divides a set of data points x_i into two clusters (in training, data are clustered depending on whether their labels y_i are set to $+1$ or -1). In a two-dimensional case, this division would be like a perceptron – a straight line dividing the two clusters of points. However, SVM has a trick to deal with clusters that are overlapping (i.e. not separable by a hyperplane), which is to map the data into a higher-dimensional representation in which the clusters are hopefully no longer overlapping. This is done by means of a kernel function $k(x, y)$, which can be chosen to match the data.

Mathematically speaking, SVM acts to assign an input vector of dimension (length) D, $\mathbf{x} = [x_1, x_2, ..., x_D]^T$, where $\mathbf{x} \in \mathbb{R}^D$, to one of K classes, $\mathbf{y} = [y_1, y_2, ..., y_K]^T$, where

$\mathbf{y} \in \{1, -1\}^K$. Since there are many variants and extensions of SVM, we will describe a solution to a particular problem called *machine hearing*, which will be discussed further in Section 8.5 and presented in full online. The same example application will also be used to illustrate deep neural networks (DNNs) in Section 8.4.7.

When classified using SVM, this application uses the so-called primal form with a soft margin [101], where the equation to be solved is:

$$\min_{w,b,\xi} \frac{1}{2} \mathbf{w}^T \mathbf{w} + c \sum_{i=1}^{V} \xi_i. \tag{8.2}$$

The idea is that this defines a hyperplane represented by $\mathbf{w}$, and is based on a regularisation constant $c > 0$ and slack variables ξ, which define an acceptable tolerance:

$$y_i(\mathbf{w}^T \psi(\mathbf{x}_i) + b) \geq 1 - \xi_i, \quad \xi_i \geq 0, \quad i = 1, \ldots, D. \tag{8.3}$$

In this equation, $\psi(\mathbf{x}_i)$ implements the kernel function by mapping $\mathbf{x}_i$ into a higher-dimensional space. Since $\mathbf{w}$ tends to have high dimensionality [102], the following related problem is solved instead:

$$\min_{\alpha} \frac{1}{2} \alpha^T Q \alpha + \mathbf{e}^T \alpha, \tag{8.4}$$

where $\mathbf{e} = [1 \ldots D]^T$ is a D-dimensional vector of all ones, Q is a $D \times D$ positive semi-definite matrix such that $Q_{ij} \equiv y_i y_j k(\mathbf{x}_i, \mathbf{x}_j)$ and $k(\mathbf{x}_i, \mathbf{x}_j) \equiv \psi(\mathbf{x}_i)^T \psi(\mathbf{x}_j)$ is the kernel function. Note that Equation (8.4) is subject to:

$$\mathbf{y}^T \alpha = 0, \quad 0 \leq \alpha_i \leq c, \quad i = 1, \ldots, D.$$

After solving Equation (8.4), using this primal–dual relationship, the optimal $\mathbf{w}$ that has been found satisfies:

$$\mathbf{w} = \sum_{i=1}^{D} \alpha_i y_i \psi(\mathbf{x}_i),$$

and the decision function is the sign of $\mathbf{w}^T \psi(\mathbf{x}_i) + b$ from Equation (8.3), which can conveniently be computed as:

$$\text{sign}\left(\sum_{i=1}^{D} y_i \alpha_i k(\mathbf{x}_i, \mathbf{x}) + b\right). \tag{8.5}$$

For the machine hearing application [97], a linear kernel works well, $\mathbf{x}_i^T \mathbf{x}_j$. Common alternatives are the nth-order polynomial $(\gamma \mathbf{x}_i^T \mathbf{x}_j)^n$, the radial basis function $e^{-\gamma \|\mathbf{x}_i - \mathbf{x}_j\|^2}$ and the sigmoid, $\tanh(\gamma \mathbf{x}_i^T \mathbf{x}_j)$, where the domain-dependent constant $\gamma > 0$.

Since SVM is a binary classifier, if the original problem has more than two classes, i.e. $K > 2$, then a one-against-one multi-class method is required, which requires $K(K-1)/2$ separate binary classifiers to represent all class divisions.

Also, it should be noted that input scaling is important, so the raw SVM input feature vectors $\mathbf{x}'$ are usually scaled before training (or classification), using a function such as:

$$x(i) = \frac{(\text{MAX} - \text{MIN}) \cdot (x'(i) - \min(\mathbf{x}'))}{\{(\max(\mathbf{x}') - \min(\mathbf{x}')) - \text{MIN}\}}, \qquad (8.6)$$

for $i = 1, \ldots, D$, and, conventionally, $\text{MIN} = -1$ and $\text{MAX} = +1$.

As mentioned above, the SVM of this form will be applied for the machine hearing classification task in Section 8.5, using the LIBSVM [102] toolkit.[4]

Working in MATLAB, assuming that a set of training feature vectors `train_data` and testing feature vectors `test_data` have been constructed, these would first be normalised by application of Equation (8.6) above:

```
train_data=(train_data-min(train_data))/
           (max(train_data)-min(train_data));
test_data=(test_data-min(test_data))/
          (max(test_data)-min(test_data));
```

The LIBSVM code is then configured and a model, `M`, can be trained from the `train_data` vector and its corresponding $+1$ or -1 labels `train_label`:

```
[train_sc,test_sc,ps] = scaleForSVM(train_data,test_
    data,-1,1);
opts='-c 32 -g 0.0009234 -t 0';
M=svmtrain(train_label,train_scale,opts);
```

The `opts` string defines the regularisation constant and γ (which can be estimated by the tool itself from the data), and kernel type `t` is set to zero for a linear kernel (since several alternatives are available). To evaluate model `M`, `test_data` is used to perform a classification which is compared against the ground truth of the correct testing labels in `test_label`:

```
[lab,acc]=svmpredict(test_label,test_scale,M);
```

yielding a set of predicted labels (`lab`) and an estimate of overall accuracy, `acc`.

SVM and LIBSVM are far more powerful than this simple single example shows; the code is given purely to demonstrate how easy it is to use SVM in practice with the assistance of such tools. Among topics not discussed here are the choice of alternative kernels and the use of support vector regression (SVR)

---

[4] LIBSVM is an open source library of SVM code, compatible with many languages and packages including both MATLAB and Octave. It is available along with excellent tutorials from www.csie.ntu.edu.tw/~cjlin/libsvm/

8.4.4 Principal and independent component analysis

Without becoming enmeshed in too much mathematics, principal component analysis (PCA) is a method of determining the relationship between the information carried by different variables – their statistical independence – and possibly transforming this relationship. In the speech and audio domains, PCA is usually performed on high-dimensionality feature vectors to shorten them by discarding information in the vector that is dependent or repeated. This is not the same as discarding elements of the vector (which would be feature *selection*), it is modifying the elements themselves into a smaller set of numbers that concentrate the information better (it is thus feature *transformation*). To do this, PCA modifies the underlying data into a set of representative components ordered in terms of their importance to the information in the original set of components. In many textbooks which describe PCA, this process is represented geometrically.

Independent component analysis (ICA) is a related technique that is applied to data which is assumed to contain a mixture of components that were generated from independent Gaussian processes.[5] The technique identifies, and is then able to separate, those independent processes. A typical application is in separating speech from background sounds, or even in separating two speakers from a recording which contains a mixture of their speech. This is called blind source separation (blind because we do not know in advance what the original sources were).

8.4.4.1 Principal component analysis

PCA is commonly used to reduce the dimensionality of something like a feature vector which has a high dimensionality, prior to a classifier (because training a high-dimensionality classifier is a lot more demanding – and much slower – than training one with a lower dimensionality). It is data-driven, meaning that it does not need to *know* anything about the features or what they mean. State-of-the-art speech systems tend to make use of a technique called linear discriminant analysis (LDA) instead of PCA, which is a kind of sub-classification method. LDA is beyond the scope of this book, but can be found in [103].

Using PCA for dimensionality reduction is not too difficult in practice. For example, given a set of N feature vectors of dimension L in the variable fvec, these can be reduced to N training vectors with a smaller dimensionality L', where $L' < L$. It may not be possible in advance to determine what the optimal value for L' is, but in practice several alternatives would commonly be tested.

MATLAB code to perform this is shown below, first standardising the data by subtracting its mean:

```
st_d=bsxfun(@minus, fvec, mean(fvec,1));
```

---

[5] Fortunately many natural processes, including most of those in the speech and audio domains, are Gaussian distributed – which is why it is also known as the *normal* distribution.

where the MATLAB function `bsxfun()` is a convenient way of repeating the given operation on all elements of the specified array (in this case `mean(fvec,1)` is subtracted, specified by '@minus', from each row of `fvec`). The alternative to using `bsxfun()` is either to use `repmat()` to repeat the mean vector enough times to make it the same size as the `fvec` matrix and then subtract directly, or to use a loop and subtract a row or column at a time (and we should avoid that, because looping in MATLAB, as an interpreted language, is considered to be inefficient and slow).

The next step after standardisation is to compute a covariance matrix, which holds the variance between each element and every other element (with the diagonal being the variance):

```
covmat = (st_d'*st_d)/(size(st_d,1)-1);
```

from which the best `Lprime` integer number of principal components are found as eigenvalues:

```
[V D] = eigs(covmat, Lprime);
```

Finally, the reduced-dimensionality feature vector, which we will call `fvec_prime`, is obtained by simply multiplying the V matrix by the original data:

```
fvec_prime = st_d*V;
```

This is essentially working with dimensions $[N][L'] = [N][L] \times [L][L']$. Furthermore, once the PCA matrix V has been defined in this way on this data, it can likewise be used as a dimensionality reduction for future data, such as for another set of data of the same dimension (but possibly different length). Thus, another set of feature vectors `fvec2` would be reduced in dimension to `fvec_prime` as follows:

```
std_fvec2=bsxfun(@minus,fvec2,mean(fvec,1));
fvec2_prime = std_fvec2*V;
```

In terms of dimensions, if there are R elements in the new data set, this is $[R][L'] = [R][L] \times [L][L']$, hence the dimensionality of the data has been reduced ($L \Rightarrow L'$), but not the number of items of data (R) – which remains the same.

This has been just a simple illustration of PCA. It is not particularly difficult, although the covariance matrix can become quite large when the data dimension is high.

8.4.4.2 Independent component analysis

A typical ICA scenario is attempting to recover the original speech from each person in a meeting which has been recorded with multiple microphones. The input to the algorithm is a set of M microphone signals, from which it is hoped to recover N voices, where $M \geq N$. If the microphones are located in different places in the room, they each receive different mixtures of the speech produced by the people in the room – and these mixtures are assumed to be linear, which is highly likely for free field microphones in the room (i.e. those that are not embedded in a corner, right up against a wall or encased in something). ICA also assumes that the original source signals (i.e. speakers' voices) are uncorrelated – which is very reasonable in practice – so that any correlation that is found in the mixed signals is due to the mixing process.

Without considering the mathematical proof (for more detail, please refer to [104]), we can create an iterative method of decreasing the mutual information between signals.

ICA is working with M received signals (such as those microphone signals), each of length T, which are placed into a mixture matrix, $\mathbf{X}$, for convenience. Matrix $\mathbf{X}$ has dimension $[M][T]$ to store the M separate signals. The assumption is that these originated from N separate sources which have been received as a linear mixture. If $\mathbf{A}$ represents the mixing matrix and $\mathbf{S}$ represents the source matrix then:

$$\mathbf{X} = \mathbf{AS}. \tag{8.7}$$

It should follow that $\mathbf{S}$, which stores N source signals, has dimension $[N][T]$, and thus the mixing matrix $\mathbf{A}$ has dimension $[M][N]$. We have already stated above that the unknown matrix $\mathbf{S}$ can only be recovered using this technique where $M \geq N$ (and where several other conditions hold true, such as linear mixing, Gaussian distribution and so on). The reader should also note that several more advanced blind source separation algorithms have been proposed for cases where $M < N$, using predictive techniques to 'guess' at the missing information.

We will now proceed to attempt to estimate $\mathbf{S}$. In reality we can only ever approximate this, and even then we cannot recover the original amplitude scaling of the rows in the matrix – and will find that the row ordering is random (i.e. we cannot know in advance which is the first original signal, which is the second original signal, and so on).

Define matrix $\mathbf{W}$ as the $[N][M]$ dimension 'unmixing' matrix to estimate the source from the received signals:

$$\mathbf{S}' = \mathbf{WX}. \tag{8.8}$$

We will use gradient descent to iteratively estimate this using the following sequence, given that $\mathbf{W}$ is randomly initialised to begin with, and $g(x)$ is a non-linear distance function (which could be various functions such as log cosh but in our case will be defined as $g(x) = 1/(1 + e^{-x/2})$), and given learning rate η:

1. Compute $\mathbf{S}' = \mathbf{WX}$, an estimate of the unknown source signals.
2. Find the overall distance, $Z_{n,t} = g(y_{n,t})$ for $n = 1, \ldots, N$ and $t = 1, \ldots, T$.
3. Define $\Delta \mathbf{W} = \eta (\mathbf{I} + (1 - 2\mathbf{Z})\mathbf{S}'^\mathrm{T})\mathbf{W}$.
4. Update $\mathbf{W} = \mathbf{W} + \Delta \mathbf{W}$.
5. Repeat from step 1 until either the estimate has converged (which can be found from the magnitude of $\Delta \mathbf{W}$) or a maximum number of iterations has been reached.

8.4 Summary of techniques

A simple demonstration of this using MATLAB is given below. First we will load some speech, create an equally long array of interfering tones, and mix these together:

```
[sp,fs]=audioread('SA1.WAV'); %load speech
sp=sp'./max(abs(sp)); %normalise
N=2;
T=length(sp);
tone1=tonegen(440,fs,T/fs); %create a tone
M=2;   %create 2 mixtures of these sounds
mix1=sp*0.23+tone1*0.98;
mix2=sp*0.81+tone1*0.19;
X=[mix1 ; mix2]; %mixture matrix
%dimension of X is [M][T]
```

Next we create an ICA loop beginning with randomly initialised **W** matrix, then carry out step-by-step refining this until either the maximum number of iterations has been reached or the convergence criterion has been met:

```
nu=0.01;    %learning rate
conv=1e-09;  %convergence criterion
%---create an [N][M] sized matrix W
W=rand(N,M)*0.1;
%---estimate the source, Sp, dim [N][t]
iter=1;
max_iter=500000;
while(iter < max_iter)
  Sp=W*X;
  Z=zeros(N,T);
  for n=1:N
    Z(n,:)=1./(1+exp(-Sp(n,:)/2));
  end
  %----calculate DW (deltaW)
  DW=nu*(eye(N) +   (1-2*Z)*transpose(Sp))*W;
  %---terminate if converged
  delta=abs(sum(sum(DW)));
  if(delta < conv)
    break;
  end
  %print out (so we know what's happening)
  if mod(iter,100)==0
    fprintf('Iteration %d, sum=%g\n',iter,delta);
  end
```

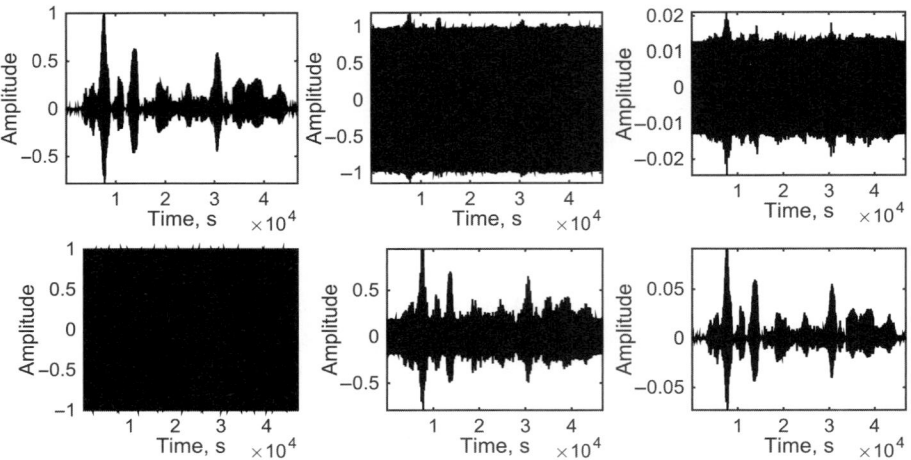

Figure 8.6 Illustration of ICA used to separate the two mixtures in the centre column. The separated signals are given on the right, as estimates of the original signals on the left.

```
    %---apply the update
    W=W+DW;
    iter=iter+1;
end
```

A plot of the output of this code is given in Figure 8.6, which shows the original speech and tone signals in the left column, the received mixture in the centre column and the estimate of the original signals $\mathbf{S}'$ in the right-hand column. In fact the speech has been very well estimated in terms of shape – although the tone estimation is much less perfect. However, listening to either of these (with `soundsc(Sp(1,:),fs)` or `soundsc(Sp(2,:),fs)`) can reveal good separation (although it does depend to some extent upon the initial random estimate of $\mathbf{W}$).

ICA can be quite easy to apply in practice, although the topic contains the potential for a lot of deep research and complex mathematics for those who wish to pursue it further.

For those who simply wish to make use of ICA to solve a practical separation problem, the best approach is probably to employ a ready-made package. There are several alternatives for MATLAB, including in some of the toolboxes, but an efficient and popular open source alternative is FastICA from Aalto University, Finland.[6] FastICA is well named, and gains its speed advantage through clever algorithmic design as well as the use of fixed point data processing.

---

[6] FastICA can be downloaded for MATLAB and Octave from http://research.ics.aalto.fi/ica/fastica, which also contains links to other information concerning ICA and its use as well as implementations in several different computer languages.

8.4.5 k-means

Some data is collected which is assumed to belong to normally distributed processes. Assuming that we can deduce or guess the number of processes that are involved, we can use the k-means clustering algorithm to find k cluster centres and then assign each item of data to the appropriate cluster. In particular, the sum of the squared distances of the data from the cluster centres is minimised.

The k-means clustering algorithm takes an iterative approach, typically beginning with random cluster start positions, then looping through a process that assigns each data point to the nearest cluster centre before adjusting the cluster centres for the next iteration based on the mean position of all data points assigned to that cluster.

This is evidently not the only method of clustering. Alternatives include the Gaussian mixture model (Section 8.4.6), agglomerative clustering and divisive clustering (which we will not consider, but is described in [105] as well as in many other machine learning texts).

Given a data set $\mathbf{x}$ (which can be of any dimensionality), k centroids, $\mathbf{m}$, which are the best possible representation of the data are found, with each centroid having the same dimensionality as the data:

$$||\mathbf{x} - \mathbf{m}_i|| = \min_j ||\mathbf{x} - \mathbf{m}_j||. \tag{8.9}$$

The method defines a global objective function which is to be minimised each iteration:

$$R = \begin{cases} \sum_t \sum_i ||\mathbf{x} - \mathbf{m}_i||^2 & \text{if } i = \text{argmin}_j ||\mathbf{x} - \mathbf{m}_j||, \\ 0 & \text{otherwise.} \end{cases} \tag{8.10}$$

It is possible to find the best set of $\mathbf{m}$ that minimises R by simply differentiating R with respect to $\mathbf{m}$ and equating to 0.

Various MATLAB functions for k-means and similar algorithms are available, but a simple implementation which can be used to cluster two-dimensional $[x, y]$ coordinate data is given in Listing 8.1.

Listing 8.1 kmeans.m

```
function centroids = kmeans(X,k)
%set error threshold
min_thresh=1e-7;
%set maximum iterations
max_iter=10000000;
%centroids
centroids = zeros(k, 2);
len = size(X,2);
nearest_c = zeros(len);
%initialise to random points
rand_i = ceil(len*rand(k, 1));
```

```
12  for i = 1:k
13      centroids(i,:) = X(:,rand_i(i));
14  end
15  %Iteration loop
16  for i=1:max_iter
17      %updated means
18      new_c = zeros(size(centroids));
19      %no, of points assigned to each mean
20      assigned2c = zeros(k, 1);
21      %Go through all data points
22      for n=1:len
23          % Calculate nearest mean
24          x = X(1, n);
25          y = X(2, n);
26          diff = ones(k,1)*X(:,n)' - centroids;
27          dist = sum(diff.^2, 2);
28
29          [~,indx] = min(dist);
30          nearest_c(n) = indx;
31          new_c(indx, 1) = new_c(indx, 1) + x;
32          new_c(indx, 2) = new_c(indx, 2) + y;
33          assigned2c(indx) = assigned2c(indx) + 1;
34      end
35
36      %Compute new centroids
37      for i = 1:k
38          %Only if a centroid has data assigned
39          if (assigned2c(i) > 0)
40              new_c(i,:) = new_c(i,:) ./ assigned2c(i);
41          end
42      end
43
44      %Early exit if error is small
45      d = sum(sqrt(sum((new_c - centroids).^2, 2)));
46      if d < min_thresh
47          break;
48      end
49      centroids = new_c;
50  end
```

To demonstrate the k-means code above, we can create three random processes, combine the data from them all, and attempt to cluster the data into three regions. For this, we need three centroids and so $k = 3$:

8.4 Summary of techniques

```
M=40;
% Create some 2D data in 3 classes
c1=([1.9*ones(M,1) 0.9*ones(M,1)])+randn(M,2);
c2=([0.9*ones(M,1) 0.1*ones(M,1)])+randn(M,2);
c3=([0.1*ones(M,1) 1.9*ones(M,1)])+randn(M,2);
% Plot the data
plot(c1(:,1),c1(:,2),'r+')
hold on
plot(c2(:,1),c2(:,2),'bx')
plot(c3(:,1),c3(:,2),'g*')

X=[c1;c2;c3];
C=kmeans(X',3);

plot(C(1,1),C(1,2),'ko')
plot(C(2,1),C(2,2),'ko')
plot(C(3,1),C(3,2),'ko')
hold off
```

Figure 8.7 plots the output from this test script for different iteration numbers, after 1, 10, 100 and 10 000 iterations respectively, clockwise from the top left. The centroids are shown as dark circles, and it is evident that these gradually move between iterations.

The k-means clustering algorithm is very useful in a number of machine learning applications for speech and other domains. It is simple to implement and relatively efficient in operations. However, it iterates towards a local maximum (there is no guarantee

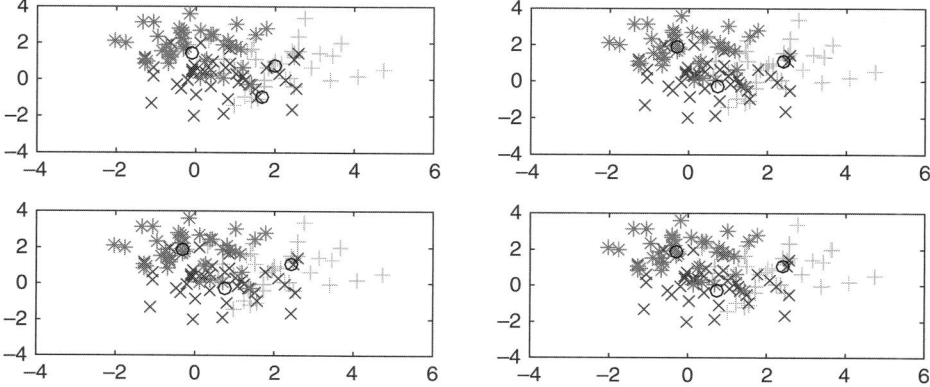

Figure 8.7 Illustration of using k-means to cluster random two-dimensional data from three processes into three classes, after 1, 10, 100 and 10 000 iterations (clockwise from top left).

that it terminates at a global maximum) and uses a fixed variance for each cluster – whereas it is quite likely that the variance of each process will be different for many applications. To get around that problem, we need to turn instead to the technique of the next section, Gaussian mixture models (GMMs).

8.4.6 Gaussian mixture models

In GMMs, we assume that data is generated from several continuous functions that might have different variances as well as means. This data will then be modelled by a mixture of Gaussians, with each point allocated to a mixture component with defined probability.

Just as with k-means, GMMs begin with an initial estimate, and then iteratively assign data points to mixture components before re-estimating the mixture components based on the data point allocation, working towards a locally optimum solution.

In this case we have K Gaussians, $\mathcal{N}_k$, each defined by their mean, μ_k, and covariance matrix, Σ_k:

$$\mathcal{N}_k(x) = \frac{1}{\sqrt{(2\pi)^n |\Sigma_k|}} e^{-\frac{1}{2}(x-\mu_k)^T \Sigma_k^{-1}(x-\mu_l)}, \tag{8.11}$$

where n is the dimension of the data vector, x, and there are $k = 1, \ldots, K$ mixtures. Using this, we can compute the probability that data item j belongs to cluster k as w_k^j and use this as a soft assignment of points to clusters (just as we did with k-means):

$$w_k^j = \frac{\mathcal{N}_k(x)\phi_k}{\sum_{j=1}^{K} \mathcal{N}_j(x)\phi_j}, \tag{8.12}$$

where ϕ_k is the prior probability of component k (the likelihood that a particular data item is generated from component k, normalised by all K components).

Given the soft assignment, w_k^j, the next step is a re-estimation of the Gaussian components, $(\phi'_k, \mu'_k, \Sigma'_k)$, for Gaussian $\mathcal{N}_k$ ready for the next iteration:

$$\begin{aligned}\phi'_k &= \frac{1}{m}\sum_{i=1}^{m} w_k^i, \\ \mu'_k &= \frac{\sum_{i=1}^{m} w_k^i x^i}{\sum_{i=1}^{m} w_k^i}, \\ \Sigma'_k &= \frac{\sum_{i=1}^{m} w_k^i (x^i - \mu_k)(x^i - \mu_k)^T}{\sum_{i=1}^{m} w_k^i}.\end{aligned} \tag{8.13}$$

The newly estimated Gaussians are then used for the soft assignment in the subsequent iteration, followed by another re-estimation and so on, until either the system converges (i.e. total error is small) or a maximum number of iterations has been reached. Again, this is very similar to the process for k-means above.

There are a number of GMM solutions for MATLAB, including those targeted for speech applications in HTK, Voicebox and so on. A simple GMM implementation is given below, starting with a definition for the Gaussian function $\mathcal{N}_k$:

8.4 Summary of techniques

```
function pdf = gaussian(x, mu, sigma)
for n=1:size(sigma,1)/size(mu,1)
   diff(:,n)=x(:,n)-mu(n);
end %ends with n set to the size difference
pdf = 1 / sqrt((2*pi)^n * det(sigma))* exp(-1/2 * sum((
    diff*inv(sigma) .* diff), 2));
```

This is then called by the GMM learning function given in Listing 8.1.

Listing 8.2 GMMlearn.m

```matlab
function [mu,sigma]=GMMlearn(X,k)
%set error threshold
min_thresh=1e-12;
%set maximum iterations
max_iter=1000;
% number of data points.
m = size(X, 1);
% dimension
n = size(X, 2);
%same initial probabilities
phi = ones(1, k) * (1 / k);

%random data points as initial mu's
indeces = randperm(m);
mu = X(indeces(1:k), :);
sigma = [];

%cov of all data as initial sigma's
for (j = 1 : k)
    sigma{j} = cov(X);
end

%initial probabilities
w = zeros(m, k);

for (iter = 1:max_iter)
    pdf = zeros(m, k);

    %for k mixtures, prob of data point per mixture
    for (j = 1 :k)
        pdf(:, j) = gaussian(X, mu(j, :), sigma{j});
    end
```

```matlab
    %calculate pdf_w
    pdf_w=pdf.*repmat(phi,m,1);
    %calculate w
    w=pdf_w./repmat(sum(pdf_w, 2),1,k);

    %store previous means
    old_mu = mu;

    %for k mixtures
    for (j = 1 : k)
        %prior prob. for component j
        phi(j) = mean(w(:, j), 1);
        %new mean for component j
        mu(j, :) = (w(:, j)'*X)./sum(w(:, j),1);
        %new cov matrix (for current component)
        sigma_k = zeros(n, n);
        %subtract cluster mean from all data points
        Xm = bsxfun(@minus, X, mu(j, :));
        %contrib of each data vector to the cov matrix
        for (i = 1 : m)
            sigma_k = sigma_k + (w(i, j) .* (Xm(i, :)' ...
                * Xm(i, :)));
        end
        %normalise
        sigma{j} = sigma_k ./ sum(w(:, j));
    end
    %early termination on convergence
    if abs(mu - old_mu) < min_thresh
        break
    end
end
```

This function has identical syntax to the kmeans.m function in Section 8.4.5. Thus, to test this, we can simply repeat the same class generation and learning code as was used to evaluate the k-means function, just replacing the call:

```matlab
C=kmeans(X,3);
```

with a call to the GMM learning function instead:

```matlab
[C,~]=GMMlearn(X,3);
```

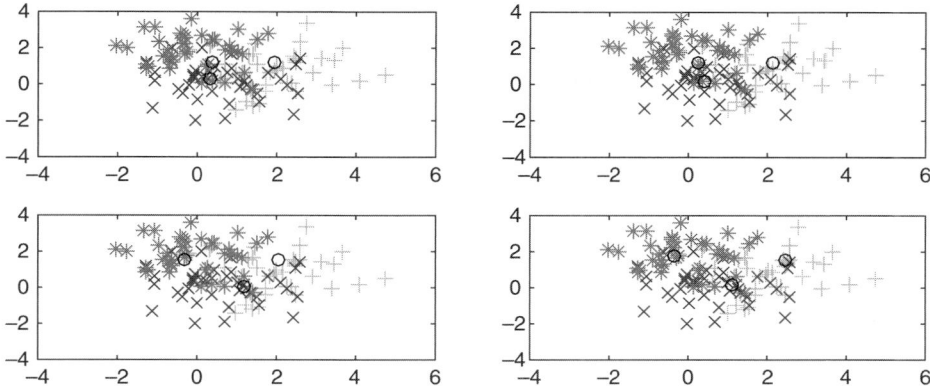

Figure 8.8 Illustration of using GMM to cluster random two-dimensional data from three processes into three classes, using identical data and conditions to the k-means example in Figure 8.7.

yielding results as shown in Figure 8.8. Note that GMM estimation is slightly more complex and thus slower than k-means, but achieves better performance. It also tends to converge slower (meaning that more iterations, and usually more data, are required). Another point to note is that the actual performance achieved depends upon how the initial points are chosen – so k-means is often used as an initialisation technique for the GMM mixtures to enable them to begin with nearly accurate data.

8.4.7 Deep neural networks

A trained DNN is very similar in structure to an MLP, with the main difference being that it is trained layer-wise from input to output layer and then fine-tuned in the reverse direction using back propagation. The layer-wise training in the forwards direction, followed by fine tuning in the backwards direction, allows deep structures to be constructed and trained. Generally speaking, DNNs tend to perform well at picking out fine details from representative feature vectors, meaning that they work well when coupled with high-dimensionality input features.

An L-layer DNN classifier is constructed to eventuate in an output layer of K classes and begin with an input layer fed with the input feature vector. The DNN is built up layer-wise from the input side to the output from individual pre-trained pairs of restricted Boltzmann machines (RBMs), each of which consists of V visible and H hidden stochastic nodes, $\mathbf{v} = [v_1, v_2, ..., v_V]^\top$ and $\mathbf{h} = [h_1, h_2, ..., h_H]^\top$. Two different types of RBM are typically used in the construction process. These are called Bernoulli–Bernoulli and Gaussian–Bernoulli respectively. The former restricts all nodes to be binary (i.e. $\mathbf{v}_{bb} \in \{0,1\}^V$ and $\mathbf{h}_{bb} \in \{0,1\}^H$), whereas the latter has real-visible nodes but binary hidden nodes (i.e. $\mathbf{v}_{gb} \in \mathbb{R}^V$ and $\mathbf{h}_{gb} \in \{0,1\}^H$).

The reason for the two types of RBM is that the input to the DNN typically comprises a vector of real-valued data, whereas the output layer is a binary classification. Hence the intermediate and final layers are Bernoulli–Bernoulli, whereas the input layer is Gaussian–Bernoulli (although many different arrangements do exist).

Like a perceptron (and similar networks), nodes have weights and biases, and there is full connectivity between the visible and hidden nodes. Thus an entire RBM layer pair is defined by the set of weights and biases.

In the Bernouli–Bernoulli RBM, this set of data is denoted by model parameters $\theta_{bb} = \{\mathbf{W}, \mathbf{b^h}, \mathbf{b^v}\}$, where $\mathbf{W} = \{w_{ij}\}_{V \times H}$ is the weight matrix and the biases are $\mathbf{b^h} = [b_1^h, b_2^h, ..., b_H^h]^\top$ and $\mathbf{b^v} = [b_1^v, b_2^v, ..., b_V^v]^\top$ respectively.

The energy function of state $E_{bb}(\mathbf{v}, \mathbf{h})$ is just the summation of these:

$$E_{bb}(\mathbf{v}, \mathbf{h}) = -\sum_{i=1}^{V}\sum_{j=1}^{H} v_i h_j w_{ji} - \sum_{i=1}^{V} v_i b_i^v - \sum_{j=1}^{H} h_j b_j^h, \tag{8.14}$$

where w_{ji} is the weight between the ith visible unit and the jth hidden unit, and b_i^v and b_j^h are respective real-valued biases for those layers.

The Gaussian–Bernoulli RBM differs only in that an extra term is required to represent the Gaussian variance in each node, σ^2, hence the model parameters are $\theta_{gb} = \{\mathbf{W}, \mathbf{b^h}, \mathbf{b^v}, \sigma^2\}$. The variance parameter σ_i^2 is pre-determined rather than learnt from training data.

The energy function for the Gaussian–Bernoulli case is thus slightly more complex:

$$E_{gb}(\mathbf{v}, \mathbf{h}) = -\sum_{i=1}^{V}\sum_{j=1}^{H} \frac{v_i}{\sigma_i} h_j w_{ji} + \sum_{i=1}^{V} \frac{(v_i - b_i^v)^2}{2\sigma_i^2} - \sum_{j=1}^{H} h_j b_j^h. \tag{8.15}$$

Every visible unit v_i adds a parabolic offset to the energy function, governed by σ_i.

Putting all of this together, an energy function $E(\mathbf{v}, \mathbf{h})$ is defined either as in Equation (8.14) or as in Equation (8.15). The joint probability associated with the configuration $(\mathbf{v}, \mathbf{h})$ for this is defined as:

$$p(\mathbf{v}, \mathbf{h}; \theta) = \frac{1}{Z} e^{\{-E(\mathbf{v}, \mathbf{h}; \theta)\}}, \tag{8.16}$$

where Z is a partition function:

$$Z = \sum_{\mathbf{v}}\sum_{\mathbf{h}} e^{\{-E(\mathbf{v}, \mathbf{h}; \theta)\}}. \tag{8.17}$$

8.4.7.1 Pre-training

RBM model parameters θ are estimated by maximum likelihood learning over a training set. The main enabler for success of the RBM was the definition of an algorithm that is able to do this conveniently and accurately (just as the back-propagation algorithm was key to the success of the MLP). For DNNs, the contrastive divergence (CD) algorithm [106] runs through a limited number of steps in a Gibbs Markov chain to update hidden nodes $\mathbf{h}$ given visible nodes $\mathbf{v}$ and then update $\mathbf{v}$ given the previously updated $\mathbf{h}$.

The input layer is trained first (i.e. the feature vector input to layer 1 $\mathbf{v}_{gb}$). After training, the inferred states of its hidden units $\mathbf{h_1}$ become the visible data for training the next RBM layer's visible units $\mathbf{v_2}$. The process repeats to stack up multiple trained

8.4 Summary of techniques

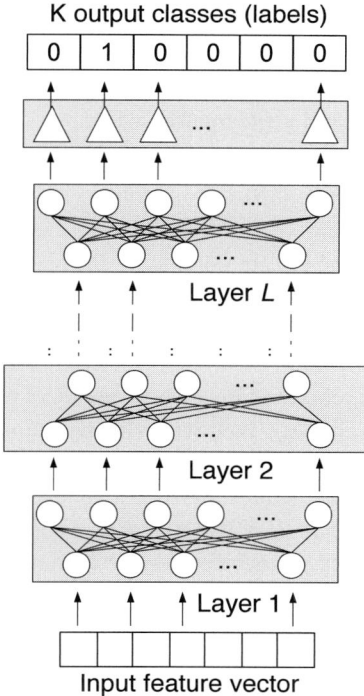

Figure 8.9 A DNN is constructed from a deep stack of Gaussian–Bernoulli and Bernoulli–Bernoulli RBMs, and is trained from the bottom up using the CD algorithm.

layers of RBMs to the required depth. Once complete, the RBMs are stacked to produce the DNN, as shown in Figure 8.9.

Configuration options, set prior to training, include the number of layers, L, the dimensionality of each layer and several training related parameters (such as learning rate, batch size and so on).

8.4.7.2 Fine-tuning

Many types of DNN have been invented for different domains, but here we will assume the same machine hearing application example [97] that was used for the SVM illustration in Section 8.4.3. In this case, the final output of the network is used to classify what type of sound (out of K classes) has been heard at the input, assuming only one input sound at a time. Thus the output is formed by a size K softmax output labelling layer to scale the network output to yield output probabilities for each class k, denoted by $p(k|\mathbf{h}; \theta_L)$ i.e. this output probability is dependent upon the model parameters of layer L and the input to layer L [107] as in:

$$p(k|\mathbf{h_L}; \theta_\mathbf{L}) = \frac{\phi(k, \theta_L)}{\sum_{p=1}^{K} \phi(p, \theta_L)} \quad \text{for} \quad k = 1, \ldots, K, \tag{8.18}$$

where $\phi(k, \theta_L) = e^{\{\sum_{i=1}^{H} w_{ki} h_i + b_k\}}$, making the output scaling a normalised exponential function.

Back propagation (BP) is then applied to train the stacked network, including the softmax class layer, to minimise the cross-entropy error between the ground truth class label for each training datum c and the probability that has been predicted at the softmax layer output. The cross-entropy cost function, $\mathcal{C}$, is computed as:

$$\mathcal{C} = -\sum_{k=1}^{K} c_k \log p(k|\mathbf{h}; \theta_L). \tag{8.19}$$

Note that, during training, various parameters are commonly introduced, including dropout (defined as the proportion of weights fixed during each training batch, instead of having all weights updated all the time – this is a pragmatic measure introduced to prevent over-training). Given that there is often a large amount of training data (needed because the DNN is *deep*), this is processed in batches, and there may be many run-throughs of the data (called training epochs), each with slightly adjusted training rates, starting large and gradually decreasing. It should also be pointed out that, during classification, the output probabilities are usually thresholded to give just a single output class (normally the one with the largest probability), but many other possibilities exist, such as (i) choosing the one with the highest probability for the current classification result compared with its long-time average, (ii) applying a sliding window and summing the probabilities of each class over that window, picking the largest (which corrects for high variance in the output probabilities), and (iii) removing the softmax entirely and, instead of doing an exponential normalisation for the current classification result as in Equation (8.18), normalising over time for each individual class. This may be useful in dealing with instances of overlapping class outputs, or when classes are themselves not entirely orthogonal (i.e. where there is information leakage between classes – as in classifying sounds that contain similarities).

There are many possibilities.

As with SVM in Section 8.4.3, we will illustrate the use of the DNN by employing a toolkit. While several alternative toolkits are publicly available, we will use the **DeepLearnToolbox** [107],[7] which is relatively straightforward to use, and can achieve excellent results.

Assuming, as with SVM (Section 8.4.3), that sets of training and testing feature vectors are available in arrays `train_data` and `test_data` respectively (and that training labels are in array `train_label` while labels for verifying the test classifications are in `test_label`), the input data first needs to be normalised and scaled:

```
mi=min(min(train_data));
train_x=train_data-mi;
ma=max(max(train_x));
train_x=train_x/ma;
```

---

[7] The DeepLearnToolbox is an open source MATLAB library for deep learning – as well as for convolutional and other neural networks. It is available from https://github.com/rasmusbergpalm/DeepLearnToolbox, along with several simple usage examples.

8.4 Summary of techniques

We now proceed to set up a DNN. In this case we will simply use most of the default values of the toolbox and a similar structure to one of the tutorial examples in [107] (which was also very effectively used in [97]):

```
nnsize=210;
dropout=0.10;
nt=size(train_x,1);
rand('state',0) %seed random number generator
%arrange input data into batches of size 100
for i=1:(ceil(nt/100)*100-nt)
    np=ceil(rand()*nt);
    train_x=[train_x;train_x(np,:)];
    train_y=[train_y;train_y(np,:)];
end
```

Now we start to create and stack RBM layers – as many as we want in fact, to create a deep structure (but this example is not particularly deep), using a number of 'dbn' functions from the DeepLearnToolbox library (DBN being an abbreviation for 'deep belief network'):

```
% train a 100 hidden unit RBM
dbn.sizes = [nnsize];
opts.numepochs = 1;
opts.batchsize = 100;
opts.momentum = 0;
opts.alpha = 1;
dbn = dbnsetup(dbn, train_x, opts);
dbn = dbntrain(dbn, train_x, opts);
% train a 100-100 hidden unit DBN
% use its weights to create a feed-forward NN
dbn.sizes = [nnsize nnsize];
opts.numepochs = 1;
opts.batchsize = b;
opts.momentum = 0;
opts.alpha = 1;
dbn = dbnsetup(dbn, train_x, opts);
dbn = dbntrain(dbn, train_x, opts);
%unfold DBN into the nn
nn = dbnunfoldtonn(dbn, 50);
nn.activation_function = 'sigm';
```

Instead of a set of RBM layers, we now have all the weights and biases, and have transformed those into a plain neural network structure. Thus it is ready to begin back propagation using the following code, where the function `nntrain()` is provided by the DeepLearnToolbox library:

```
%train NN by backprop
opts.numepochs = 1;
opts.batchsize = b;
nn.dropoutFraction = dropout;
nn.learningRate = 10; %start big
%loop through many epochs
for i=1:1000
    fprintf('Epoch=%d\n',i);
    nn = nntrain(nn, train_x, train_y, opts);
    %adapt the learning rate
    if i==100
        nn.learningRate = 5;
    end;
    if i==400
        nn.learningRate = 2;
    end;
    if i==800 %end small
        nn.learningRate = 1;
    end;
end;
```

The outcome of all of this training is a fairly large array in MATLAB memory called `nn` which defines the DNN structure and sizes, containing all weights and connections. This is now ready for evaluation – testing proceeds by attempting classification of each test feature vector in turn and comparing with the ground truth labels, with the final performance score obtained from the percentage of correct classifications:

```
correct=0;
test_now=0;
nfile=4000;
for file=1:nfile   %for all TESTING files
[label, prob] = nnpredict_p(nn,test_x(test_now+1:
   test_now+nFrames(file),:));
    if label==ceil(file/80)
        printf('correct\n');
    else
        printf('NOT correct\n');
    end
    test_now=test_now+nFrames(file);
end
```

The function that this calls, `nnpredict_p()`, implements whatever type of softmax or output voting strategy is to be applied on the classified output, such as the code below which simply outputs the maximum probability class:

```
function [label, maxp] = nnpredict_p(nn, x)
    nn.testing = 1;
    nn = nnff(nn, x, zeros(size(x,1), nn.size(end)));
    nn.testing = 0;
    prob = sum(nn.a{end});
    label = find(prob==max(prob));
    maxp=max(prob);
end
```

The function `nnff()` used in this code is provided by the DeepLearnToolbox library to implement the feed-forward network. Note that we will use several more of these functions to illustrate convolutional neural networks (CNNs) in the following section.

8.4.8 Convolutional neural networks

CNNs are multi-layer neural networks formed as a stack of convolution layers, sub-sampling layers and fully connected layers although, as with most topics in machine learning, there are a variety of implementation methods where details may vary. A CNN will typically [108] begin with a convolutional layer and subsampling layer pair, which may be repeated several times (i.e. the layers are interleaved). The final subsampling layer will then feed into a fully connected set of output layers which is just like an MLP. Thus the CNN's novelty could be viewed as being a front end to a basic network such as an MLP.

Convolution layers work to gather local features from a rectangular window on an input image map, and transform these into a single output point. In operation, the rectangular window steps across the full input image in the x- and y-directions, with the output features being formed into a smaller image with the x- and y-direction resolution dependent upon the rectangular window step size. There are usually a few output images generated from each input image map – with each output image created through exactly the same operation, but with different weights and bias settings that have been learned. Meanwhile the subsampling process simply reduces the spatial resolution of the image maps using a process such as max pooling (which represents an entire rectangular window by the maximum element within that window), averaging (where the output is the mean value from the input data) or some similar process.

While the network complexity is high due to the large amount of connectivity, CNNs make use of shared weights within layers (i.e. to form each output image map in a convolutional layer), which assists in reducing the number of parameters that need to be trained. However, CNNs share with deep neural networks (DNNs) the need to consume large amounts of data in order to become well-trained. In general, convolutional layer l forms layer output maps from the outputs of the previous layer acted upon by a scaled kernel with bias. The jth output map of layer l, $\mathbf{x}_j^l$, is defined as:

$$\mathbf{x}_j^l = f\left(\sum_{i \in M_j} \mathbf{x}_i^{l-1} * \mathbf{k}_{ij}^l + b_j^l\right), \tag{8.20}$$

where $\mathbf{x}_i^{l-1}$ is the ith input map of layer l and $\mathbf{k}_{ij}^l$ denotes the kernel that is being applied. Meanwhile M_j represents a selection of input maps [109].

The subsampling layer is simpler than this, being defined as:

$$\mathbf{x}_j^l = f(\beta_j^l \downarrow (\mathbf{x}_i^{l-1}) + b_j^l), \tag{8.21}$$

with $\downarrow$ (.) representing whatever kind of subsampling is being used (for example, representing a group of pixels by their mean), and β and b are biases [109].

The fully connected output layer is effectively a multi-layer perceptron (MLP) network, with its input layer size defined by the total number of points being output by the final CNN subsampling layer, and its output size defined by the number of classes being classified. Otherwise it is formed as a typical MLP, and, as with any MLP, it can be learned by gradient descent using the back-propagation algorithm. Since units in the same feature map share the same parameters, the gradient of a shared weight is simply computed from the sum of the shared parameter gradients at the end of a mini-batch (i.e. it is usually not adjusted each iteration).

The CNN is widely used in image processing [108, 110], where it has demonstrated good performance in a number of tasks. It has been applied to the ASR and speech processing fields [111, 112], and has been shown to out-perform traditional networks for many different tasks – especially those which have image-like input features.

A diagram of a CNN is shown in Figure 8.10, which is derived from an attempt to use a CNN classifier [113] for the machine hearing application example that was described in Sections 8.4.3 (SVM) and 8.4.7 (DNN). The diagram illustrates the input feature map (which in this case is a spectrogram) and two sets of convolution and subsampling layers prior to a fully connected output layer. The dimensions of each image for this particular application example are given below each layer, starting with a 40×52 spectrogram, to six separate 36×48 output maps which are then down sampled to 18×24 and so on. The final output is 12 output maps of resolution 7×10, which means 840 MLP input nodes for the fully connected output layer, and an output of 50 classes. All of these figures are related directly to the machine hearing application example that will be discussed further in Section 8.5, but illustrate a very typical kind of CNN arrangement.

Implementation is significantly simplified by making use of one of several open source packages, and as with the DNN example we shall choose DeepLearnToolbox [107], working in MATLAB. Let us assume in this case that all data is available in matrix `all_x` with corresponding labels in `all_y`.[8] We can thus condition data, set up

---

[8] We further assume that all data has been normalised and scaled prior to this point – incorrectly normalised data can have significantly detrimental effects on the performance of a CNN.

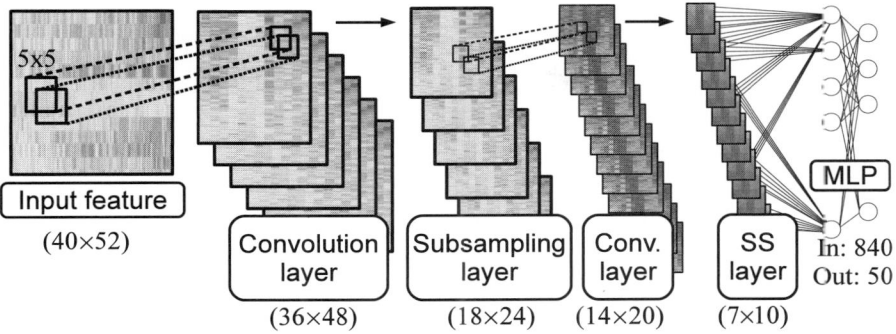

Figure 8.10 A six-layer CNN designed for a machine hearing classifier, comprising two sets of convolution and subsampling layers plus a single fully interconnected output layer. Dimensions are shown for each layer for the specific application example [113].

batches and distribute the data with 80% reserved for training (trainportion=0.8) and the remaining 20% for testing. Training data will then be stored in matrix train_x with corresponding labels in train_y and test material in test_x and test_y:

```
datalen=length(all_y);
trainportion=0.8;
batchsize=50;  %batch size for CNN training
databatches=floor(datalen/batchsize);
trainbatches=floor(databatches*trainportion);
trainbit=trainbatches*batchsize;

test_x=all_x(:,:,trainbit+1:end);
train_x=all_x(:,:,1:trainbit);
test_y=all_y(:,trainbit+1:end);
train_y=all_y(:,1:trainbit);
```

The parameters of the CNN are then set up into a structure as follows, in this case for a network with six CNN layers prior to the final MLP:

```
%Set up the CNN
rand('state',0)
cnn.layers = {
  %input layer
struct('type', 'i')
  %convolution layer
struct('type', 'c', 'outputmaps', 6, 'kernelsize', 5)
  %sub sampling layer
struct('type', 's', 'scale', 2)
```

```
    %convolution layer
struct('type', 'c', 'outputmaps', 3, 'kernelsize', 5)
    %sub sampling layer
struct('type', 's', 'scale', 2)
    %convolution layer
struct('type', 'c', 'outputmaps', 2, 'kernelsize', 3)
    %sub sampling layer
struct('type', 's', 'scale', 1)
};
%set up the network and training data
cnn = cnnsetup(cnn, train_x, train_y);
```

The final line above sets up the CNN structure along with the training data ready to initiate the training. This is launched as follows, where we have selected just a small number of epochs (four) since this is sufficient for convergence in the current application:

```
opts.alpha = 1;
opts.batchsize = batchsize;
opts.numepochs = 4;
%do the training
cnn = cnntrain(cnn, train_x, train_y, opts);
```

The trained CNN model is now available in the cnn matrix, which can be used in a feedforward fixed fashion for testing:

```
cnn = cnnff(cnn, test_x);
%cnn now contains output results
[~, h] = max(cnn.o);
[~, a] = max(test_y);
bad = find(h ~= a);
errors = numel(bad) / size(test_y, 2);
```

The final output class for each testing data point is in cnn.o, and the output variable errors holds the final proportion of errors over the testing set.

8.4.9 Future methods

The chapter has, up to now, surveyed a number of useful machine learning algorithms that are useful to speech and audio processing. In the modern world these algorithms work together with the existence of big data to learn good models of observed processes.

If a reader is to refer back to the original invention of each one of these methods, whether in a research article or book, it is highly likely that the authors will justify their algorithms in terms of being 'brain-like' or 'eye-like' (for image processing), or 'ear-like' (for machine hearing). The idea is that they are inspired by cognitive processes in animals which learn from the world around them. However, brains are complex living organisms which, in practice, are extremely difficult to replicate artificially, and thus the algorithm authors must choose which aspects of the brain (or cognitive process) they wish to model, and in how much detail. The resulting models are therefore all compromises with different strengths and weaknesses, as well as occupying different positions on the trade-off between fidelity and efficiency. One of the most important attributes of the machine learning models that we do use is that they can be trained – and that generally requires that they are backed up by a mathematical proof that the models will converge during training on a solution which is at least locally optimum. Put another way, anyone can propose a new cognitive model, but unless it can be trained (and demonstrated to be able to be trained), it may not be very useful in practice! This is a highly active research area, so maybe the only constant is that there will be continual improvement and change.

As for any future machine learning methods, it is highly likely that more and more 'brain-like' systems will be developed that incorporate more and more aspects of the human cognitive system. The trend – as we found at the beginning of this chapter – is also towards larger and larger data sets, i.e. towards bigger and bigger big data.

8.5 Big data applications

Big data, as an approach to understanding the world around us, is inextricably linked with modern machine learning, and is finding applications in a wide variety of areas touching virtually all domains of life. Within the audio, speech and hearing fields, it has been crucial to recent performance advances across the entire research domain, with the possible exception of speech compression.

We have briefly mentioned some extremely capable big data-driven speech applications such as Apple's Siri, Google's voice assistant and iFlytek's speech cloud already. However, these are commercially secret applications that are hidden from examination by academic researchers. Thus we will present an open example which is a complete approach to machine learning. This is available along with full MATLAB code,[9] employing machine learning techniques (specificatally DNNs) to implement a machine hearing system. Rather than printing several pages of raw MATLAB code here, the software is provided on the accompanying website for download from http://mcloughlin.eu/machine_hearing along with full instructions and example data. Readers are encouraged to download, view, edit and make use of this software. It is likely that readers may recognise some of the code listed there, because snippets were included in DNN and SVM examples above.

---

[9] This machine hearing example also works well in Octave, with minor code changes.

8.6 Summary

This chapter has run through a fast tour of the main machine learning methods that are commonly applied to the speech, audio and hearing research fields. Our tour has been necessarily short, because there is far more to the fascinating field of machine learning and big data than can be presented here (the reader is referred to the bibliography below to source deeper information about this topic). Although we touched on many topics in this chapter, we have not discussed much about the mechanics of automatic speech recognition (ASR) – which will be the subject of our next chapter.

Before moving on, however, a brief word of caution. Big data is often presented as a standalone research field, and similarly some machine learning experts claim to be data-agnostic (meaning that they will work with any data, without needing to know exactly what the data contains). To some extent these views do truly represent the fact that the same machine learning techniques are being applied across a vast array of research fields – from video interpretation through text processing, social networking monitoring, financial performance analysis, medical sensors, brain–computer interfaces and so on. These views also reflect the fact that (machine) learning systems will learn whatever is presented to them, without necessarily having pre-conceived notions about the data being presented. However, the author believes that it is extremely important that the underlying *meaning* of the data should never be forgotten. Not knowing the meaning of data is almost guaranteed to result in less efficient analysis and processing, as well as in less relevant outputs. Conversely, knowledge of the data characteristics will often allow better insights into the underlying processes involved in shaping it, and can be leveraged to improve processing efficiency and performance.

Bibliography

- *Pattern Recognition and Machine Learning*
  C. Bishop (Springer-Verlag, 2006)
  Probably the de-facto standard text for machine learning (the underpinning technology area for 'big data'), this 700-plus page book serves as the most complete and best-presented resource for further study in this field. This is widely considered to be the machine learning bible.

- *VLFeat*
  www.VLFeat.org
  An excellent set of machine learning toolboxes, primarily designed for visual learning, but many of which are equally applicable to speech and audio data. In addition to providing a fast, efficient and easy-to-use toolbox which works in MATLAB and Octave and can be linked to from C programs (as well as from other languages), the authors have provided tutorial examples showing how to use the programs. For example, consider the GMM tutorial at www.vlfeat.org/overview/gmm.html and the SVM tutorial at www.vlfeat.org/overview/svm.html#tut.svm.

This resource is highly recommended for those who simply want to get up and running with some of these machine learning algorithms quickly yet efficiently.

- *Multi-layer perceptron explanation and example*
  http://3options.net/brainannex/multilayer-perceptron-in-matlab-octave/
  This website contains working code for MATLAB and Octave.

- *Multi-layer perceptron explanation and tutorial*
  http://users.ics.aalto.fi/ahonkela/dippa/node41.html
  Mathematical background and explanation for MLPs, including a good presentation of the back-propagation algorithm.

Questions

Q8.1 Provide three reasons why raw data is generally transformed to a smaller set of features prior to being used for machine learning training.

Q8.2 By monitoring the error rate of a general machine learning system, how might (a) overtraining, (b) insufficient training data and (c) mis-labelled data be diagnosed?

Q8.3 Explain why, in machine learning, a data set might be split into two parts – one for training and one for testing. What not simply use all data for training, and then all data for testing?

Q8.4 Sometimes, the data is split into three parts to make a separate development dataset. Explain what this might be used for.

Q8.5 Briefly identify the difference between supervised and unsupervised learning.

Q8.5 With the aid of a simple scatterplot (a two-dimensional x–y plot containing lots of individual data points), draw shapes to illustrate and explain the difference between the concepts of classification and clustering.

Q8.6 How can a binary classifier such as SVM be used to classify data into four classes instead of into two classes?

Q8.7 Working in MATLAB, repeat the ICA example of Section 8.4.4.2, but use two arrays of speech (instead of one array of speech and one of noise). You may need to reduce the learning rate and increase the convergence criterion. Try two recordings from the same speaker, and then try recordings from different speakers. Can you see which works best, and speculate why that might be?

Q8.8 Try plotting two Gaussian-shaped distributions from MATLAB function `gaussian()` in Section 8.4.6 on the same axis, generated with different means and sigma. Now plot three separate histograms of the data `c1`, `c2`, `c3` from the previous section (Section 8.4.5). Are these histograms roughly Gaussian in shape?

Q8.9 In what way are deep neural networks (DNN) described as being 'deep', and why might that be considered an advantage?

Q8.10 What is the difference between a Gaussian–Bernoulli and a Bernoulli–Bernoulli restricted Boltzmann machine, and where are these both typically used to construct DNNs?

Q8.11 Identify the major difference in the format of the input features in DNN and convolutional neural network (CNN) implementation.

Q8.12 What feature of the CNN internal arrangement is used to simplify its training requirements compared with the complexity of a fully interconnected network of roughly the same size and shape?

Q8.13 If an industrial engineer is asked to use a 'big data' approach to solve a new problem, list some of the likely technical questions she would be asking herself and others in order to accomplish her task.

Q8.14 With regard to something like MFCC features, explain what ΔMFCCs and $\Delta\Delta$MFCCs are.

Q8.15 Consider an audio-based speaker identification (SID) system. Apart from MFCCs, what are some of the other types of common input features that researchers might consider using?

9 Speech recognition

Having considered big data in the previous chapter, we now turn our attention to speech recognition – probably the one area of speech research that has gained the most from machine learning techniques. In fact, as discussed in the introduction to Chapter 8, it was only through the application of well-trained machine learning methods that automatic speech recognition (ASR) technology was able to advance beyond a decades long plateau that limited performance, and hence the spread of further applications.

9.1 What is speech recognition?

Entire texts have been written on the subject of speech recognition, and this topic alone probably accounts for more than half of the recent research literature and computational development effort in the fields of speech and audio processing. There are good reasons for this interest, primarily driven by the wish to be able to communicate more naturally with a computer (i.e. without the use of a keyboard and mouse). This is a wish which has been around for almost as long as electronic computers have been with us. From a historical perspective we might consider identifying a hierarchy of mainstream human–computer interaction steps as follows:

> **Hardwired:** The computer designer (i.e. engineer) 'reprograms' a computer, and provides input by reconnecting wires and circuits.
> **Card:** Punched cards are used as input, printed tape as output.
> **Paper:** Teletype input is used directly, and printed paper as output.
> **Alphanumeric:** Electronic keyboards and monitors (visual display units), alphanumeric data.
> **Graphical:** Mice and graphical displays enable the rise of graphical user interfaces (GUIs).
> **WIMP:** Standardised methods of windows, icons, mouse and pointer (WIMP) interaction become predominant.
> **Touch:** Touch-sensitive displays, particularly on smaller devices.
> **Speech commands:** Nascent speech commands (such as voice dialling, voice commands, speech alerts), plus workable dictation capabilities and the ability to read back selected text.
> **Natural language:** We speak to the computer in a similar way to a person, it responds similarly.

Anticipatory: The computer understands when we speak to it just like a close friend, husband or wife would, often anticipating what we will say, understanding the implied context as well as shared references or memories of past events.

Mind control: Our thoughts are the interface between ourselves and our computers.

At the time of writing, researchers are firmly embedded between the **speech commands** and **natural language** stages, despite some notable successes in the latter field. Similarly, researchers working on future digital assistants have made great progress on systems able to understand context and shared references – although such systems are far from being able to appreciate something like an 'in' joke. While **mind control** remains the remit of science fiction, researchers working with fMRI (functional magnetic resonance imaging) and sensitive EEG (electroencephalogram) equipment have isolated brain activation patterns associated with fixed concepts, activities, single words or phrases: even today there are methods of sensing information from the brain to predict which word a person is currently thinking of. There is a great deal of hard work and difficult research between recognising a single word and continuous thoughts; however, please remember that speech technology was at the 'single word' stage 30 years ago but today handles continuous dialogue with relative ease.

Within this chapter, we will mainly consider speech recognition. In particular, the task of deciding what words have been spoken – thus avoiding the issue of understanding the meaning of whatever was spoken.

However, the scope is slightly wider than just speech recognition. Eventually we will have considered the linked questions of detecting the presence of speech, identifying who is speaking and in what language, as well as how many people are speaking, and finally what is being spoken (as well as what is the meaning of the speech).

9.1.1 Types of speech recogniser

Speech recognition technology itself covers a wide field, normally categorised by a few key descriptive phrases:

Automatic speech recognition (ASR) describes a system that can recognise what has been spoken, without requiring additional user input.[1]

Keyword spotting means looking out for particular words or phrases (e.g. 'attack' or 'plant a bomb'), as well as listening out for a particular phrase that signals the user's intention to begin speech communications (e.g. 'Okay Google' or 'Computer:'). This technology is used primarily for initiating vocal commands, which means recognition of single words delimited by pauses, as well as monitoring continuous telephone conversations for public security purposes.

Continuous speech recognition is recognition of full sentences or paragraphs of speech rather than simple phrases or words. These applications do not require a user to pause when speaking (so the system must be quick enough to cope with

---

[1] *Speech recognition* is the task of recognising speech, by both human and computer listeners, whereas *automatic* speech recognition implies that it is a computer doing the listening.

their fastest rate of speaking – a real problem with current technology), and would encompass dictation and transcription systems.

Natural language processing (NLP), whilst not strictly limited to speech, describes the computational methods needed for a computer to understand the meaning of what is being said, rather than simply recognising what words have been spoken. Unlike automated transcription systems, where the meaning may be irrelevant as long as the words have been captured correctly, NLP enables the creation of a virtual butler who is able to cater to human needs, and so the semantics or meaning of what is said would be of primary importance.

As we have mentioned previously, this chapter will concentrate mainly on the task of word recognition (which in turn depends on phoneme recognition). This technology forms the foundation of most of the application areas above, but does not require the higher-level interpretation that would be needed by NLP, and does not need to worry about the real-time issues of continuous recognition or automated transcription of live events.

9.1.2 Speech recognition performance

Established researcher Victor Zue and colleagues wrote a book several years ago that identified various parameters that can be used to characterise speech recognition systems and their performance. Based upon this work, Table 9.1 lists a few of the characteristic parameters of these systems (this information was mainly derived from Section 1.2 of [114]). This interesting qualitative analysis of important factors provides some useful insights into the potential variability of such systems. Overall, we can safely say that ASR becomes much easier when the smaller end of the range is chosen for each parameter in the table. For example, a single-user system that attempts to recognise two words spoken by a known voice is a much simpler proposition than a system that attempts to recognise any word from a 100 000 word vocabulary, spoken by any person. Let us now examine some of those factors in turn.

9.1.2.1 Vocabulary and SNR

It is reasonable to assume that the larger the vocabulary, the more difficulty any system will have in accurately detecting a word, because there are more opportunities to

Table 9.1 Speech recognition system parameters.

Parameter	Typical range (easier	–	more difficult)
Vocabulary	small	–	large
Users	single	–	open access
Speech type	single words	–	continuous sentences
Training	in advance	–	continuous
SNR	high	–	low
Transducer	restricted	–	unrestricted

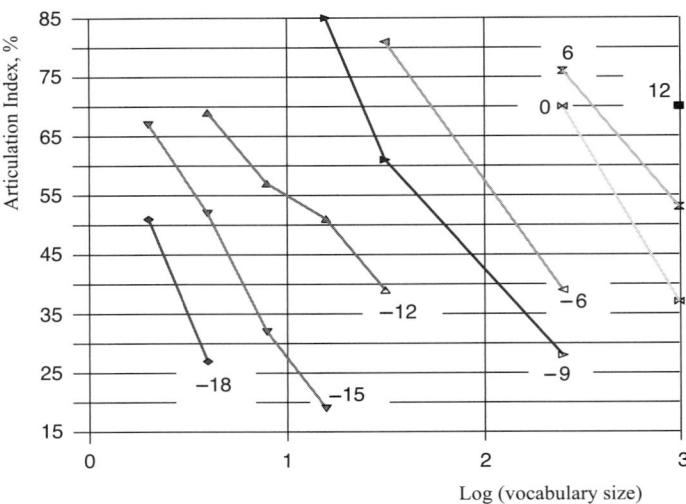

Figure 9.1 Plot of articulation index (a measure of recognition accuracy) versus the logarithm of vocabulary size for speech recognised at various levels of additive white noise, with SNR ranging from −18 dB to +12 dB.

mis-recognise that word and a greater probability of there being similar-sounding words. In fact, the same is true for human speech and this relationship between vocabulary and recognition accuracy is explored in Figure 9.1. The plot is generated from the same data that was used to plot Figure 3.10 in Section 3.4.4, but excludes the extreme accuracy points above 85% or below 20% (when other effects such as saturation come into play). The original data came from the results of experiments conducted over half a century ago by Miller et al. [34]. The figure plots the percentage recognition accuracy (articulation index) of human listeners for different vocabulary sizes. The vocabulary size is plotted using a log scale, spanning 2 to 1000 words. There are different curves for different signal-to-noise ratios (SNR) of the speech. This graph thus illustrates two parameter ranges of Table 9.1. The first is vocabulary size – the slope of all of the lines shows the same reduction in performance as vocabulary size increases. The second is SNR – for a given vocabulary size, the performance for higher SNR is always better than that for lower SNR.

In fact, the logarithmic relationship has been shown to be present in some of the largest variable-vocabulary recognisers, such as the famous Bellcore telephone directory assistance system [115].

The US Government's National Institute Of Standards and Technology (NIST) has run ASR-related competitions for a number of years. An evaluation of these in 2007 led to the historical performance data plotted in Figure 9.2. This graph shows the word error rate (WER) achieved by different tasks and NIST-related competitions over the years, compared with a WER of 4% which is the approximate upper error rate of average human listeners. Many of the results plotted on the graph are from famous research systems, some of which we have discussed elsewhere.

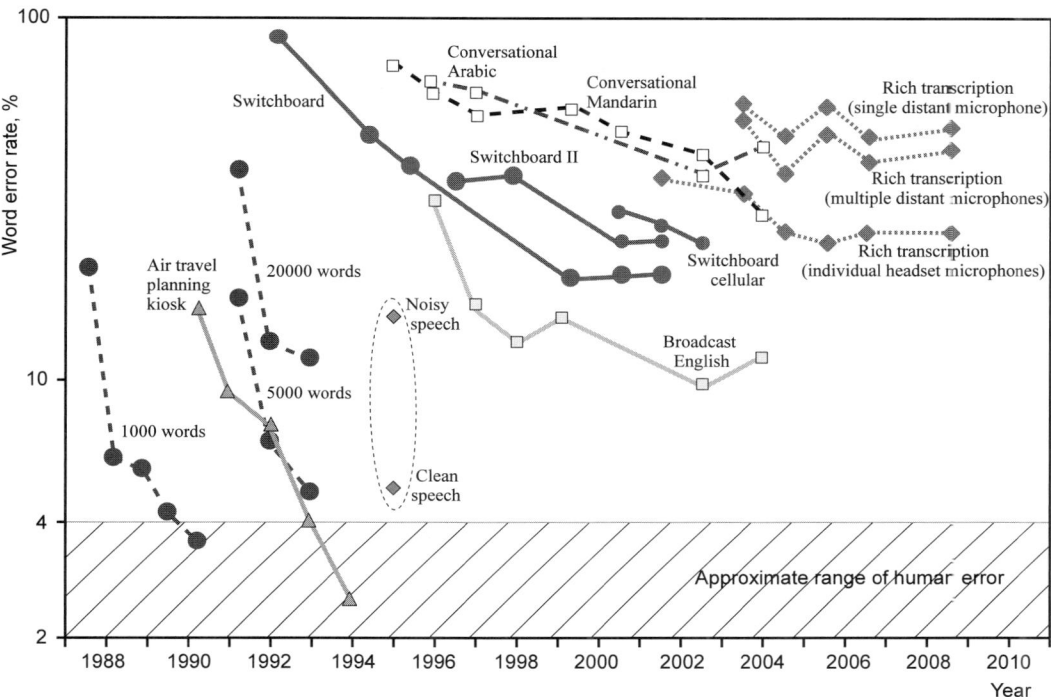

Figure 9.2 ASR word error rate achieved over time. Reproduced using the data presented in Figure 4 of [116], showing the historic results of various NIST ASR competitions and evaluations.

We note some interesting trends from this data. The obvious trend is that tasks (lines) have a slope that indicates how performance gradually improves over time. Unfortunately it appears that the slopes flatten from left to right, indicating that the recent year-on-year progress is slower than that of thirty years ago.

A quick glance at the plot might leave the impression that actual WER is getting worse – the more recent lines have higher average error rate than the older ones. However, such a conclusion would be incorrect, because the more recent competitions are a lot more difficult than the earlier ones. Some are more difficult in terms of vocabulary, as discussed earlier (the effect of which can easily be seen in Figure 9.2 by comparing the 1000, 5000 and 20 000 word vocabulary results at the bottom left of the figure). Switchboard (which contains more than 3 million words of spontaneous telephone conversations by more than 500 American English speakers of both genders, with an overall vocabulary of about 6000 words) is characterised by a large amount of speaker variability, noise and real-life quality issues. This saw continual WER improvements over the years, plateauing at around the turn of the century.

Some other interesting points are visible from the data. Firstly, a 1995 study showed the large detrimental effect of noise on WER (circled). A similar effect may be at work in separating the three lines at the top right, which are results from the transcription of meetings using a single distant microphone, multiple distant microphones and headset microphones. Obviously the further away the microphone is from the speaker, the more noise is introduced, so the two sets of results seem to be consistent. Other results in Figure 9.2 are for unrestricted broadcast English speech, and for conversational Arabic and Mandarin. Each of these follows a very similar trend, although perhaps we can conclude that the performance of English ASR is currently better than for those other languages.

Finally, it is interesting to contrast these kinds of results with those achieved through the use of big data and machine learning in a cloud environment (for example, see the ASR performance plot for iFlytek systems in Figure 8.1, which was primarily derived from Chinese ASR performance).

9.1.2.2 Training

Most modern ASR systems require some form of training to acclimatise the system, either to the speech of a particular individual or to a group of similar individuals.[2] Training is usually accomplished in advance by using a large dataset of similar speech (or similar speakers), and the larger the variability in conditions that system is required to cope with, the harder it is to train or to achieve a given error rate, as well as the larger the training set must be. There should also be roughly equal amounts of different types of training material to match the different types of speech the system is designed for; it cannot be expected to recognise an input that it has not been trained for. In terms of ASR, for example, this probably means that all of the words that we wish to recognise should be in the training set (or alternatively, all of the phonemes within all of those words).

Apart from quantity, the quality of the training set matters too – it needs to be representative of the actual speech that the system will encounter (although there have been some recent advanced research techniques developed to adapt systems trained in one domain to work in another domain, often referred to as *model adaptation*). This also means that the words in the training set should ideally be spoken in a similar way (e.g. pronunciation) to the speech that will be recognised.

9.1.2.3 Restricted topic

It should be noted that the continuous speech recognition research field also includes topic-specific recognisers, which are trained with a vocabulary of subject-specific words. The idea is that restricting the type of words – the vocabulary choice and perhaps the grammar – can improve performance. One example in Figure 9.2 was the famous air travel planning kiosk, which achieved excellent accuracy (maybe better than many humans) partially by being confined to a single application area. Other researchers have similarly created systems that are restricted to understanding medical terminology, as well as some trained on legal terminology (which some would argue is often incomprehensible to most humans as well). Since these niche ASR systems

---

[2] Note that some machine learning systems support *online training* which adapts the system to new information during operation. The more common alternative is to go through a training process and then fix the system parameters during operation, although the choice of training frequency and batch size are completely configurable.

can perform so well, topic recognisers have also been developed – used as front-end analysers which can determine which back-end ASR system should be used to handle a specific sentence based on the topic it refers to. In more general terms, such front-end analysis can be used to switch which ASR vocabulary is currently being used. For example, should the next sentence use a general vocabulary, or should it use a medical vocabulary or perhaps a legal one?

9.1.2.4 Transducer

The process of capturing speech input for a recogniser is extremely important, and can affect not only the signal-to-noise ratio of the input speech, but also other characteristics, such as distortion, which influences the quality of the information being captured. For ASR, the two main physical attributes of the capture process are (i) the placement of the microphone with respect to the sound source (i.e. mouth) and interfering background noise source(s), and (ii) the transfer function between input sound and the electrical signal output from the microphone.[3] The former influences the signal-to-noise ratio by determining how much of the wanted signal (speech) is captured compared with unwanted sound, whereas the latter influences signal distortion. The entire process from source to processing (in this case ASR) is shown diagrammatically in Figure 9.3, including the transducer technology, the analogue connection and amplification of its signal, digitisation and storage prior to processing – which assumes the use of MATLAB or Octave in this diagram.

In general, a transducer that is located in such a position as to make the sound path shorter will perform better by improving the channel. However, note that microphones and loudspeakers have different patterns of sensitivity, ranging from omnidirectional (sensitive in all directions) to unidirectional (sensitive in one direction). Some specialised devices such as shotgun microphones, and those paired with a parabolic dish, are extremely directional, and are designed to capture sound over a distance. Interestingly, loudspeakers can be similarly directional, with perhaps the most directional ones being those used for sonic weapons like LRADs (long range acoustic devices), popularly known as sound cannons.

Use of multiple microphones can allow some sophisticated processing to perform beam steering, which allows some fixed sources to be emphasised at the expense of others. This effectively means electronically steering the captured signal towards the wanted mouth while cancelling other mouths and nearby interfering sources. Some examples of this will be shown later in Section 10.2.4, both for loudspeakers and for microphones. The benefit of multiple microphones was also seen in Figure 9.2 in the three sets of rich transcription results (top right). Directional microphones which can 'steer' themselves towards a speaker to avoid background sounds are better than single microphones.

Even with a well-trained ASR system tuned to a particular voice that can achieve an accuracy level of over 90%, an issue as simple as changing microphone placement,

---

[3] Since we are primarily working in the digital domain within this book, we would often include in the transfer function everything that lies between the sound and the digital signal.

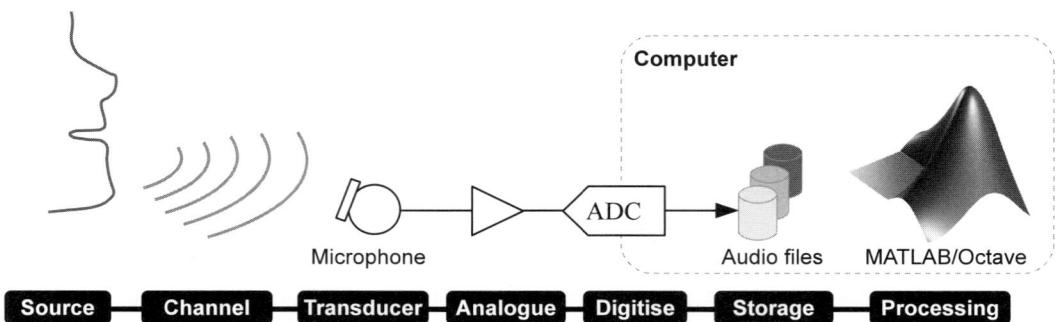

Figure 9.3 A diagram illustrating the various phases in the capture and processing of speech to perform ASR.

using a different type of microphone or speaking in the presence of some light background noise could reduce accuracy by 20% or even more.

The presence or absence of background noise is a critical operational factor (see Figure 9.2 noisy/clean speech results), and recognition systems designed to work with headset-mounted microphones or similar (including electronically steered microphone arrays) will naturally perform better than those capturing speech from a single transducer located far away from the mouth of a speaker. Again, this is seen in Figure 9.2 for the rich transcription results.

Some ASR research systems have made use of non-speech cues to improve recognition performance, including: video images of a speaking mouth, low-frequency ultrasonic echoes of the mouth (we will discuss this in Section 10.8), body gestures, facial expressions, and even nerve impulses in the neck. Each of these requires some kind of transducer to capture the information, and all have different characteristics

9.1.3 Some basic difficulties

Although we have looked at the main parameters related to speech recognition, there are several issues that speech recognition systems in general need to cope with. These may include:

Voice activity detector (VAD), a device able to detect the presence of speech. It would serve no purpose for an ASR system to attempt the computationally intensive task of trying to recognise what is being said when no speech is present, and thus having the ability to accurately detect when speech is present is important. However, this is not a trivial task, and is in fact a research area in its own right. We will discuss this topic more in Section 9.2.

Segmentation of speech into smaller units is often required in processing systems. Whilst this is generally based on fixed-size analysis frames when performing general audio processing (see Section 2.4), in ASR systems, segmentation into words, or even into phonemes, may be required. Again, this is non-trivial, and is not simply a matter of searching for gaps within continuous speech, since the gaps *between* words or sentences may on occasion be shorter than the gaps *within* words.

Word stress can be very important in determining the meaning of a sentence, and, although it is rarely captured in the written word, it is widely used during vocal communications. As an example, note the written sentence 'He said he did not eat this' and consider the variations in meaning represented by stressing different words:

He *said he did not eat this* : indicating that someone else said so
He **said** *he did not eat this* : indicating that you probably disbelieve him
He said **he** *did not eat this* : indicating that someone else ate it
He said he **did** *not eat this* : indicating that he is, or will be eating this
He said he did **not** *eat this* : indicating an emphatic negative
He said he did not **eat** *this* : indicating that he did something else with it
He said he did not eat **this** : indicating that he ate something, but not this

Of course several of the above stresses could be used in combination to create an even wider variety of secondary meanings.

Context was discussed in Section 3.4.4, and may include phoneme context, word context, phrase context, sentence context and beyond. Although word context cannot be judged reliably in all cases (especially in automated systems), it can often be used to strengthen recognition accuracy. Phoneme context – the sequence of two or three neighbouring recognised phonemes – is extremely important to the performance of modern ASR systems.

9.2 Voice activity detection and segmentation

VAD, sometimes known by the unhappy term 'speech activity detection' (SAD), aims to detect the presence of speech in audio. This might include a scenario of identifying when the signal from a hidden microphone contains speech so that a voice recorder can operate (also known as a voice operated switch or VOS). Another example would be a mobile phone using a VAD to decide when one person in a call is speaking so it transmits only when speech is present (by not transmitting frames that do not contain speech, the device might save over 50% of radio bandwidth and operating power during a typical conversation). Turning to speech recognition, we know (and we will definitely see later in this chapter) that ASR is very computationally demanding, requiring significant amounts of computing power and energy. For such applications, a VAD means the difference between the ASR system working hard to understand every audio frame, and the ASR system being triggered and consuming energy only when speech is known to be present.

In these ways, VAD turns out to be extremely important when *implementing* speech systems. In other words, when translating most speech technologies from the laboratory into a product, a VAD is often needed.

When designing a VAD, as with ASR, if the situation is more constrained, better performance is possible. Thus, a VAD operating in known types of background noise, with a pre-determined microphone, detecting continuous speech from a known range of speakers, is much simpler to design (and can perform much better) than a VAD designed to capture any speech – even of short single words – spoken by any person in any type of noise, captured using any microphone.

9.2.1 Methods of voice activity detection

There are, of course, a myriad of ways to design a VAD, with researchers inventing and publishing new techniques and variations on techniques, continuously (a survey of several methods is in [117]). Like most speech-based systems, different techniques have benefits for various operating scenarios, leading to a wide range of trade-offs – in particular between performance and computational complexity. In general, VAD performance (which we will learn a little more about in Section 9.2.2) depends strongly upon the quality of the audio signal, and degrades rapidly with increasing levels of background noise. It also depends upon the type of background noise because, even for humans, some kinds of noise can mask speech much better than other kinds. In particular, some noise is inherently 'speech-like', meaning it has similar characteristics to speech.

The earliest and simplest VAD techniques were energy detectors. These were predominantly analogue systems, but can easily be replicated in MATLAB. In operation, input speech is divided into analysis frames, and an energy score obtained for each frame. The energy of each frame is then compared with a threshold. If it exceeds the threshold, then speech is judged to be present. This simple technique works well when there is little or no background noise, and the noise is not speech-like. It is illustrated in Figure 9.4, which actually plots a section of a speech by Winston Churchill corrupted by (recently recorded) office background noise at four different levels from 20 dB to −10 dB SNR. In each case, the energy of the speech plus noise mixture is analysed in 0.1 s non-overlapping frames. A fixed threshold of 10% of maximum energy is used: frames whose energy exceeds this level are judged to contain speech, those that do not are judged to contain non-speech. The VAD decisions are shown on the lower energy plots using small circles (a circle at the top of the energy plot indicates that the frame below it is judged to contain speech). Consider first the 20 dB SNR plot (top left). Churchill is speaking for about 13 seconds in these clips ('I feared it would be my hard lot to announce the greatest military disaster in our long history, I thought . . .'), and the quality of the June 1940 recording leaves much to be desired. However the VAD works very well in this case: whenever the waveform shows the obvious presence of speech, the VAD has detected it.

Given that the same speech and the same noise is used in each case (the only difference is that the noise amplitude is progressively increased from top left to bottom right), there should be the same amount of speech detected in each test shown in Figure 9.4 if the VAD method works correctly. While the 10 dB SNR plot (top right) seems to show the VAD working fairly well, the 0 dB SNR condition (bottom left) shows noise being mis-detected as speech at around the 11 s mark. However, the −10 dB SNR plot (bottom right) reveals the opposite problem: most of the speech is ignored.

9.2 Voice activity detection and segmentation

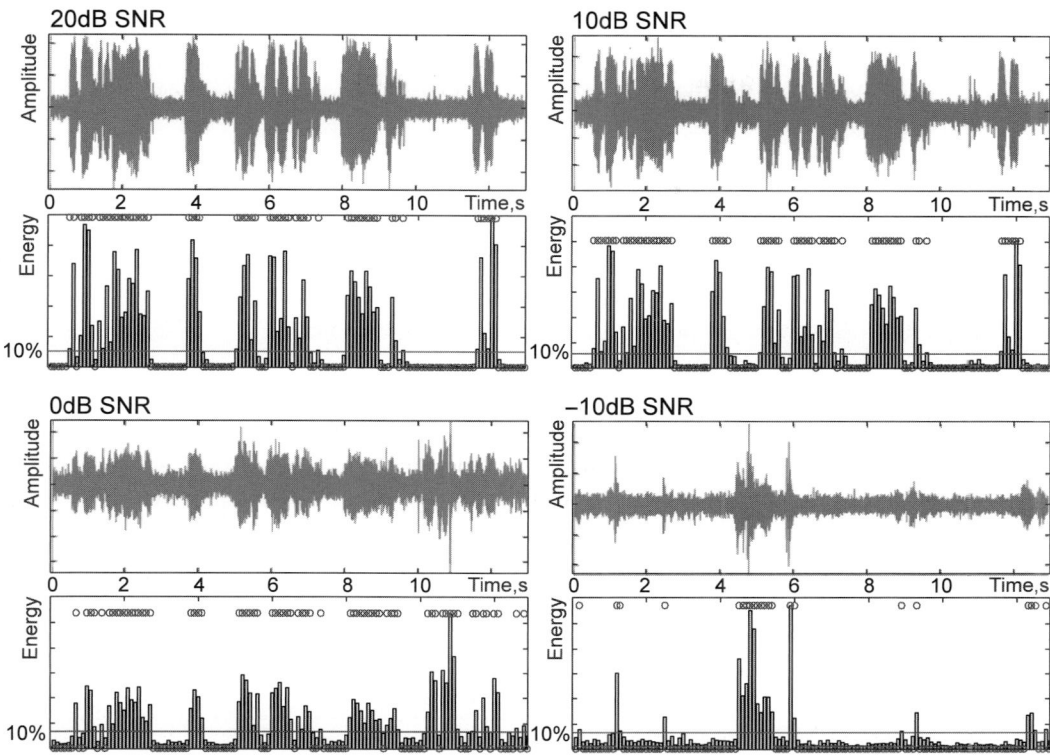

Figure 9.4 Illustration of using an energy detector to perform VAD, using a 10% level on a 13 s sample of broadcast speech mixed with office noise at levels of 20, 10, 0 and −10 dB SNR. In each case the waveform is plotted above bars representing the frame-by-frame energy. A horizontal line indicates the 10% energy threshold, and a simple frame-wise VAD decision is made by selecting frames with energy above the threshold, plotted as circles.

Code to perform this VAD in MATLAB is as follows:

```
%the noisy speech is in array nspeech
%fs is the sample rate
L=length(nspeech);
frame=0.1; %frame size in seconds
Ws=floor(fs*frame); %length
Nf=floor(L/Ws); %no. of frames
energy=[];
%plot the noisy speech waveform
subplot(2,1,1)
plot([0:L-1]/fs,nspeech);axis tight
xlabel('Time,s'); ylabel('Amplitude');
%divide into frames, get energy
```

```
for n=1:Nf
  seg=nspeech(1+(n-1)*Ws:n*Ws);
  energy(n)=sum(seg.^2);
end
%plot the energy
subplot(2,1,2)
bar([1:Nf]*frame,energy,'y');
A=axis; A(2)=(Nf-1)*frame; axis(A)
xlabel('Time,s'); ylabel('Energy');
%find the maximum energy, and threshold
emax=max(energy);
emin=min(energy);
e10=emin+0.1*(emax-emin);
%draw the threshold on the graph
line([0 Nf-1]*frame,[e10 e10])
%plot the decision (frames > 10%)
hold on
plot([1:Nf]*frame,(energy>e10)*(emax),'ro')
hold off
```

Evidently, this simple VAD technique is easily defeated by noise. It may work well in a setting such as a broadcast studio, but in a home, vehicle or office it would constantly mis-detect speech, either being triggered when there was no speech (called a false positive) or not being triggered when it should have been (called a false negative).

Another significant problem is that it interprets *any* sound that occurs as potentially being speech, with the only criterion being the energy of the sound. Instead, it should try to detect particular types of sound as being speech-like and reject other types of sound which are not speech-like, or which are noise-like.

To make such decisions, the detection measure should really relate to known features of speech, which could be from either the time domain or the frequency domain. We will look briefly at both, starting with the frequency domain, because, in practice, we often assess the time-domain characteristics of frequency-domain measures before making a VAD decision.

9.2.1.1 Frequency-domain features

From Chapter 3 we know that speech is comprised of a sequence of voiced (V) and unvoiced (UV) phonemes, and that these two classes have different characteristics due to their production mechanisms, which could potentially be detected.

V phonemes contain clear formants which can be detected in a short-time spectrum as a peak. For most speakers, the first three formants lie in well-known ranges. Thus a spectral peak detector might identify phonemes in the correct ranges and infer that speech is present. However, many sounds in nature, including musical instruments and animal calls, contain spectral peaks that can have the same approximate harmonic relationship as formants do in speech. Consider UV phonemes then: some, such as fricative /s/, contain predominantly wideband high-frequency information. We could detect this in the frequency domain by comparing the energy in different spectral regions. However, energy in the high-frequency region does not only indicate an /s/; it might be the sound of gas escaping from a pipe instead, or wind rustling through the leaves on a tree. It turns out that many types of noise are characterised by wideband noise with energy in the high-frequency region.

Pitch is important in speech too. We know that $f0$ lies within a narrow range for most speakers, and is present in V phonemes. It also varies more slowly in frequency than formants do. So we could detect spectral energy in the range of pitch, or detect spectral peaks due to pitch. However, this would not work in practice, because many sounds and background noises have pitch-like characteristics, especially rotating machinery (like a lot of noise-making components in a vehicle).

It might seem that this subsection is not very useful – after all we have considered several characteristics of speech, and found that none of them is discriminative enough in practice to use for VAD. Each of them is susceptible to some common types of noise. However, a *combination* of these techniques is a better proposition. While one detector can be mis-triggered by a particular type of noise, it is unlikely that a second technique is also triggered. It is even less likely that three techniques are triggered simultaneously – although it does happen, which is why we need to combine even more information. In particular, combining the frequency-domain measures with the time domain can yield the best results.

9.2.1.2 Time-domain features

Energy alone, as we have seen, is not a good enough VAD metric. Part of that is because speech energy is so variable: V speech tends to have high energy, UV speech tends to have low energy, so how can an energy criterion be reliable? The answer is that we must factor in exactly that kind of variability. Just having a few frames of high energy does not mean that speech is present, but a specific speech-like pattern of high-, medium- and low-energy frames could tell us more.

We already know from Chapter 3 that speech is spoken with certain patterns. Some patterns are limited by our equipment (i.e. lungs, glottis, throat, lips, the need to breathe and so on), others are constrained by language, content and our own particular way of speaking. There are defined patterns in speech which include syllabic rate, phoneme rate, V/UV switching (in most languages), patterns of silence and changes in amplitude. Some of the patterns can be picked up by analysing how the energy varies, but even better is to look at how the frequency-domain features that we discussed above change over time.

This now becomes a matter of statistics: first creating a model of the variation in these parameters that is found in normal speech and then obtaining the statistics of some audio being tested, comparing these observed statistics with the model for speech to form a hypothesis that speech is present (or not). Depending on how much the statistics differ between the model and the observed signal, a probability value can be assigned to each feature. Then the feature probabilities are summed to obtain an overall probability that speech is present. Our statistics should at least include the mean, but also the first-order measure of the spread of frames around the mean (the variance, or standard deviation), and probably the second-order measure too, or the spread of observed variances around the overall average variance.

Remember that different features vary at different rates – so some features (like syllabic rate) may need to be examined over a few seconds to determine their mean, while others may need to be examined within just a few frames (e.g. formant peaks). Most features themselves are computed over a single frame, whereas some such as pitch are generally computed over multiple analysis frames, but can be overlapped to yield a single output per frame. Having obtained various feature measures for a sequence of frames, each of those features is then examined over a longer time duration to obtain their statistics. If you stop and think about this process, it means that some measures take a significant duration to compute accurately, and therefore may not pick up shorter utterances such as single isolated words.

9.2.2 VAD performance

Normally, VADs include something called a *hang time* (or *hang over*) and occasionally a *hang before* which extends the detection period after and before the detected speech period respectively. This is because many words end with a low-energy phoneme, such as a fricative, which is difficult to detect. Early VAD algorithms would frequently cut off the last phoneme within their detection region (or miss the initial phoneme in words that begin with a quiet unvoiced sound), and hence the idea of widening the decision region, as illustrated in Figure 9.5. A hang time of around 0.5 s to 1.0 s would ensure

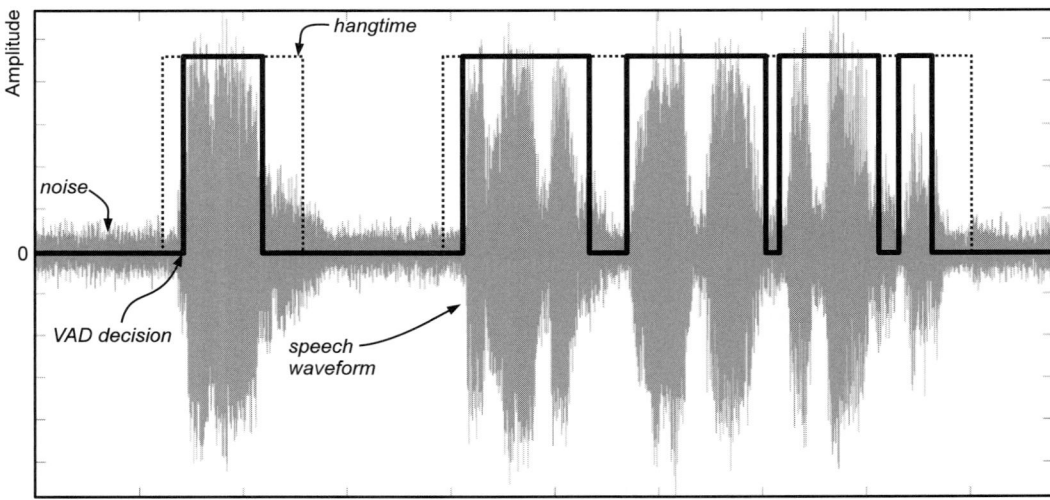

Figure 9.5 An illustration of VAD output (solid line) corresponding to the waveform of a speech recording. Often the VAD decision is extended in time to include a section of audio immediately before and immediately after a detected section, to ensure that the start and end portions of the utterance are captured.

that these possible quiet phonemes are not lost, at the small expense of often including a short section of non-speech at the beginning and end of the detection region.

The subject of this section, how to assess VAD performance, is not as easy as it might at first appear. A simplistic approach might be to count the proportion of times the VAD triggers during regions in which speech is present – but this would not account for when the VAD completely misses a segment of speech or when it triggers within a speech period and misses the beginning and end. An equally simplistic approach would be to count the number of speech utterances that are detected by the VAD, but again that does not account for false positives – when the VAD is triggered by noise – and neither does it take any account of when part of an utterance is missed.

An additional difficulty is that natural speech contains pauses between sentences, paragraphs and words, and sometimes within words (consider a long glottal stop). Are these gaps considered to be part of the speech? The solution to this is to slice the speech into frames of a convenient length for speech analysis, which one hopes are longer than most of the pauses in continuous speech.

So-called *ground truth*[4] indicates whether each frame contains speech or not. A good VAD system should classify each frame to match the ground truth. Note that the best ground truth in many cases is that obtained from experienced human listeners

Now, given a frame-by-frame analysis, we can start to identify VAD performance in terms of the types of errors, rather than just the number of errors. At least two scores are required to truly characterise the VAD performance. The same basic information can be presented in several ways, for example (i) the percentage of analysis frames correctly detected as being speech and (ii) the percentage of speech frames that are incorrectly identified as containing noise.

In fact, we can often understand VAD performance even better by looking deeper at the types of errors. Thankfully, we can use some standard performance criteria developed by Beritelli *et al.* [118] to analyse and report results:

- *Front-end clipping* (FEC), the percentage of erroneous frames when transitioning from a non-speech region to a speech region.
- *Mid-speech clipping* (MSC), the proportion of speech frames erroneously classified as being non-speech.
- *OVER* (or overrun), the percentage of erroneous frames when transitioning from a speech region to a non-speech region (i.e. the end-of-speech equivalent of the FEC metric).
- *Noise detected as speech* (NDS), the proportion of non-speech frames which are misclassified as being speech.

In operation, these are mutually exclusive counters that are incremented for each erroneous frame, and the sum over all frames yields the total number of errors. These scores

---

[4] This term comes from the remote sensing community, who devise algorithms to analyse images of land areas obtained from radar in space and from other distant sensors. The task is to determine the nature of regions in the image. The ground truth, usually obtained by someone travelling to that region and inspecting it, is the knowledge of what is actually there in reality. The aim is to develop algorithms that output detection results that match the ground truth.

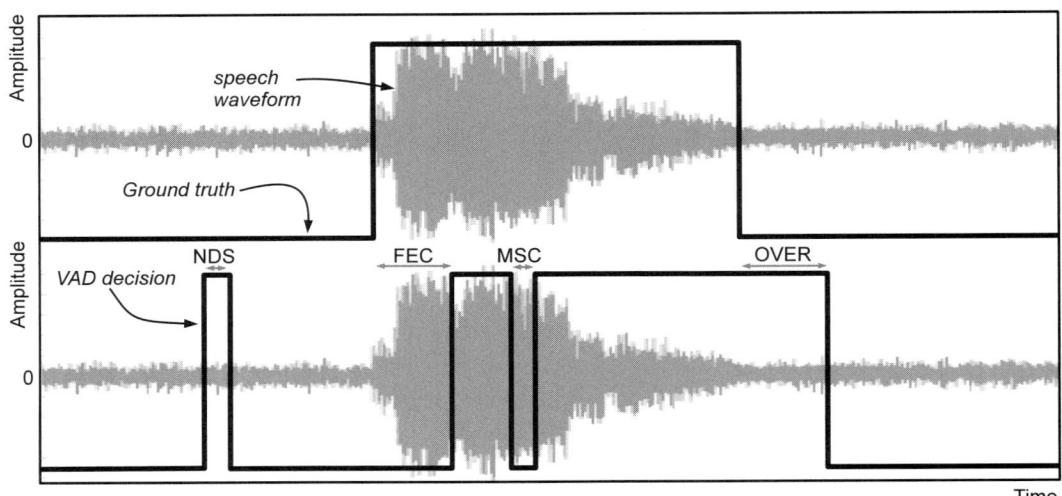

Figure 9.6 The lower plot illustrates four types of error used to score VAD performance (NDS, FEC, MSC, OVER). Ground truth is shown on the upper plot.

are each divided by the total number of analysis frames and reported as proportions. Figure 9.6 illustrates each of these error types.

9.3 Current speech recognition research

The state-of-the-art in ASR research is quite fast moving, with new techniques and approaches being introduced continually by researchers worldwide. While we cannot hope to present a comprehensive overview of this deep and dynamic topic, we can at least consider some of the mainstay machine learning-based techniques.

9.3.1 ASR training and testing

As we discussed in the last chapter (see Figure 8.2), machine learning systems have several phases of operation. First is the training phase, which uses a data set specifically for training a model within the heart of the system, there may then be a refinement phase, and a testing phase, both of which operate on a separate data set. For ASR specifically, we can see in Figure 9.7 how features extracted from training data are used to train a model of speech. For example, in a system with phoneme outputs, the model will examine the features (e.g. MFCCs) from a segment of speech and predict which phoneme is being spoken. Ground truth information accompanying the training data – in this case a list of phonemes along with start and stop (or duration) information – is compared with the output of the model. Training means that the system internal weights are then either reinforced or adjusted based upon whether the phoneme identification is correct or incorrect. While testing, another set of data is used to feed features into

9.3 Current speech recognition research

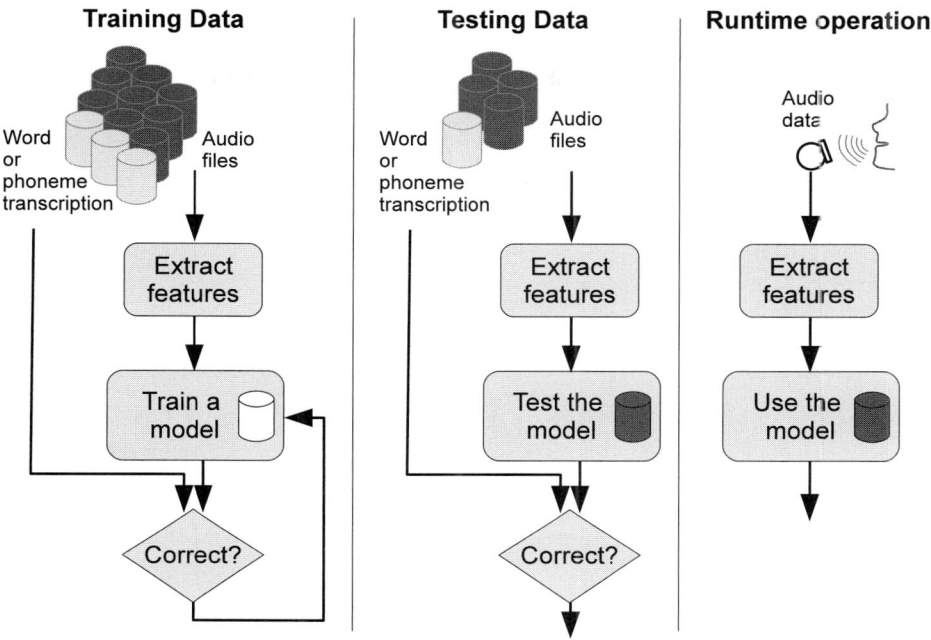

Figure 9.7 ASR phases.

the trained model. The output is compared with the ground truth to determine the error rate. During runtime operation, speech waveforms are input to the system for the model to predict which phoneme is being spoken. There is no ground truth, and hence no comparison or model adjustment in this phase of operation.

Systems can be built that output words (and are thus trained with individual words), phonemes (and thus are trained with phonemes) or smaller items, but whatever level of operation is chosen, the ground truth needs to match the information that is being trained – or rather the system will gradually be trained to resemble the ground truth as much as possible. Some of the highest performance is available to systems that use sub-phonetic modelling. Something called *senones* [119], or tied phoneme states, allows multiple states to represent each combination of phonemes (e.g. start, middle and end) and performs well in practice.

9.3.2 Universal background model

Gaussian mixture models (GMMs, see Section 8.4.6) are heavily used in modern ASR systems, particularly in being used to form something called a universal background model (UBM), i.e. into a GMM-UBM. This approach models each segment of speech though two components. In effect, the first part of the model includes all speech and channel effects and can be trained using a huge amount of data from many different speakers. This can be done just once, and the trained GMM-UBM then used in different speech-based systems. The second part is speaker- and channel-specific. This uses

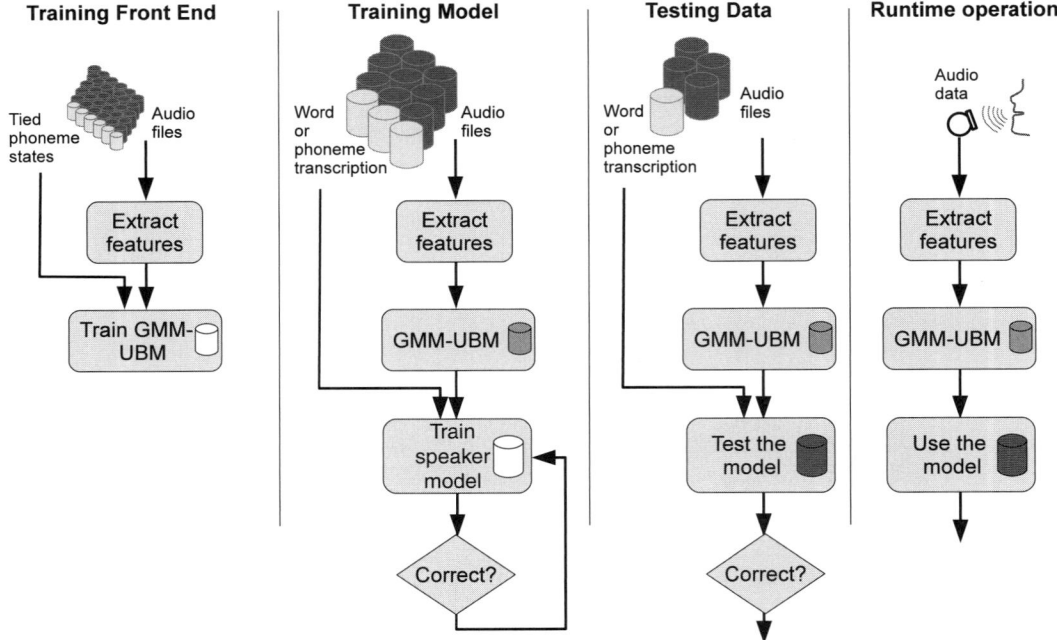

Figure 9.8 Many systems require four or more phases – where different parts of the system need to be trained using different data.

training data that one hopes is very similar to the testing data or operating scenario, to create a model that is much more specific than the background model. While the UBM is 'train-once and then use many times', the second part is trained specifically for the current user and operating scenario (e.g. microphone type, background noise, unique speaker characteristics). It is effectively applying to the UBM an adaptation towards the current speaker. Two-pass training is illustrated in Figure 9.8. Two, three or even four training passes can be found in the research literature in the search to achieve better performance.

The UBM is trained using different data from that used for the speaker model. This means that there needs to be a lot of multiple-speaker training data for the UBM (which is relatively easy to obtain), but little training is needed for the target speaker. Now we know that, to some extent, performance is dependent upon the amount of training data. Thus the UBM method means, for example, that one particular user can have an ASR system that performs as well as if he or she had spoken into it for hundreds of hours of training, but this is achieved with only a small amount of training data.

9.3.3 Features

We have already discussed a large range of input features in Section 8.3.2, with MFCC-based features being by far the most common, almost always augmented by their delta and delta-delta (i.e. the frame-by-frame change and frame-by-frame rate of change).

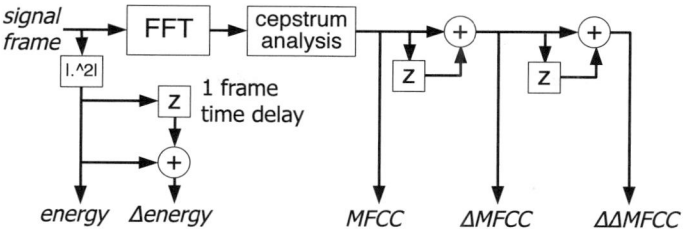

Figure 9.9 A block diagram showing the feature extraction process for energy and MFCC features, and their delays, prior to ASR input.

Their generation process is shown in Figure 9.9, along with energy and delta-energy. In practice there would probably be 13 MFCC coefficients used, yielding a total feature vector size of $\{13 + 13 + 13\} + \{1 + 1\} = 41$ per frame. Mel filterbank features may be used instead, or spectrogram (and spectral envelope) features, along with pitch. Each of these yields a different feature vector size and has different performance characteristics.

ASR systems often have highly overlapped and short analysis frames, perhaps 10 ms long, overlapped by 8 ms per frame. We can see that each frame then yields one feature vector. Since the frames are short, not much time-domain information is available there for recognition. Hence researchers normally add context to the feature vector. This means concatenating the previous few and subsequent few frames together. If the current frame index in time is i, and this is represented by feature vector $\mathbf{v}^i$ of length L and containing elements $v_0^i, v_1^i, v_2^i, \ldots, v_L^i$, then the feature vector with context c is:

$$\left[\mathbf{v}^{i-c}, \mathbf{v}^{i-c+1}, \ldots, \mathbf{v}^{i-1}, \mathbf{v}^i, \mathbf{v}^{i+1}, \ldots, \mathbf{v}^{i+c-1}, \mathbf{v}^{i+c} \right]. \tag{9.1}$$

So the feature vector length now becomes $(2c+1)L$, which is starting to become large and unwieldy, probably resulting in quite a slow ASR system. Hence researchers use a number of methods to reduce the size of the feature vector, such as through principal component analysis (PCA – see Section 8.4.4).

A linked question is the mismatch between the content of the underlying speech and the feature vector's context. For a phoneme state recogniser, one phoneme in the speech might span tens (or even hundreds) of analysis frames. A word recogniser is even worse, possibly spanning thousands of analysis frames. This is where the i-vector approach comes in, with 'i' standing for 'intermediate' (in size) vector, because the vector is longer than the MFCC parameters, but shorter than the output from a GMM-UBM (called a super vector).

The important feature of an i-vector is that it is one common-sized feature vector that is used to represent a part of speech (i.e. a word, phoneme, senone or other unit), irrespective of how many analysis frames spanned across the feature in the original waveform. Thus an i-vector from a slowly spoken phoneme is constructed from more analysis frames than an i-vector which is formed from a quickly spoken phoneme. However, the important point is that there is a single i-vector corresponding to a single speech state in each case. This makes classification much more convenient.

The same could also be said for other input representation methods, including the use of vector quantisation, which we met in a different context back in Section 6.2.6. When used to constrain feature vectors, it is effectively a system that maintains a codebook of allowed feature vectors. Whatever potentially 'unruly' set of input data is received is quantised (or mapped) into one (or possibly more) codebook entries. Normally the codebook entries are learned during a training phase, and the mapping or quantisation process uses some kind of distance measure to ensure that the input vector is mapped to the 'nearest' codebook vector. The decision as to which is nearest could be the Euclidean distance (minimising the sum of the squared difference between each vector element), or might employ some other criterion, such as a perceptual weighting.

9.3.4 Variations on a theme

There is a wide variety of research approaches to forming the feature vectors in ASR systems, using almost every conceivable type of information representation, as well as a vast array of analysis methods. We will just examine one variation more closely here as an example of a *transfer learning* system.

Transfer learning is the idea of using very good training data to create a very well-trained machine learning system, and then transferring this to another nearby domain. This is extremely important because there are some application domains that do not have sufficient training data, whereas other application domains have vast amounts of good quality training data. Transfer learning allows the weaker domain to share training data with the stronger domain.

One of the best examples where transfer learning is useful is in building ASR systems for minority languages. If we take a less widely spoken language such as Cornish, it is highly unlikely that there will be large amounts of training data available to develop a working Cornish language recogniser. However, the Cornish language shares many phonemes with English, and thus it might be useful to create a phoneme recogniser for English – which can be trained very well using the vast array of training materials in English – and then adapt this for use with the Cornish language. This might work perfectly if all of the Cornish phonemes are found in English, but could clearly become confused by any phonemes that are important in Cornish but are not present in English. Hence the English-trained phoneme recogniser probably cannot be used directly with the Cornish language.

Instead, we use a technique like the deep bottleneck network (DBN) [120, 121] which is illustrated in Figure 9.10. The DBN acts as a feature extractor, and is well trained using a large-scale speech dataset to extract phonetic features. This is shown in the top part of Figure 9.10: MFCCs or other kinds of standard features with context are used to train a multi-layer deep neural network (DNN) to output senones. Typically the input dimension would be something like 43×15 in size, and the output dimension around 6000. However, an internal constriction called the bottleneck layer forces all of the discriminative phonetic information to be squeezed through a much lower-dimensionality vector of dimension around 50. Given a good quality, well-labelled and large-scale speech database, the DNN can be trained with very high accuracy, i.e. it

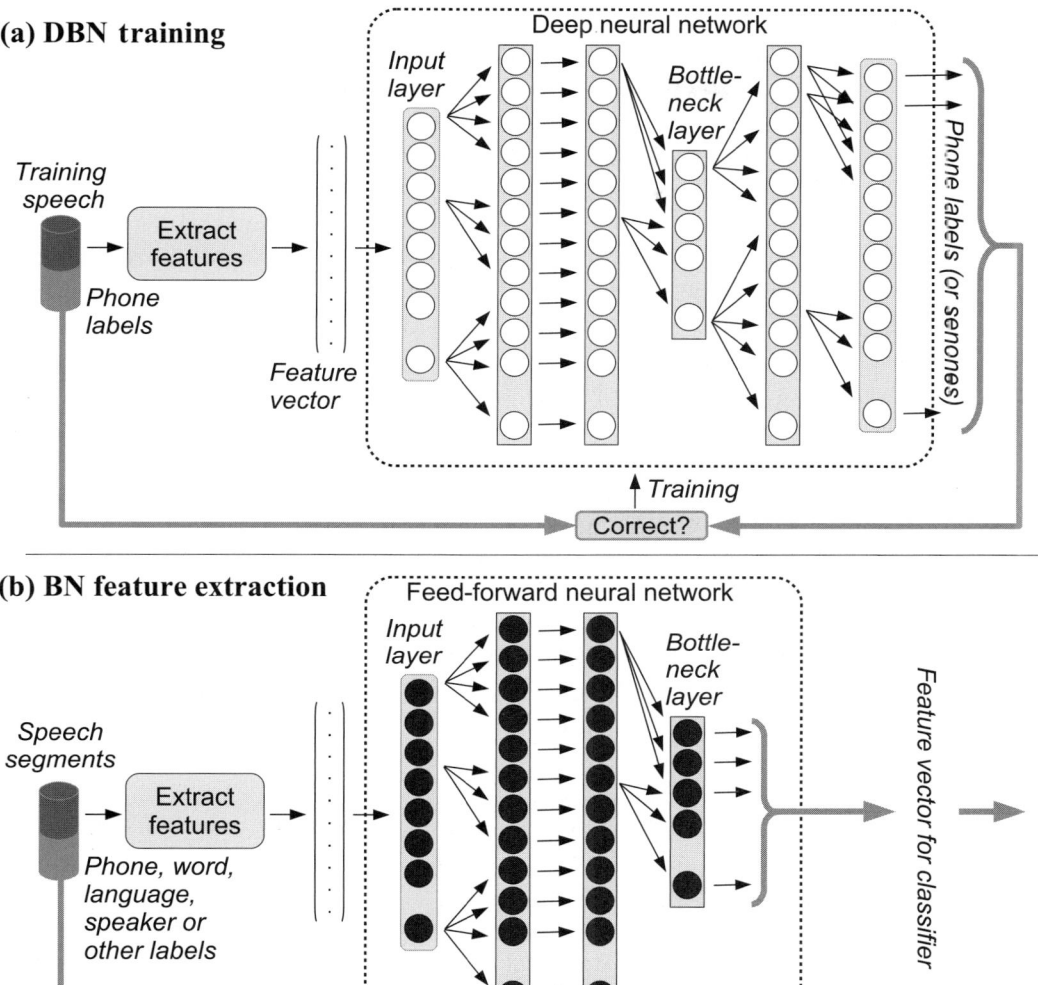

Figure 9.10 The initial training phase (above) and operating as a feature extractor (below) for a deep bottleneck network front end.

can very reliably predict the phonetic state of input speech features. The layer-by-layer nature of the DNN means that all of the classification information for a given analysis frame is being squeezed through the bottleneck layer; that single low-dimensionality layer vector thus forms a compact and discriminative representation of the phonetic information in the much larger input feature vector. However, the bottleneck layer is clearly *not* a phonetic representation, and tends to incorporate much of the acoustic information that is present in the input domain (in addition to the phonetic information found in the eventual output domain).

Once the DNN has been well trained, the layers after the bottleneck are removed, and the remainder of the trained network is used – without changing the internal weights – as a front-end feature extractor. For our example scenario, having been well-trained on English speech, it would then be applied to Cornish speech. Frames of Cornish (represented by MFCCs, moments and context) would enter the DBN. The output vector from the bottleneck layer would then be used instead of the MFCCs as the input to a Cornish ASR system.

Since there is not enough labelled Cornish speech to train an ASR system directly, we are thus able to leverage the materials available for English to create a Cornish ASR front end. What little Cornish information can be found may then be used to train the ASR back end, which will be significantly easier given the much more discriminative input features that are now available thanks to the DBN.

In fact this approach has not been used to create a Cornish language recogniser, to the author's knowledge, but has been used very successfully for training speaker and language recognition systems. These achieve excellent performance by using good quality English and Chinese ASR training materials. In fact, at the time of writing, the best performance for recognising highly confusable language pairs like Farsi/Dari, Ukrainian/Russian and Urdu/Hindi is achieved by a DBN system trained using over 500 hours of phonetically labelled conversational Mandarin speech [121, 122], which certainly has little in common with those languages.

The DBN is not the only transfer learning system. There are several very new methods of adapting DNNs in particular speaker adaptations, such as the speaker code system [123] (which has even been combined with DBN in [124], yielding excellent results).

9.4   Hidden Markov models

A Markov model is a kind of state machine that steps through a sequence of states, each of which might cause some kind of output. Every time unit provides an opportunity to change state and produce an output. When in a particular state, the model does not 'remember' where it came from (i.e. it is memoryless), and the choice of which state to jump to next is described by a probability. So a particular state i might be able to jump to several states next (and one of these jumps might be back to itself). The sum of all *transition* probabilities is 1.0, but usually some transitions are much more likely than others. The hidden Markov model (HMM) is *hidden* because the external observer (i.e. us, looking at it) does not really know what state it is in; all we know is the output of the model, called its *emission*. Each and every state is able to emit something, and again each possible emission (output) is described by a probability.

Given a particular model, we can look at a sequence of outputs and try to guess what sequence of states was responsible for them. We (usually) cannot be certain about the state sequence, but can only know it to a certain likelihood. Or, if we have a particular sequence of observations (outputs) and a set of models, we can decide which model was the most likely producer of that sequence.

9.4 Hidden Markov models

Maybe all of this sounds a little strange, so we shall look at an illustrative example. First, however, we should discuss why this might be useful, and relate our knowledge to how it is used in ASR (and other speech domains).

In ASR, HMMs are used to model phonemes, syllables, words or even higher units of speech. Sometimes, three different sets of HMMs will be used to model different parts of speech in one ASR system: one set of HMMs for tri-phone states, one set for allowed words and one set for higher-level grammar. These respectively implement acoustic models, lexical models and syntactic models. Since this type of system becomes complex to discuss, we will instead consider an isolated word recogniser.

For isolated word recognition, a separate HMM can be constructed for every word that is included in the training data set (and it is hoped that this will cover every word in the actual test or operating data). So if there are ten words in the dictionary we can create ten HMMs, one for each word. A sequence of feature vectors extracted from speech, just like those shown in Figure 9.9, is the observation sequence, and the recognition task is to decide which HMM was most likely to have produced this observation. Since there is one HMM for each word, deciding which is the most likely HMM is akin to deciding which is the most likely word. In practice, this could work reasonably well for a single speaker and a small vocabulary (assuming a reasonably sized training set); however, a separate HMM is needed not only for each word, but also for each pronunciation of each word (so we would not do it precisely this way in a modern ASR system – which we will discuss later).

9.4.1 HMM example

Rather than thinking about speech, let us assume that there are three types of weather, which we will call sunny, rainy and overcast, and each type of weather lasts for an entire day. These are thus three states, S, R and C, and we will assume that the weather can only be in one of those states each day, but can change from day to day, and it is probably more likely that we would experience a change from C to R rather than going directly from S to R. Furthermore, we know that the chance of strong winds is somehow related to the states – less likely in the S state, more likely in the C and R states.

We can make a state diagram of this, shown in Figure 9.11, where the probabilities of transitions between states (based on a long history of weather observations) are shown. This type of example is called a first-order ergodic Markov model.

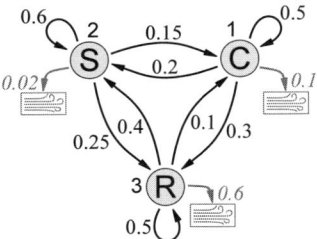

Figure 9.11 A fully trained HMM model (since the transition and output probabilities are given) for three states of sunny (S), cloudy (C) and rainy (R), with each state having different probabilities of generating high winds.

It is easy to read off the probabilities from this example: if today is sunny (S), the probability of tomorrow being sunny also is 0.6, whereas the probability of rain (R) is 0.25. These add up to 0.85, but tomorrow must be something, so the transition probabilities should add up to 1.0. Hence the remaining 0.15 gives the probability that tomorrow is cloudy (C).

If today is Friday and it happens to be cloudy, and we are planning a trip to the beach, we could easily compute the probability that the coming three days will be sunny (i.e. states C–S–S–S). This is simply the product of the following probabilities, where day 0 is today:

- P(day 1 is S | day 0 is C)
- P(day 2 is S | day 1 is S ∩ day 0 is C)
- P(day 3 is S | day 2 is S ∩ day 1 is S, day 0 is C)

where we read P(X | Y) as being 'the probability of X given (or assuming) Y', and we multiply them because we want to know when *all of them* happen (we would add them if we were deciding whether just one of them happens). When we feed in the probabilities from Figure 9.11, this equates to $0.2 \times 0.6 \times 0.6 = 0.0432$, or a 4% chance: I would probably pack an umbrella!

However, the HMM goes one step further because it can also encode output probabilities. For example, it can model the probability in each state that the wind is strong enough to blow off my hat. If we assume that the probability of strong wind in states C, S and R is 0.1, 0.02 and 0.6 respectively, then it is easy to compute the chance of my hat being blown off today or tomorrow:

- P(wind on day 0 | day 0 is C)
- P(wind on day 1 | day 1 is C ∩ day 0 is C)
- P(wind on day 1 | day 1 is S ∩ day 0 is C)
- P(wind on day 1 | day 1 is R ∩ day 0 is C)

where we read (X ∩ Y) as being 'X and Y'. The answer is now the sum of all terms (because once is enough), so this is $0.1 + 0.5 \times 0.1 + 0.2 \times 0.02 + 0.3 \times 0.6 = 0.334$, meaning that it should be fairly safe to wear my hat.

I hope this is all well and good so far; it is simple probability. Let us take another step to illustrate the power of the HMM at work. Imagine a scenario where I telephone my friend in New Zealand who knows nothing about the weather where I am currently located, except that the average probability of the weather for states C, S and R is 0.7, 0.1 and 0.2 (unlike in New Zealand, where the S state probability would obviously be a lot higher) and he has a copy of my HMM. I tell him that the only time my hat blew off during the past week was two days ago. On the basis of this information, do you think he is able to advise me whether today is likely to be rainy? To answer this, we will look more at HMMs, and revisit the question later in Section 9.4.3.

9.4.2 How HMMs work

There are a number of tutorial guides on HMMs for ASR, not least in the famous HTK book [125] as well as an excellent (but dated) tutorial by Rabiner [126], but we shall instead describe general HMMs here without proof, followed by a separate discussion of their application to ASR in Section 9.4.4.

A single HMM, $\mathcal{H}$, is described by five items of information (S, V, π, A, B), which together define:

- N separate states denoted by $S = \{s_1, \ldots, s_N\}$.
- M possible output symbols (called a vocabulary), $V = \{v_1, \ldots, v_M\}$, each of which may be emitted by the HMM.
- N initial state probabilities π, $\{\pi_i, \ldots, \pi_N\}$, each bounded in the range $[0, 1]$, summing to 1 (to ensure that the model does actually start in some state, i.e. $\sum_{i=1}^{N} \pi_i = 1$).
- A transition probability matrix, A, comprising elements $a_{ij}, i \in S, j \in S$, which defines the probability of a transition from state j to state i in the next cycle, again bounded in the range $[0, 1]$. Since there *must* be a transition each cycle, $\sum_i a_{ij} = 1$ for each current state j.
- An emission probability matrix, B, comprising elements $b_{ij}, i \in V, j \in S$, which defines the probability of emitting symbol i when in state j.

In general, S and V are defined structurally, based on the understanding of the problem and the content of the data, whereas A, B and π need to be learned from data. During operation, the HMM transitions from one state S to the next each cycle, simultaneously emitting an output symbol from V. Unlike the kinds of state machines taught in digital logic, both the transitions and the outputs are defined by probability distributions: this system does not progress through a fixed sequence of states, instead the actual sequence it follows is a likelihood.

9.4.2.1 Use of a trained HMM

Trained HMMs can be used for tasks such as classifying or decoding. In both cases the task is to explain an observation, $X = <x_1 \ldots x_T>$, which is a length T sequence of output symbols (from set V). This sequence is something that we have measured or detected, and are now using an HMM to try and understand. In effect, this is making a hypothesis that the system which was responsible for sequence X can be modelled by HMM $\mathcal{H}$. Classification means having several trained models $\mathcal{H}_1, \ldots, \mathcal{H}_K$, from which we need to find a model, $\mathcal{H}_k$, which has the highest probability of being able to generate the sequence X. Mathematically speaking, the probability of observation sequence X being generated by model $\mathcal{H}$ is:

$$P(<x_1 \ldots x_T> | \mathcal{H}), \tag{9.2}$$

and we wish to find the maximum likelihood, which is the highest probability that sequence X came from one of the K models, by examining all of them:

$$\text{class}(k) = \underset{K}{\text{argmax}}\, P(X | \mathcal{H}_k). \tag{9.3}$$

Decoding, by contrast, uses a single trained HMM, $\mathcal{H}$, to attempt to find the highest probability state sequence $\hat{S} = <\hat{s}_1, \hat{s}_2 \ldots \hat{s}_T>$ that could have caused output sequence X.

We will now look separately at each of those tasks before we progress on to how the HMM is trained: i.e. the following subsections assume that K HMMs $\mathcal{H}_k$ have been previously trained, and are therefore presumably good descriptions of the system that caused the observation X.

9.4.2.2 Classification

Working from the perspective of the observation, X (since the internal states S are hidden), it is necessary to scan through the sequence of observed symbols $X(1)$, $X(2)$ and so on, to compute the probability that they came from the current model being examined. In fact there are only two relationships for each symbol, because the states are memoryless. These are the relationship to the previous symbol and the relationship to the next symbol. Assessing the probabilities of these two relationships leads to separate scanning processes in the forwards and backwards directions.

Working in the forwards direction, $s(t)$ is the state at time t, and the observation sequence up to this state has been $< X(1), X(2), \ldots, X(t) >$. The probability of being in state i at time t having observed this sequence is:

$$\alpha(t, i) = P\{X(1, \ldots, t) \cap s(t) = i \mid \mathcal{H}\}, \tag{9.4}$$

and this can be computed by starting at the initial state, $P(s(1) = i)$, which is the definition of π_i, and $P(X(1) \mid s(1) = i)$, which is the definition of B, so that $\alpha(1, i) = b[X(1), i].\pi_i$.

Moving to step 2 and beyond, we simply need to find the probability of the current state, assuming that the previous state is known (and of course that the observations are known), i.e. $P\{X(1, \ldots, t+1) \cap s(t+1) = i\}$, which is equivalent to $P\{X(1, \ldots, t) \cap s(t+1) = i \cap X(t+1)\}$. From this we sum up to the end of the state sequence:

$$\alpha(t+1, i) = b[X(t+1), i]. \left\{ \sum_{j=1}^{N} a_{ij}.\alpha(t, j) \right\}. \tag{9.5}$$

Remember that we have multiple models, $\mathcal{H}_k$, so we need to compute which model is most probable, by summing over all state sequences in each model:

$$P(X(1, \ldots, T) \mid \mathcal{H}) = \sum_{j=1}^{N} \alpha(T, j). \tag{9.6}$$

In practice, not all state transitions are allowed (perhaps they are restricted by grammar rules in a particular application scenario), and thus we would only need to sum over allowed transitions (i.e. those for which $a_{ij} \neq 0$). This would be accomplished by computing the set of $\alpha(t, j)$ for each model, using Equation (9.6), and then finding the maximum to derive the most likely model $\hat{H}_k$:

$$\hat{H}_k = \underset{k}{\mathrm{argmax}}\, P(X \mid \mathcal{H}_k). \tag{9.7}$$

In fact, we could obtain an equivalent probability computation by working in the backwards direction, by defining $\beta(t, i)$ as the probability of observing symbols

$< x_{t+1}, x_{t+2}, \ldots, x_T >$ from state $s(t) = i$. In other words, this is $P\{X(t+1, \ldots, T) \mid \mathcal{H} \cap s(t) = i\}$.

Again, the boundary state, $t = T$, is simple since $\beta(T, i) = 1$. The iterative case again relies upon the fact that the next state is known, but the previous state probability based on this needs to be found. Hence,

$$\beta(t-1, j) = \sum_{i=1}^{N} b[X(t), i].a_{ij}.\beta(t, i), \quad (9.8)$$

and as with the forward case, disallowed transitions (i.e. $a_{ij} \neq 0$) can be excluded from the summation to save time, and the overall probability is defined as:

$$P(X(1, \ldots, T) \mid \mathcal{H}) = \sum_{i=1}^{N} \pi_i.b[x_1, i].\beta(1, i). \quad (9.9)$$

In any practical application of these HMM calculations, both α and β variables would need to be normalised to prevent them from rapidly overflowing or underflowing as the number of states becomes large. Usually they are also transformed to the logarithmic domain.

9.4.2.3 Decoding

The overall probability of being in state i at time t is defined as $\gamma(t, i)$, which is based on forward and backwards sequences, thanks to Bayes' theorem:[5]

$$\gamma(t, i) = \frac{\alpha(t, i).\beta(t, i)}{\sum_{j=1}^{N} \alpha(t, j).\beta(t, j)}. \quad (9.10)$$

In practice, because of the scaling issue mentioned above, this would normally be computed using log probabilities instead.

Then the most likely state at time t is the one with the greatest probability:

$$\hat{s}(t) = \underset{1 \leq i \leq N}{\operatorname{argmax}} \gamma(t, i). \quad (9.11)$$

This is all very well, but it assumes that the state probabilities are independent, when what we really want is the most probable sequence of states, *not* the sequence of most probable states!

The time-honoured solution to this problem is to apply the Viterbi algorithm, invented by Qualcomm founder Andrew Viterbi. It is actually very similar to the forwards–backwards computations above, but with some refinements.

We start by defining two variables, $\delta(t, i)$, which is the highest probability of being in state i at time t, and $\psi(t, i)$, which records which transition was the most likely one to have reached the current state. Both take into account all of the sequences that have

---

[5] Thomas Bayes (1702–1761) was a Presbyterian Church minister and statistician who developed one of the most important statistical theories used in speech processing today. In modern notation, he stated that $P(B \mid A) = \frac{P(A \mid B).P(B)}{P(A)}$.

occurred up to time t, i.e. $\delta(t,i)$ is roughly equivalent to the forward calculation, α:

$$\delta(t,i) = \max_{\hat{s}(1,\ldots,t-1)} P\{s(1,\ldots,t-1) = \hat{s}(1,\ldots,t-1) \cap s(t) = i \cap X(1,\ldots,t) \mid H\}.$$

(9.12)

This computation accounts for both the probability of being in the current state, as well as the probability of having reached the current state. Computation of this quantity begins from an initial case for time $t = 1$, $\delta(1,i) = b[X(1),i].\pi_i$, which is the same as we saw for $\alpha(1,i)$, while the time indices beyond this are given by:

$$\delta(t+1,i) = b[X(t+1),i] \cdot \left\{ \max_{j=1}^{N} a_{ij}.\delta(t,j) \right\}.$$

(9.13)

Again this is conceptually similar to the formulation of $\alpha(t+1,i)$ except that we are interested in the maximum probability instead of the summation.

Meanwhile the initial state for $\psi(t,i)$ is $\psi(1,i) = 0$ since no transition was needed to reach the first state. Beyond this, the step-wise computation is:

$$\psi(t-1,i) = \operatorname*{argmax}_{j=1}^{N} a_{ij}.\delta(t,j).$$

(9.14)

Then the final combined probability for a sequence of states is equivalent to the highest-probability at the final state, which can be computed using $\delta(T,i)$ alone:

$$\hat{P} = \max_{j=1}^{N} \delta(T,j).$$

(9.15)

Similarly, from this we can determine the highest-probability terminal state precursor:

$$\hat{\psi}(T) = \operatorname*{argmax}_{j=1}^{N} \delta(T,j),$$

(9.16)

and then backtrack along the sequence used to arrive there:

$$\hat{\psi}(t-1) = \psi(t-1,\hat{\psi}(t)) \text{ for } t = T, T-1, \ldots, 1.$$

(9.17)

This sequence $\hat{\psi}$ is the most probable state sequence in the model that gave rise to the observed output sequence.

9.4.2.4 Training

Both the decoding and classification methods described above assumed that there was an HMM model, $\mathcal{H}$, that described the system (or that there were several potential models). Since we know that an HMM is defined by five items of information (S, V, π, A, B), it follows that each of these needs to be accurately defined in order for the HMM behaviour to be accurate. In fact, we normally pick the number of states S in advance through knowledge of the problem domain, and we can see the vocabulary V through analysis of the training data.

Given that S and V are set in advance, the action of *training* an HMM is to find appropriate values for π, A and B. There is apparently no analytical method of finding globally optimal parameters [126], but the Baum–Welch and expectation maximisation (EM) methods can iteratively find local maxima for parameters $\lambda = (\pi, A, B)$.

The approach is that we start with 'educated guesses' about the parameters and then iteratively evaluate their probable effectiveness before adjusting them in some way and repeating.

To do this we introduce another probability variable, $\xi(t,i,j)$, which is the probability of being in state i at time t and state j at time $t+1$, given the model λ and observation sequence X, such that $\xi(t,i,j) = P\{s(t) = i \cap s(t+1) = j \mid X, \lambda\}$.

In fact this comprises several probability components that we already know. The first is the probability of observing $X(1,\ldots,t)$, ending in state i (which is $\alpha(t,i)$). The second is the probability of a transition from state i to state j (which is a_{ij}), while the third is the probability of emitting symbol $X(t+1)$ while in state j (which is $b_{jX(t+1)}$). Other components are the probability of observing symbols $X(t+2,\ldots,T)$ given that $s(t+1) = j$ (which is defined as $\beta(t+1,j)$). Since we already have definitions for everything, we just need to combine them together:

$$\xi(t,i,j) = \frac{\alpha(t,i).a_{ij}.b_{jX(t+1)}.\beta(t+1,j)}{\sum_{i=1}^{N}\sum_{j=1}^{N}\alpha(t,i).a_{ij}.b_{jX(t+1)}.\beta(t+1,j)}. \quad (9.18)$$

Note that the numerator and denominator are identical, except that the denominator is summed over all possible states, thus we are dividing $P\{s(t) = i \cap s(t+1) = j \cap X \mid \lambda\}$ by $P(X \mid \lambda)$.

Now, we combine this with $\gamma(t,i)$ from Equation (9.10) to create a re-estimate for the A parameters, A^R, by first estimating new values for a_{ij}, which we will call A':

$$a'_{ij} = \frac{\sum_{t-1}^{T-1} \xi(t,i,j)}{\sum_{t-1}^{T-1} \gamma(t,i)}. \quad (9.19)$$

Rather than using these new parameters directly, a learning rate, η, is normally applied, so that the parameters used for the next iteration are a linear combination of the old and newly estimated values, $A^R = \eta A' + (1-\eta)A$. In this way the training progresses by incrementally updating the A parameters.

The process for B^R is similar, derived from:

$$b'_{ij} = \frac{\sum_{t-1}^{T} \gamma(t,i)[\text{when } X(t) = k]}{\sum_{t-1}^{T} \gamma(t,i)}. \quad (9.20)$$

Thus $B^R = \eta B' + (1-\eta)B$, and the re-estimated initial probabilities π^R are even simpler to define as $\pi^R = \eta\gamma(1,i) + (1-\eta)\pi$. All of this, of course, being subject to the parameter scaling issues that were mentioned previously.

9.4.2.5 Continuous observation densities

While the subsections above have described HMMs, their uses and their training, there is one final refinement that is important in the speech processing field.

Do you remember that the output X was discussed in terms of a sequence of symbols from vocabulary $V = \{v_1,\ldots,v_M\}$? This is all very well for discrete outputs, but in speech the outputs are often relatively high-dimensional vectors of features. An example would be the 39-dimensional feature vector of MFCCs, with their Δ and $\Delta\Delta$. Since this does not map at all well to a vocabulary of states (unless we quantise them very heavily), researchers long ago decided to allow the HMM output probabilities to be continuous. The most common way of doing that is to model the observation sequence by a mixture of Gaussians. This method is popular because it does not overly complicate the re-estimation procedure for B.

In this way, each element in B is a mixture of Gaussians:

$$b(j, \mathbf{y}) = \sum_{m=1}^{M} c_{jm} \mathcal{N}(\mathbf{y}; \mu_{jm}, \Sigma_{jm}), \qquad (9.21)$$

where $\mathbf{y}$ is the vector that the mixture is modelling, c_{jm} is the prior probability of the mth mixture in state j and

$$\mathcal{N}(\mathbf{y}; \mu_{jm}, \Sigma_{jm}) = \frac{1}{\sqrt{2\pi \Sigma_{jm}}} \exp\left[\frac{(\mathbf{y} - \mu_{jm})^2}{2\Sigma_{jm}}\right],$$

which is a Gaussian defined by its mean (μ_{jm}) and covariance (Σ_{jm}) vectors. This is a multivariate Gaussian mixture, unless $M = 1$, when it would be univariate instead. Note that the mixture weights must sum to unity, $\sum_{m=1}^{M} c_{jm} = 1$.

The Baum–Welch re-estimation process for the mixture parameters in a multivariate Gaussian HMM is described in [126] and [127], and in practice involves three stages: (i) a random or constant initialisation leading to (ii) a so-called flat start which initialises all Gaussians with the global mean and covariance parameters, and then (iii) Viterbi alignment. It should come as no surprise to learn that the current state-of-the-art in ASR is even more complex than the situation described here. This is why most research makes use of very capable research packages that implement all of the software necessary to carry out these steps and more. The best-known packages are HTK and Kaldi, discussed further in Section 9.5.

9.4.3 Revisiting our HMM example

Knowing what we now know, let us refer back to our HMM weather example of Section 9.4.1.

To recap, we had a model of the weather in Figure 9.11. The model had evidently been trained, because probabilities were given for the various state transition probabilities, as well as emission probabilities for the observables. In addition, my friend in New Zealand knew the average probabilities for being in each state, as well as the recent observed history. Putting all of this together, and enumerating states C, S and R from 1 to 3, we get an HMM $\mathcal{H}$ defined from:

$$\pi = [0.7, 0.1, 0.2],$$

$$A = \{a_{ij}\} = \begin{bmatrix} 0.5 & 0.2 & 0.3 \\ 0.15 & 0.6 & 0.25 \\ 0.1 & 0.4 & 0.5 \end{bmatrix}, \qquad (9.22)$$

$$B = \{b_{ij}\} = [0.1, 0.02, 0.6].$$

9.4 Hidden Markov models

The information available leads to π (i.e. my friend knows what the overall chance of starting in one of those states is) and an observation sequence, X. This can be populated based on the sentence 'the only time my hat blew off during the past week was two days ago', hence $X_T = <0,0,0,0,1,0,0>$, where $T = 7$ and a 1 in the sequence of days indicates wind of sufficient strength to remove a hat on that day.

What we are interested in knowing is whether my friend can compute the probability that today is likely to be rainy, based on the observation sequence but without knowing anything about the actual weather on any of the days. Hence the actual weather state is hidden.

To find this, we need the variable $\alpha(t, i)$, which is the probability of being in state i at time t, given the model and observation sequence, i.e. $P(X = <0,0,0,0,1,0,0> \cap$ day 7 is R $\mid \mathcal{H})$.

Starting with $\alpha(t, i)$ in the initial state at time $t = 1$, $\alpha(1, i) = b[X(1), i].\pi_i$, the subsequent steps were defined back in Equation (9.5). Rather than doing this longhand, it may be easier to use MATLAB.

First we set up the HMM model $\mathcal{H}$, the observed sequence X and the lengths:

```
Pi=[0.7, 0.1, 0.2];
B=[0.1, 0.02, 0.6];
A=[0.5 0.2 0.3
   0.15 0.6 0.25
   0.1 0.4 0.5];
N=length(Pi);

X=[0 0 0 0 1 0 0];
T=length(X);
```

Next we define an empty array for $\alpha(j, i)$ and compute the initial state:

```
alpha=zeros(T,N);
%initial state
alpha(1,1:N)=B(:).*Pi(:);
```

Finally, iterating forwards through the observations:

```
for t=1:T-1
  for Pi=1:3
alpha(t+1,Pi)=B(Pi)*sum(A(Pi,:)*alpha(t,:)');
  end
end
```

Running this code, the output probabilities are quickly computed to give the following α matrix:

$$\{\alpha_{ij}\} = \begin{bmatrix} 0.0700 & 0.0020 & 0.1200 \\ 0.0071 & 0.0008 & 0.0407 \\ 0.0016 & 0.0002 & 0.0128 \\ 0.0005 & 0.0001 & 0.0040 \\ 0.0001 & 0.0000 & 0.0012 \\ 0.0000 & 0.0000 & 0.0004 \\ 0.0000 & 0.0000 & 0.0001 \end{bmatrix}.$$

The rows in this matrix are the time sequence, starting with day 1 and ending at day 7. The columns are the probabilities of the three states, C, S and R, for the given times.

Looking at row 7 at the bottom, we can read off the probabilities for today; however, we note that they are all nearly zero. This very clearly illustrates the scaling problem: as the sequence becomes longer, $\alpha(j,i)$ rapidly decreases towards zero. In practice we would need to scale these values for each step.

In fact, the probabilities for today are small but have not yet reached zero, namely $[1.4 \times 10^{-5}, 2.2 \times 10^{-6}, 1.2 \times 10^{-4}]$. From this we realise that my friend will not really know what the weather is today with any degree of certainty (because the probabilities are so small). However, the probability of rain is almost 10 times the probability of it being clear and almost 60 times the probability of it being sunny. Therefore he would be justified in reminding me to wear my raincoat for the rest of the day because it will probably rain.

9.4.4 More about HMMs in ASR

Our brief treatment of big data in Chapter 8 should have made clear that, generally speaking, the fewer features that can be extracted from a given unit of input data, the more testing material is required. Similarly, the more complex the model (or input feature), the more training is required. The huge flexibility of machine learning methods such as HMMs means that there is a vast array of trade-off parameters that can be adjusted when implementing HMMs in ASR. We saw that, while we already divide a dataset into training and testing parts, a third division is often necessary – the development set, which is used to fine-tune a machine learning algorithm by tweaking the various parameters.

9.5 ASR in practice

This book is not really aimed at the general public (something which the reader should hopefully be aware of by now) and thus the term 'practical' for this book's audience is likely to mean 'a research system that can be implemented on computer and modified for experimentation'. We will address exactly this point below in Section 9.5.1; however, we also acknowledge that some readers may not be particularly interested in low-level modifications to an ASR research framework, and simply require something to

9.5 ASR in practice

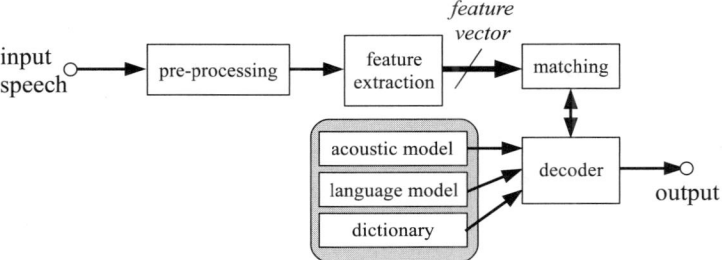

Figure 9.12 Block diagram of a generic speech recognition system, showing input speech cleaned up and filtered in a pre-processing block, feature extraction and then the matching and decoding processes driven from predefined models of the sounds, language and words being recognised.

be implemented quickly and efficiently for use as part of a larger project. This will be addressed in Section 9.5.2 below.

Practical ASR systems tend to share a similar generic processing structure, although the actual implementation details can vary widely. Starting at the mouth of a human speaker, we have already considered the channel and transducer (see Figure 9.3 on page 274), and followed the sequence through to stored audio files on computer (or streamed audio data on computer). Working now in the digital domain, a block diagram of a generic system, shown in Figure 9.12, shows how input speech is first cleaned up by a pre-processing system before a feature vector is extracted (of which we discussed many in Section 8.3.2). The pre-processing may take the form of filtering, probably windowing and normalisation, and some method of segmentation. It often includes de-noising where the input speech is known to be noise-corrupted.

Following pre-processing, features are extracted from the speech. There are many possible features which can be used, including LPCs, LSPs, cepstral coefficients spectral coefficients and so on, although, as we have mentioned, mel-frequency cepstral coefficients (MFCCs) are probably the most popular at present, augmented with pitch or energy information, as in Figure 9.9.

There is, of course, no reason why the vector needs to contain just one feature. Each feature may thus include several tens of coefficients, and is typically computed over a 20 to 30 ms analysis window which is updated (shifted) 10 ms between frames.

The features are then classified by a trained HMM to obtain posterior probabilities of particular phoneme outputs (or, more normally these days, tri-phone state probability outputs[6]). A dictionary of all words and pronunciations supported by the system is used to constrain phoneme sequences into known words and a language model weighs the probabilities of the top few word matches based upon their adherence to learned language rules. For example, if the highest matching word is found to be something

---

[6] It is interesting to note that in the commonly used 300-hour Switchboard English language dataset there are about 9000 tri-phone states, compared with around 6000 for a similar-sized Mandarin spoken news database.

disallowed based on the previous word in the language being spoken, then it probably should be rejected in favour of the second highest matching word.

The complexity of this process is encoded in the number of degrees of freedom, and the vast range of possibilities inherent in the language. This is one reason why the vocabulary should be restricted where possible in practical systems, but also why the size of the feature vector should be minimised where possible. However, more restrictive grammars can actually be helpful.

The language model, as described, considers the probability that the current speech is correctly matched given knowledge of the previous unit of matched speech. In general this history can extend back further than to just the previous sound. An *n-gram language model* looks back at the past n speech units, and uses these to compute the probability of the next unit out of a pre-selected set of a few best matches from the acoustic model. Of course, this again increases computational complexity, but significantly improves performance (especially in more-regular languages such as Mandarin Chinese). The units under consideration in the n-gram language model could be phonemes, words or similar, depending upon the application, vocabulary size and so on. In a less-regular language such as English, the number of phonemes in different words can vary significantly, so again a dictionary can be used to adjust the parameter n in the n-gram language model.

The output of an ASR system could be given as a string of phonemes, but is more usefully delivered as a sequence of recognised words, although again this depends upon the particular application and configuration of the system. The models of speech themselves, namely the acoustic model, language model and dictionary as shown in Figure 9.12, are particularly important to ASR: a sound missing from the acoustic model, language features not covered by the language model and words not in the dictionary cannot be recognised.

Although the dictionary is often created from predefined word lists, the two models are usually the result of training. Whilst it is theoretically possible to define language and acoustic rules by hand, it is far easier and more accurate to train a system using representative speech to build up these models statistically. For a system operating with different but known speaking subjects, it is possible to detect who is speaking (see Section 9.6) and then switch to an individual acoustic model for that speaker, which can be better than having one big model to cover everyone. However, large vocabulary continuous speech recognition (LVCSR) systems of today are primarily statistical and use one huge model for all speakers, possibly with some speaker adaptation (one such common adaptation is vocal tract length normalisation (VTLN) [128], which is useful for working with different age or gender speakers).

Similarly, for systems encompassing several different languages or dialects (see Section 9.7) it can be better to detect these and switch language models appropriately.

Up to this point, we have discussed ASR systems in general. However, it is instructive to briefly turn our attention to some examples. As mentioned, we will first look at two prominent research systems, and then a prominent implementation system.

9.5.1 ASR Research systems

HTK (Hidden Markov Model Toolkit) is a freely available software package originally developed by the Cambridge University Engineering Department. This highly configurable framework, available as easily modifiable C source code, contains everything necessary to implement, test, modify and experiment with HMMs, particularly for speech processing. The latest version (at the time of writing) is 3.4.1, which has been stable for over 7 years, is widely used worldwide for ASR research (and indeed for many other branches of speech research too) and has underpinned a large amount of research progress in speech technology. The software structure and its use are fully documented in the HTK book [125].

More recent is Kaldi [129], which describes itself as 'a toolkit for speech recognition written in C++ and ... intended for use by speech recognition researchers'. This is also freely available, and is open source software written by an international group of collaborators. It is well documented and maintained, and is able to replicate most of the HTK functionality while allowing many more modern extensions (including compilation support for GPU computation, which is required, to speed up operation, which can stretch into weeks of computing time on large datasets – despite the acknowledged speed and efficiency of this software).

While both HTK and Kaldi are excellent resources, at the time of writing the former is regarded as a gentler introduction to speech processing (although it is still very complex for a beginner), whereas the latter is able to achieve better performance on the most demanding tasks, at the cost of a significantly steeper learning curve than for HTK. Note that both HTK and Kaldi require a full featured Unix-like computer operating system such as Linux or Mac OS-X for full functionality, although parts of the software can be compiled on lower-quality operating systems such as Microsoft Windows, with some difficulty.

9.5.2 ASR Implementations

The open source Sphinx recogniser, originally developed at Carnegie Mellon University in the USA, is one of the best implementation examples of a flexible modern speech recognition system. It can be used for single word recognition or expanded up to large vocabularies of tens of thousands of words, and it can run on a tiny embedded system (PocketSphinx) or on a large and powerful server (which could run the Java language Sphinx-4). CMU Sphinx is constantly updated and evaluated within the speech recognition research field.

Sphinx-4, in common with most current ASR implementations, relies upon hidden Markov models to match speech features to stored patterns. It is highly configurable, and incredibly flexible – the actual feature used can be selected as required. However, the one that is most commonly extracted and used for pattern matching purposes is the mel-frequency cepstral coefficient (MFCC) [130].

This flexibility extends to the pre-processing sections, the 'FrontEnd' where a selection of several different filters and operations can be performed singly or chained together, and also to the so-called 'Linguist', which is a configurable module containing a language model, acoustic model and dictionary. The linguist is responsible for consulting these based upon a particular feature vector, and determining which subset of stored patterns are compared with a particular feature vector under analysis.

Sphinx-4 has been tested extensively using industry-standard databases of recorded speech, which are commonly used by ASR researchers to compare the performance of systems. Accuracy rates of over 98% are possible for very small vocabularies (with a response time of 20 ms), over 97% for a 1000-word vocabulary (in 400 ms) and approximately 81% for a 64 000-word vocabulary (below 4 s) [130]. These figures are assumed to be for high-SNR cases.

9.6 Speaker identification

Speaker identification (SID) or speaker classification means the automated determination of *who* is speaking. This is related to, and overlaps with, the very similar research area of speaker verification, which ensures that the person speaking is indeed the person they claim to be. At the time of writing, neither technology is particularly common, although voice-unlocking systems exist for smartphones and computers, and the related area of diarization (see Section 9.8) relies upon speaker identification methods. However, there are indications that speaker identification and verification technologies are about to become significantly more important with the probable introduction of vocally authorised financial transactions in the near future: this would involve a customer speaking into a smartphone or a computer terminal to authorise payment from their account.

In both verification and identification, a-priori information is used to determine either who a given speaker is (with a classification output) or whether a given speaker is who she claims to be (with a binary true or false output). In practice, candidate users will need to be registered with the system doing the analysis – their vocal information has to have been captured in some way and stored. During operation, vocal information from a user is compared with the stored information to check for matches. There are two further main branches to this overall research area: the first is where the material being spoken is fixed in advance and the second concerns the case in which the material being spoken is unrestricted. The former means using some kind of passphrase (which might be secret), whereas the latter means using any spoken material. In the unrestricted case the problem becomes much more difficult, with the final accuracy probably being more closely related to the amount of captured data that can be analysed than to the accuracy of the analysis method which is employed.

Whichever branch is being considered, the methodology relies upon the way in which the speech of two speakers differs. If there are instances of both speakers saying the same words, either through restricting the words being spoken or perhaps through fortuitous sample capturing, then the analysis becomes easier. Having two speakers saying the same phoneme at some time is far more likely to be achievable in the unrestricted case than speaking the same word. However, this would presuppose that an accurate method of identifying and isolating different phonemes is available – this is itself a difficult task to perform automatically. In the unrestricted case, it would be possible to use information pertaining to *what* is being said as part of the analysis. At a higher language level, grammar, pronunciation and phraseology can each help to differentiate among speakers. Some of the progress in this field has been tracked by A. P. A. Broeders of Maastricht University in two papers reviewing the period 1998 to 2001 [131] and from 2001 to 2004 [132], and a slightly more recent review is available in [133].

The restricted and unrestricted text cases mentioned above are also known as *text-dependent speaker recognition* and *text-independent speaker recognition* in the research literature. In one of the classical reviews of this field, S. Furui not only subdivides the research field similarly to the way we have discussed above, but separately discusses the ability of several processing methods [134]:

Dynamic time warping (DTW) where strings of features matched from speech under analysis are shifted in time, compressed and expanded to match to stored templates of those features from each candidate speaker.

Hidden Markov model (HMM) to match statistical property similarities between a speaker and candidates, as discussed in Section 9.4.

Long-term statistical methods to consider how the feature tendencies of one speaker match those of the other candidates (see Section 9.2.1.2 for similar arguments applied to VAD).

Vector quantisation (VQ) methods to compile a simplified feature vector based upon the measured features from one speaker, and to match this against the stored codebook from candidates.

Speech recognition-based methods which can, in principle, be used to detect phonemes, or phoneme classes, either for analysis of their distribution or potentially to improve the accuracy of any of the methods already listed.

Even assuming that no background noise is present, the same microphone is used (with the same distance and orientation to the lips) and the spoken words are identical, the speech of two people may differ for many reasons, including:

Physical characteristics such as length of vocal tract, size of nasal cavity, tooth position and so on. The effect of vocal tract length is one reason why the voices of children change, becoming less 'squeaky' as they grow older. It is most noticeable during puberty when a boy's voice may break. This is caused by his small larynx and short, thin vocal chords starting to lengthen and thicken, while simultaneously facial bones grow to create larger nasal and buccal cavities. The change does not happen overnight (although it may seem to), but shows how even relatively gradual changes in physiology can cause much more dramatic changes in speech. Unfortunately changes can also happen as many people enter old age, causing further differences in vocal characteristics.

Physiological characteristics such as the speed of tongue movement, size of mouth opening, placement of tongue, lip shaping and so on that are the features of a particular person when speaking.

Behavioural characteristics include the loudness and expressiveness of a person's voice. Expressiveness may well include pitch changes, and may be for linguistic reasons (accent, pronunciation and so on). Some people speak faster than others, and would thus have a higher syllabic rate. Other people *flatten* their vowels, produce rolling /r/ sounds guttural /g/ sounds and so on. Many of these changes are context dependent.

However, in the computational domain, we have simply a one-dimensional signal to consider: the audio vector. The process for almost any SID task will begin similarly by performing an analysis upon the input signal to extract features. In Chapter 7 we discussed many such features. It is very likely that some of those features are more suitable for use in classifying speakers, and there are probably as many different features to choose from as there are researchers working in this field [135], although, as with other domains, MFCC-based approaches predominate.

Before leaving the discussion on SID, it must be remembered that there can be significant variation in the recorded voice of a person. We have already alluded to issues such as microphone response, distance to the mouth of the speaker, orientation to the speaker and the presence of background noise. But even assuming perfectly repeatable recording, the voice may change. Some reasons for this might include:

- Long-term effects due to physiological changes such as growth, ageing or disease, or, as another example, smoking- or pollution-induced cell damage to the throat lining.
- Long-term behavioural changes. To take one example, it has been observed by the author that those living in Japan tend to speak quietly while those living in Hong Kong tend to speak loudly, from a UK English perspective. Would I learn to speak more quietly after living in Japan for a few years, or more loudly after living in Hong Kong for a few years?
- The emotional state of the speaker, including anger, but also stress and even gentleness.
- Illness, such as influenza, sinusitis, inflamed throat, blocked nose, mouth ulcers and so on. Similarly, inner ear infections of various forms will alter the feedback path between voice and brain that is part of the speech production regulatory mechanism, thus affecting speech. Lung complaints influence temporal and loudness patterns whereas dental complaints can prevent some phonemes from being produced accurately.
- Time of day – again the author observes that people tend to speak more slowly in the early morning if they are unused to waking early. Waiting for the 'brain to warm up', as an explanation, may not satisfy our medical colleagues, but at least explains the effect adequately enough from a speech perspective.
- Fatigue or tiredness, whether mental or physical (vocal).
- Auditory accommodation – coming from a loud machine room into a quiet area often causes speakers to initially misjudge the volume of their voice. Imagine a user walking up to a microphone at a voiceprint-based door entry system, extracting the in-ear headphones from their iPod and then speaking their secret pass phrase (LOUDLY!).
- Coffee, curry and many other beverages or foods seem to adhere to the lining of the mouth and throat, affecting the qualities of a person's voice for some time after consumption, or perhaps cause production of mucous and saliva which can affect speaking.

To highlight the problems that these *intra-voice* changes can cause, accomplished speaker recognition researcher S. Furui admits that so far (at least up until he wrote this in 1997) no system has succeeded in modelling these changes [134], the problem being that relaxing the accuracy requirements of a speaker recognition system to allow for variations in a user's voice will naturally tend to increase the percentage of incorrect classifications. In the vocabulary of the researchers, allowing more 'sheep' (valid users that are correctly identified) and fewer 'goats' (valid users that are not correctly identified) also causes more 'wolves' (invalid users that can impersonate the sheep). In addition, researchers sometimes refer to 'lambs' – the innocent sheep who are often impersonated by the big bad wolves [136].

As with ASR technology, the best SID performance is obtained from systems employing deep neural networks (DNNs), with MFCC feature inputs (augmented with deltas and accelerations), and context.

9.6.1 SID research

Current speaker identification research tends to use hybrid approaches of deep neural network (DNN) analysis of acoustic features (including deep bottleneck features, DBFs), fused with a GMM-UBM approach at a phonetic level. The reason is that speaker identity is bound up with both the choice of speaking unit (such as pronunciation and favourite ways of saying things) and the way it is spoken (primarily defined by the voice production apparatus). Standard evaluation tasks for SID aim to identify speakers from recordings of unconstrained speech. Unsurprisingly, acoustic methods have an advantage for short duration recordings (since there is insufficient phonetic-level information within the data), while phonetic approaches are more useful for longer duration recordings. Even where DBFs are used, these are analysed alongside standard features such as three orders of MFCC plus context (or, alternatively, using PLP features, which seem to have a slight advantage in noisy conditions).

A US government agency called the National Institute of Standards and Technology (NIST) runs a worldwide open competition called the Speaker Recognition Evaluation (SRE) to provide a baseline score for research systems. The competitions include training, development and evaluation datasets and standard tasks on different length recordings, as well as defining scoring methodologies. SRE has been run aperiodically since 1997 (the most recent being 2008, 2010 and 2012). The latest exercise, SRE12, comprises nine tests for automated SID systems on English speech recorded by telephone and by higher-quality microphone sources under one of five conditions. A complicated scoring mechanism is used to derive an equal-error rate performance measure from results (which reports the trade-off between false positive and false negative errors).

9.7 Language identification

Automatic language classification or language identification (LID), by analysis of recorded speech, has much in common with the automatic speaker classification task of Section 9.6. It can be subdivided in a similar way – namely whether there is any constraint upon what is being said and whether there is any constraint upon the identities and number of persons speaking. More importantly, the base set of analysis features and processing techniques is common between the two tasks.

In an extensive review of language identification research, Zissman and Berkling [137] cite four auditory cues that can be used to distinguish between languages:

Phonology (see Section 3.2) generally differs in that not all languages comprise the same set of phonemes, and undoubtedly they are used in different sequences and arrangements between languages.

Morphology, meaning that languages tend to have different, but often similar, lexicons. By and large, languages derived from the same root will share a more common morphology. However, imported or shared words blur this distinction.

Syntax differs in style, sequence and choice of framing words. For example, some languages tend to prefix nouns with prepositions, others do not. Some languages, such as Malay, have far more word repetitions than others, such as English.

Prosody is the rate, spacing and duration of language features.

Although the researchers do not normally discuss the issue, the research field is confronted by a rather difficult problem: those brought up speaking one language may retain the prosody and even syntax when speaking another language. Anyone who has travelled to Pakistan or India and heard locals speaking English would be struck by how the prosody of the language had become localised. Similarly, syntactic differences lead many Mandarin Chinese speakers who are starting to learn English to ignore rules of plurality and tense. Greeks speaking English may well include too many definite articles 'having the knowledge is the good thing'. Conversely the author has been accused at times of speaking both Chinese and Greek like an Englishman (i.e. not particularly well).

The difficulty of speakers bringing the prosody from their native tongue to another language is analogous in many ways to the issue of one speaker impersonating another to a speech classification system. It is a source of inaccuracy, but is not necessarily a major problem. Imagine a language recognition system built into a call centre that routes incoming callers to customer service operators able to speak their language. That a Greek speaking English with Greek syntax would be misinterpreted as speaking Greek is not a major problem. However, if he were misinterpreted as being a Japanese speaker, that may well be a major problem.

Continuing on to the techniques used for the classification, Zissman and Berkling [137] provide an overview of the various techniques, and also go on to provide a comparative evaluation of each of these. Their techniques are:

Spectral-similarity approaches based upon the different spectral characteristics of languages or upon the statistics of the spectral changes during speech for each language.

Prosody-based approaches which look at the timings and duration of captured linguistic features.

Phone-recognition approaches, since, as with the speech recognition approach in the speaker classification task, there are differences in the relative frequency, time-distribution and sequencing of phone features.

Multilingual speech units called *poly-phones*, *mono-phonemes* or *key-phones* are the widest possible set of phonemes represented by all languages under consideration (poly-phones), or just the most important ones in terms of relative frequency or importance to classification (mono-phonemes/key-phones). These can then be detected by an automatic speech recognition system, with the output analysed statistically.

Word level approaches can search for the occurrence of key words. These systems can even be trained, without having prior knowledge, to identify and select the key words.

Continuous speech recognition where, at the extreme, parallel speech recognisers for each of the candidate languages compete to provide the best (most sensible) transcription. In many ways this is a brute-force approach, which would be sensible only for a relatively small set of candidate languages.

As alluded to above, the driving forces behind automated language classification are predominantly telephone call centre-related at the present time. The advent of a Star Trek universal translator (or, if you prefer, the in-ear Babelfish of THHGTTG[7]) is still some years away, and the present bias of the Internet is toward visual media rather than auditory communications. Having said that, for single-digit sized groups of languages, classification accuracy can exceed 90%, indicating that the technology is maturing quickly.

While neither automatic speaker recognition nor automatic speech recognition can approach the abilities of the average human, automatic language recognition systems exceed average human abilities (partly because few humans can speak ten languages). In this way, despite the disadvantages of the technology developed to date, it can find ready application outside the research laboratory.

9.7.1 LID research

As with SID research, typical LID tasks are the classification of multiple languages from speech segments of different average duration – typically 3, 10 and 30 seconds. Also similar to SID research is the use of DNN systems, including DBF front-end features, fused with higher-level phonetic information. The reason why both levels are important is that, while some languages are very different at a phonetic level, there are some pairs of languages that share almost all phonemes and phoneme probabilities. Similarly, there are differences between some languages at an acoustic level, but other languages may appear similar at this level.

NIST (mentioned above in Section 9.6.1) runs what is probably the world's foremost baseline evaluation of LID systems. This is called the Language Recognition

---

[7] *The Hitchhiker's Guide to The Galaxy*, Douglas Adams.

Evaluation, an open competition that has been run in 1996, 2003, 2005, 2007, 2009 and 2011 (but not since then, at the time of writing). The competition provides a rich set of data for training and evaluation. Considering just the latest exercise, 23 target languages were included, namely Amharic, Bosnian, Cantonese, Creole (Haitian), Croatian, Dari, American English, Indian English, Farsi, French, Georgian, Hausa, Hindi, Korean, Mandarin, Pashto, Portuguese, Russian, Spanish, Turkish, Ukrainian, Urdu and Vietnamese, although many more languages were included as training data. Apart from the lack of standard English (which disadvantages most English-speaking countries), Arabic and Indonesian/Malay in the list, it is a fairly representative set of common world languages. More importantly, it contains seven highly confusable pairs of similar languages: {Cantonese–Mandarin}, {Portuguese–Spanish}, {Creole–French}, {Russian–Ukrainian}, {Hindi–Urdu}, {Farsi–Dari} and {Bosnian–Croatian}. As mentioned above, this data is divided into sets of different length recordings of average duration 3, 10 and 30 seconds.

A complicated cost detection function is defined to score the ability of any LID system competing for the NIST LRE, which combines false alarm and false error scores. Without providing the exact definition of the score here, the best error rates reported for 3, 10 and 30 s recordings in LRE2009 were about 10%, 3% and 2% respectively.

9.8 Diarization

Figure 9.13 is an analysis of part of a broadcast radio programme. The recording comprises speech from a main presenter, a second speaker and a pre-recorded sound clip containing a third speaker, in addition to a few gaps and periods with background

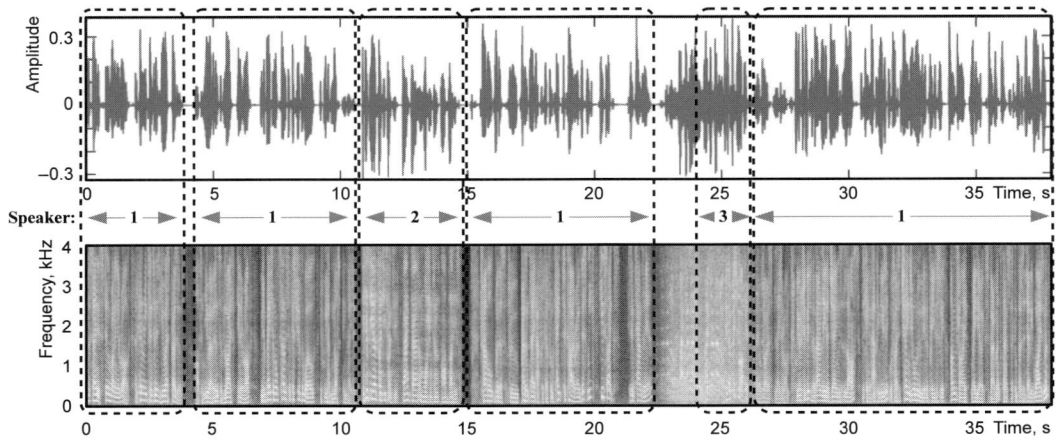

Figure 9.13 A recording from a broadcast radio programme, viewed as both a waveform (top) and a spectrogram (bottom), with the regions spoken by different people identified and enumerated (speakers are indexed from 1 to 3) along the centre of the plot.

sound effects. This type of recording is extremely common: broadcast audio and television, audio tracks from film, recorded telephone conversations, meetings, surveillance recordings and so on. Diarization is the unwieldy sounding name given to the science of interpreting such recordings (and note that it is traditionally spelt with a 'z'). In essence it is asking the question 'who spoke, and when?', with the variables being the number and identity of speakers, along with the timings of their utterances.

There are several separate, but interrelated, diarization research topics, which include simply identifying the number of speakers present, through to identifying when they spoke and what they spoke. Typically (but not always) the number of speakers is known a priori, and thus algorithms are developed to identify which periods of the recording contain speech from which speaker (or none). A further simplification is to disallow overlapping speech.

Historically, trained transcribers would need to listen to recordings of multiple speakers to decide who spoke and what they were saying, but, with the exponential growth in media recordings, there are good reasons to develop tools capable of performing this task automatically.

Diarization research is very cutting-edge and fast moving. However, we can describe a modern diarization research system in outline [138, 139]. Input audio is first segmented using a VAD to remove non-speech periods. The remainder is analysed using highly overlapped frames (typically 10 ms in size). Features such as MFCC, PLP or DBF are then extracted for each frame. A Bayesian inference criterion (BIC) is then used to split frames. Given a window containing multiple frames, the simple criterion is whether the probability distribution function of the larger frame is tighter than that when the window is split. In other words it determines whether the window contents are 'better' represented by a single window or by two windows. The output from the BIC process is a list of segments that are known to have been produced by different speakers (on the basis that the probability distribution of features from different speakers is likely to be different). However, it is significantly over-fragmented (because speech is not constant, so the probability distribution of different phonemes from one speaker is naturally going to show differences), and thus a process of clustering is required to grow sequences of frames from individual speakers. The powerful Viterbi algorithm re-emerges in this, primarily statistical, procedure. Agglomeration of frames into clusters ceases when the correct number of speakers has been reached (or proceeds until some goodness measure is met, in the case where the number of speakers was not known a priori).

9.9 Related topics

If we were to discuss all of the topics related to ASR then this chapter would become much longer than it already is, because the field is ever expanding in both scope and depth. However, the movement over recent years has primarily been application-oriented, fuelled by the improved accuracy of speech models – in particular those that make use of advanced machine learning techniques. In addition to those topics we have discussed above, there are significant research efforts ongoing (or being formed) in several areas, including:

- Topic detection – which aims to determine what a speaker is talking about, in particular as a front end to an ASR system related to a particular domain (since we have seen that restricting the topic domain and hence vocabulary and/or grammar can significantly improve recognition performance).
- Code switching is the name given to speech which switches between languages, often during a sentence (or even mid-word). This type of speech seems to be particularly prevalent in South East Asia, possibly exacerbated by the ability of many people to speak multiple languages [140]. For example, a sentence that would not be out of place in Singapore or Malaysia like 'I go makan with my pengyou' contains three languages (some English, the Malay word 'makan' which means to eat and the Chinese word 'pengyou' which means 'friend'). Often there could be several dialects of Chinese instead of Mandarin, including Hokkien, Cantonese, Dejiu and Hakka (with Hokkien seemingly well-suited to swearing).
- This brings multilingual issues to the forefront, including accents, dialects and the influence of first language on second language (L2) speakers.
- Language learning, including pronunciation training and correction.
- The use of dialect or colloquial speech – especially colloquial terms for items (an example from the public security domain is the word 'pineapple' referring to a hand grenade rather than a prickly fruit).
- ASR for speakers with abnormal speech and several related disorders, explored in Infobox 9.1 on page 311.
- ASR for children, the elderly and the sick, as well as those who are temporarily incapacitated (e.g. drunk, or in the midst of eating).
- Silent speech interfaces (SSIs) – not only used for ASR, but useful nevertheless, are different modalities of speech input used to complement audible speech input when recording voice. One example of an SSI is the ultrasonic speech system given in Chapter 10.
- Emotion detection from speech – which may be useful for applications such as the automated handling of call centre traffic, ensuring that irate customers (as detected from their speech patterns) are dealt with earlier and by more suitable customer relations staff.
- Mood detection, as distinct from emotion detection, could apply to a group of people. For example, tracking the prevailing feeling of a large crowd as it marches to the endpoint of a demonstration.
- Thought police – is it possible to detect what someone is really thinking from what they are saying? This has been done in science fiction for some time, but will it work in the real world? Perhaps researchers working on voice stress patterns will reach such an endpoint.

If nothing else, this list should convince everyone that ASR-related research is dynamic, interesting and important to both society and commerce. Hopefully readers will be able to add their own new application domains to this list in future.

> **Box 9.1** Dysarthria
>
> There are a number of terms that describe abnormal speech production, including the following:
>
> - **Dysarthria** – a disorder of speech articulation muscles caused by a nerve defect.
> - **Aphonia** – the inability to produce any voice.
> - **Apraxia of speech** – interruption in the ability of the brain to control the voice, usually by the inability to translate from thinking of saying something into actually saying it.
> - **Dysphonia** – disorder in voice production, so that whatever voice is produced is degraded (usually sounding quite unnatural).
> - **Stammer/stutter** – interruptions in the flow of speech, often caused by nervous issues.
> - **Speech disorder/impediment** – a speech communications difficulty which prevents normal speaking.
> - **Lisp** – a difficulty in producing sibilants.
>
> If ASR is truly to become a technology for all, then solutions will need to be robust enough to handle many of the above issues, as well as speech from children and the very old. A few speech researchers are working on those extremes, although assistive speech technology is not currently a priority area for many researchers.

9.10 Summary

As we can see, ASR and its related topics makes for a very diverse, fast-moving and exciting field. Unlike 20 years ago when the author was completing his PhD in speech, the topic is highly popular today, and well regarded by industry.

It does not take a great deal of imagination to envisage future commercial applications that make use of ASR in education, the automative industry, call centres, for financial transactions, in healthcare and social services, the legal profession, entertainment, public security and countless other domains.

While we cannot do justice to the entire ASR domain in a single chapter (and possibly not even in a single book these days), we can make a good start by laying down the foundations of important techniques and approaches. In particular, by explaining how the rapid performance gains of the past decade have come about, i.e. mainly thanks to the improving capabilities of machine learning.

Bibliography

- *Hidden Markov Model Toolkit (HTK)*
  Available from http://htk.eng.cam.ac.uk
  The excellent HTK book can also be obtained from the same location as the software.

- *Kaldi*
  Available from http://kaldi.sourceforge.net

- *Sphinx and Pocketsphinx*
  Available from http://cmusphinx.sourceforge.net

Questions

Q9.1 Describe the main differences between continuous speech recognition and keyword spotting tasks.

Q9.2 Using your knowledge of ASR technology and accuracy factors, consider the likely reasons behind an ASR system achieving very different word error rates for the words 'shy' and 'balloon'.

Q9.3 A simple voice activity detector (VAD) calculates the energy in each analysis frame to decide whether speech is present. Briefly describe in what everyday situations such an approach may not work well, and suggest some simple improvements.

Q9.4 A voice recorder is triggered using a VAD, turning on to capture speech, and turning off when the speech is no longer present. When listening to the recorded speech, the developer notices that sometimes the system turns off before the end of a sentence (especially when the last word ends in 's' or similar phonemes). Identify a common reason for this problem and suggest a solution.

Q9.5 An end-user purchases ASR software for her laptop. Following installation instructions, she trains the software by speaking a sequence of words and sentences. The more training she completes, the better the word recognition accuracy becomes. What elements (i.e. internal models) of the complete ASR process will have been improved through the training process? [Hint: her friend tries to use the system from time to time – he noticed that at first the error rate when he spoke was the same as that for the user, and found that it increased after she had trained the system, but by a much lower margin than the improvement in her recognition accuracy.]

Q9.6 List and briefly explain four differentiating features of speech that might be useful for performing language identification.

Q9.7 List the steps that need to be followed in the setup and training sequence for a deep neural network (DNN), and identify how a deep bottleneck network (DBN) differs from a DNN.

Q9.8 For ASR using less common languages, why might a neural network be initially trained using English (or Chinese), and then a transfer learning method implemented? What advantage does this confer?

Q9.9 What is speaker diarization, and how does it differ from basic speaker identification or ASR?

Q9.10 We discussed the use of trained hidden Markov models (HMMs) for classification and decoding. With the aid of an example, briefly describe and contrast how classification and decoding can be applied.

Q9.11 Refer to the trained HMM of Figure 9.11. My friend from New Zealand is planning a week's holiday with me, and is concerned that it will rain for the entire week. Assuming that it is raining on the first day of his holiday, what is the likelihood that the rest of the week will also experience rain, according to my HMM?

Q9.12 Explain why a phonetic dictionary is usually required for ASR systems that perform phoneme-level recognition but output recognised words.

Q9.13 Briefly describe the difference between speaker identification and speaker verification, and determine which task is likely to be the most difficult to achieve good performance in.

Q9.14 List five reasons why an enrolled user of a speaker identification system may not be recognised on one particular day, when she is normally recognised without difficulty.

Q9.15 Describe, in one sentence, the meaning of 'code switching' when referring to spoken language systems.

10 Advanced topics

The preceding chapters have, as far as possible, attempted to isolate topics into well-defined areas such as speech recognition, speech processing, the human hearing system, voice production system, big data and so on. This breakdown has allowed us to discuss the relevant factors, background research and application methodology in some depth, as well as develop many MATLAB examples that are mostly self-contained demonstrations of the sub-topics themselves. However, some modern speech and audio related products and techniques span across disciplines, while others cannot fit neatly into those subdivisions discussed earlier.

In this chapter we will progress onward from the foundation of previous chapters, discussing and describing various advanced topics that combine many of the processing elements that we had met earlier, including aspects of both speech and hearing, as well as progressing beyond hearing into the very new research domain of low-frequency ultrasound.

It is hoped that this chapter, while conveying some fascinating (and a few unusual) application examples, will inspire readers to apply the knowledge that they have gained so far in many more new and exciting ways.

10.1 Speech synthesis

Speech synthesis means creating artificial speech, which could be by mechanical, electrical or other means (although our favoured approach is using MATLAB of course). There is a long history of engineers who have attempted to synthesise speech, including the famous Austrian Wolfgang von Kempelen, who published a mechanical speech synthesiser in 1791 (although it should be noted that he also invented something called 'The Turk', a mechanical chess playing machine which apparently astounded both public and scientists alike for many years before it was revealed that a person, curled up inside, operated the mechanism). The much more sober Charles Wheatstone, one of the fathers of electrical engineering as well as being a prolific inventor, built a synthesiser based on the work of von Kempelen in 1857, proving that the original device at least was not a hoax.

These early machines used mechanical arrangements of tubes and levers to recreate a model of the human vocal tract, with air generally being pumped through using bellows and a pitch source provided by a reed or similar (i.e. just like that used in a clarinet or oboe). Different combinations of lever settings could cause the systems to create the shape – and sound – of vowels and consonants. A well-trained human operator was required to learn how to control the system, working levers and valves to form sounds which were then sequenced together to form phonemes and speech, albeit very slow and mechanical-sounding speech. In fact, much the same methodology of stringing together different phonemes is still in use today, having survived the transition to electrical systems in the 1930s through electronic systems in the 1960s and into computers. Of course, speech synthesis researchers would argue that both quality and usability have improved significantly over the intervening 300 years. However, all of these systems have been inspired by the mechanics of the human voice production system, presented at the start of Chapter 3.

In general, the aim of speech synthesis is to create speech. The actual word synthesis derives from *synthetic*, the opposite of natural, and describes the process of creating or manufacturing something which is artificial. We already know that human speech has a number of characteristics, including semantics (i.e. meaning, emotional content and so on), identity information and situational characteristics, along with timbre, tone and quality. Some or all of these can be the target of speech synthesisers for different applications. Examples include attempting to create a specific person's voice, replaying a fixed stored message and automatically reading general text (e.g. reading a story to a listener).

By far the most common applications are those that reproduce messages – either broadcast announcements, or attempts to translate textual information into speech, so-called text-to-speech (TTS) systems. The two main approaches to speech synthesis are (i) *concatenative* systems, which concatenate (string together) sequences of sounds to synthesise an output [141], and (ii) speaker models, which derive equations or descriptions for the varying parts of continuous speech (e.g. pitch, vocal tract, gain and so on) and then use these to generate speech.

As a basic technology, stored voice playback is the simplest form of speech synthesis.

10.1.1 Playback systems

Basic speech playback systems, often used for applications such as telephone voicemail menus, simply record and store entire words, phrases or sentences which are replayed upon command. An example would be systems which record spoken digits and then replay them in different sequences to generate larger numbers. Although the quality of the speech itself can be extremely high, these systems do not sound particularly natural because the intonation of various words does not always match listener expectations: in reality a human would speak a word differently depending upon its context (i.e. neighbouring words) as well as its position in a sentence. Because of this listeners will usually be able to identify an unnatural change in either pitch or cadence (see Section 7.4) when words are joined together. However, recent advances in the processing of speech can allow for post-processing of a stitched-together sentence to improve naturalness – for example, by imposing an overall pitch contour (with an end-of-sentence downward tail – or upward for a fake Australian-like accent which also appears to have taken root across the UK over the past decade).

Unfortunately systems that play back stitched-together sound recordings in this way are not particularly flexible: they cannot create new words, and require non-volatile storage (with fast access) of every word in their vocabulary. Advances in speech quantisation, in particular the CELP analysis-by-synthesis systems of Chapter 5, allow stored audio to be compressed in size, such that storage requirements, even for quite a large vocabulary, are not excessive; however, a potentially large number of words still have to be indexed, accessed and recorded in the first place.

Although even a small English dictionary contains around 90 000 words (e.g. *Oxford English Mini Dictionary*), not all of the words are in common or continuous usage. In fact something called 'Basic English' was proposed in the 1940s [142]: a restricted set of 850 English words (and restricted grammar) that can convey the meaning of more than 90% of the concepts in the 25 000-word *Oxford Pocket English Dictionary*.[1] The fact that such a restricted vocabulary exists and is usable might imply that a voice playback system capable of reproducing just 850 words would suffice for most situations. However, we can easily think of words that might be needed which exist in few, if any, printed dictionaries. Particularly proper nouns including unusual surnames or Christian names, new place names, product names, colloquial or slang phrases and exclamations. To glimpse the extent of the problem, consider some examples that begin with 'K':

- First names: Kelvonah, Kylie, Kyle
- Surnames: Kolodziejczyk, Krzeszewski, Krzycki
- Place names: Kirkleyditch, Kirkbymoorside, Knaresborough
- Product names: Kaukauna (cheese), Kombiglyze (medical), Kawaii (piano)
- Slang: kerfuffle (commotion), kiasu (Singapore, fear of losing), klutz (USA, clumsy)

What a shame that Scrabble disallows proper nouns and slang.

Clearly, while playback systems have high quality and are very simple to implement for basic systems, they are not flexible enough for anything but the simplest and most rigid interactions that do not involve too many nouns (or even involve some of the new verbs and adjectives that appear from time to time, so-called *neologisms*).

We have considered playback of word-level recordings above. When used to recreate continuous speech, this is an example of a concatenative system. However, concatenation could equally well be used at any level in the hierarchy of language. One level up in the hierarchy would be the simplistic but inflexible case of replaying stored phrases or sentences, whereas one level down is the more flexible case of phoneme concatenation, which we will discuss next.

10.1.2 Text-to-speech systems

As we have mentioned previously, text-to-speech (TTS) describes the process of turning written words into audible speech, and could include concatenation or other

---

[1] In fact Basic English has a number of proponents who have 'translated' many texts and novels over the years. Today we might consider using such an approach for learners of English, although the word choice in the restricted vocabulary has definitely moved on since the 1940s.

reproduction mechanisms. For a readable overview and survey of such systems, the reader is encouraged to refer to a paper by Dutoit [143] which captures all but the very latest approaches. The simplest TTS systems operate at a single word level (for example, using word playback as discussed in Section 10.1.1), but work only if the word is known in advance. A much more flexible approach is to build a system that is capable of producing different phonemes and then stringing these together in the correct sequence to produce different words. Although the English alphabet contains 26 letters, there are generally considered to be about 44 separate phonemes in English, so what is needed is a system that is able to produce each of those 44 sounds and can combine them flexibly and naturally in the correct sequence upon demand.

What is needed at a word level then is a dictionary or heuristic which relates each written word to a sequence of phonemes: in effect a rule specifying how to pronounce each word.

In English, pronounciation is a non-trivial task because spelling is often not phonetic: there are many words which must be pronounced in a way contrary to a phonetic reading of their spelling. Some languages benefit from more regular pronunciation rules, so translating a written word into phonemes becomes easier. In Mandarin Chinese, for example, words that are romanised (i.e. written in Western letters) make use of something called the pinyin system, which is a phonetic transcription of the approximately 56 phonemes used in the language. In other words, all Chinese characters have an official pinyin transcription which encodes how they should be pronounced. Chinese TTS is considered more in Infobox 10.1.

Moving back to English, it is common in TTS systems, including most commercial speech synthesisers, for there to be a mixture of dictionary and guesswork applied to pronunciation. Given a word to speak, the dictionary is first queried to determine whether a known word-to-phoneme mapping exists (and if so, that is used), otherwise a heuristic is applied to guess the pronunciation.

Early-years schoolchildren tend to learn their English in a similar way to this: if they know how to speak a word they will do so, otherwise they will attempt to pronounce it phonetically (with occasionally humorous results). The primary difference between toddlers and TTS systems is that the latter do not learn new pronunciations as a result of listeners laughing at their incorrect first attempts.

10.1.3 Linguistic transcription systems

Text-to-speech systems are all very well, but humans do not simply read words in isolation – and that is one reason why the word playback systems of Section 10.1.1 tend to sound unnatural. In reality humans modulate their speech over a sentence, adding stress and intonation to different words to alter the meaning of what is being said (see the sentences reproduced with different stressed words in Section 9.1.3), or to augment the meaning with emotional information such as interest, distaste, happiness and so on. All of this is in addition to the basic lexical information being conveyed by the text itself (i.e. over and above the meaning of the written words).

Most speakers are able to modulate their voice in a number of ways, including frequency-domain adjustment (pitch, formants) and time-domain changes (rate of

Advanced topics

> **Box 10.1** Chinese TTS
>
> We have discussed various aspects of Mandarin Chinese in previous chapters, including the structure of Chinese speech (Infobox 3.1 on page 63), as well as the testing of Chinese intelligibility in Chapter 7 and its features related to ASR in Chapter 9.
>
> From these discussions, we know that spoken Chinese is a tonal language with relatively few different pronunciation units (between 408 and 415 depending upon who measures it), but which are distinguished primarily by tone (and secondarily by context). There are four tones in Mandarin Chinese plus a neutral and all vowels must take one of these (Cantonese has something like eight tones, by contrast).
>
> Written Chinese meanwhile consists of tens of thousands of characters or ideograms (tiny pictures that often convey some semantic meaning and sometimes even hint at a pronunciation – but often don't) that represent a spoken word. Unfortunately many ideograms carry alternative pronunciations, which depend upon context, which tends to complicate the Chinese, TTS task. More unfortunate is that even *simplified* Chinese, as used in China, Singapore, Malaysia and among most foreign learners, comprises at least 3000 characters to convey simple information, such as in a mass-market newspaper, and over 13 000 characters for common works of literature. Each of those 13 000 conveys one or several phonetic transcriptions plus one or more tone alternatives. Thankfully we are only considering *simplified* Chinese here.
>
> An example is shown in the figure below:
>
>
>
> The Chinese ideograms (characters) ma3 guo2 ling3, representing the author's name in Mandarin Chinese.
>
> Like many translations of proper nouns, this example is primarily phonetic rather than having been chosen to convey a meaning. It is also limited to three characters (the norm for Chinese names is either two or three characters). Fortunately there are no alternative pronunciations or tones for these characters. It is read as 'ma' (meaning 'horse'), 'guo' (meaning country) and 'ling' (meaning the summit of a hill), which must be spoken with tones 3, 2 and 3 respectively: get the tones wrong and the name is unintelligible.
>
> So the TTS or character-to-speech task is (i) decide from context which pronunciation variant to use for a given character, (ii) map the pronunciation to phonemes, and (iii) decide which tone to use for each character.
>
> As if that task was not difficult enough already, written Chinese does not contain any spaces between characters: context is thus sometimes difficult to determine, as is the correct insertion of pauses between the phonemes of one word and the next. Finally, the tones corresponding to each word (character) are context-sensitive in some cases too!

speaking, pauses within and between words, and so on), or in more complex ways that change the perceived sounds. Most of the time we do not stop to think of the modulations we are using – they just occur naturally – although we may choose to override or accentuate an emphasis. Actors, in particular, learn to control their voice to order (see Infobox 10.2 on page 319).

> **Box 10.2** Actors and voice artists
>
> As with other human skills, people vary enormously in their speaking ability. At one extreme is the speaker who produces an inflexible, monotonous, emotionless and emphasis-free stream of semi-intelligible phonemes. I am reminded of one or two lecturers during my time as an undergraduate who had the seemingly amazing ability to reduce the most exciting and interesting topical content to a mind-numbingly boring endless drone. The other extreme might be those who have the ability, like the famous actor William Shatner (better known as Captain Kirk of the starship *Enterprise*), to imbue almost any textual content with dramatic pauses, changes in speed and emphasis, and pseudo-emotional overtones. These are primarily time-domain rate and amplitude variations.
>
> More accomplished actors (including Shatner's replacement, Patrick Stewart), accomplish deeper and more fundamental modulations to their voice, adjusting the tone, resonance, timbre and so on, to change accents at will.
>
> Extreme examples of this skill are voice artists who exhibit a mastery over their own voices that enables them to impersonate others as well as adopt completely different vocal personas. Many readers will likely have friends and acquaintances who are able to mimic others, in speech content (for example, on International Talk Like A Pirate day, 19 September each year), speaking pattern (for example, visit one of the many 'talk like William Shatner' websites), accent or tone. Professional voice artists who do this are hired by radio and television companies based upon their abilities to alter their voices at will.
>
> Several fascinating studies have been made of the speech produced by voice artists. Researchers conclude [144] that impersonators are able to control pitch, speaking rate, vocal fold patterns (i.e. timbre of pitch) and vowel formant distributions on a vowel-by-vowel basis. They also change the mapping from word to phoneme (i.e. pronounciation) to take account of prosody (the rhythm of speech), which they are also able to control. Studies show that the available range of voices that are physically achievable by most speakers is extremely large, limited only at the extremes by physiology. The limits include someone with a short vocal tract not being able to create all of the vowel frequencies of someone with a long vocal tract (i.e. a small female may find difficulty in mimicking the voice of a typical large male, and vice versa). However, most of the remaining factors that differentiate different speakers are potentially under the control of the speaker, and could be learned given time, patience and sufficiently good hearing.

Anyone who, like the author, has endured listening to overlong dreary monologues in lectures and academic conferences should value the ability of speakers to introduce intonation changes and modulations. As a practical demonstration of how this alters speech, try to read the sentence 'Friends Romans countrymen lend me your ears' aloud with uniform spacing between words. Now read it with the correct punctuation; 'Friends, Romans, countrymen, lend me your ears'. Try again with some emphasis on the first three words, plus longer pauses between them. Most would agree that these efforts can change the impact of the sentence quite noticeably. Such techniques are artfully employed by the best speakers, those whom it is a pleasure for us to hear. Readers who would like to improve their public speaking skills (for example, at a conference), might like to refer to the short article [145] for a few pointers.

Having demonstrated the importance of stress, intonation, pitch and pace/pauses in speech, we now need to acknowledge that these aspects are subjective elements added by a speaker, and which are not represented in the basic text. Thus even a TTS system that can reproduce individual words so well they are indistinguishable from a human speaker's would fail to produce a natural output when those words are joined together, unless the issues mentioned above are taken into account. Therefore, to achieve natural sounding speech synthesis, as many of those elements as possible should be incorporated into the synthesiser. Two alternative approaches are possible: firstly the use of a transcription system that includes the required information in the input instructions (i.e. in place of bare text input) and secondly a method of extracting or inferring the information from the text automatically. The second method is the way in which a human interprets, and augments, text that they are reading.

In fact the first system is not truly TTS since the input is not just text, but text plus linguistic markers. In a purely practical sense, stress markers can be found as accents within the international phonetic alphabet, and can be added by an expert transcriber – although it would probably require another expert transcriber to read the result. Experimental methods which can begin to add this information automatically to text do exist, and perhaps future speech recognition systems would be capable of recognising (and outputting) this information too. However, the task of determining this kind of information requires a degree of human intervention, and will probably continue to do so for the foreseeable future. For a more complete synthesis system, pitch variations, prosodic changes, volume modulation and so on would also need to be added to the text, perhaps along with precise phoneme timing information. The degree of effort and difficulty involved in doing this for a particular item of text means that systems using such linguistic marker information are better suited to those with a constrained vocabulary – however, these are precisely the kinds of situations for which stored playback solutions already work best.

The second method of improving naturalness requires a machine to 'read' a sentence without a-priori knowledge of stress and intonation patterns, extract such information from the text, and incorporate this into its reading. This is much the same task as the one a human speaker faces when given a paragraph to read out loud: scan the text, decide which words to stress, where to speed up, increase volume, pause and so on. This task is no longer considered to be within the domain of speech synthesis since it is one of interpretation, and is instead part of *natural language processing* (NLP) research.

Figure 10.1 shows a block diagram of a generic concatenative speech synthesiser of the second type mentioned above. A database of stored phonetic sounds is strung together by the system to match the phonemes of the input text. Most words will hopefully be found in a phonetic dictionary (which maps words to a sequence of phonemes). Words that are not in the dictionary are instead generated through phonetisation rules. For English speech, the dictionary needs to be quite extensive because there are many exceptions to phonetic pronunciation. This is especially true for place names, such as 'Derby', which is actually pronounced 'Darby', and 'Worcestershire', which should be pronounced something like 'Woostershur', as well as family names such as 'Cholmondeley', pronounced as 'Chumley'. Occasionally, context would dictate a different pronunciation of a word, and this would be detected through the use of an NLP system. A simple example is the word 'close' which is pronounced differently when used to indicate proximity (short 'o') or as an action (long 'o'). Apart from different contextual changes and basic pronunciations, certain forms of words and phrases are

10.1 Speech synthesis

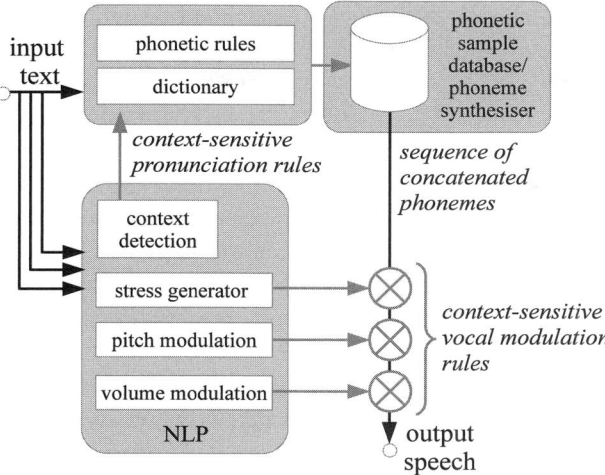

Figure 10.1 Block diagram of a generic speech synthesis system, showing input text and output of a concatenated strings of phonemes, A natural language processor (bottom left) accounts for context-sensitive pronunciation changes, adding stress, intonation, pitch and volume, and modulates output phonemes into a sentence structure with word emphasis.

never pronounced in the sequence in which they are written, or need to first be expanded into a full form. Some examples include £100 ('one hundred pounds' not 'pound one hundred' and definitely not 'pound one zero zero'), 23/04/2015 ('the twenty-third of April two thousand and fifteen'), etc. ('et cetera').

NLP is also used to identify places in the recreated phoneme string in which to add stress patterns, probably accompanied by changes in prosody, pitch variation volume variation and so on. As we have discussed above, intelligent NLP systems are being developed to identify such instances automatically from input text, but the alternative is for the input text to identify such changes. An example of the latter is the speech synthesis markup language (refer to Infobox 10.3 on page 325).

It should of course be noted that the outline presented above in this section is simply one possible method of performing text-to-speech synthesis: there are several alternative approaches represented in the research literature [143], and this remains a very active field of academic research at present.

10.1.4 Practical speech synthesis

There are many speech synthesis technologies available today. Rather than survey all of these, we will simply consider two alternatives. Both are open source software, so they can be freely downloaded, tested and modified. The first, called eSpeak, is a relatively simple, low-complexity standalone synthesiser which is ideal for rapid prototyping or for incorporating within a larger project. The second, the Festival speech synthesiser, is a highly advanced and configurable speech synthesis framework that is suitable for research in speech synthesis and capable of very good TTS quality. It can also be used for rapid prototyping if required, but is better suited as a research-based synthesis tool. We will consider each in turn, along with the emerging alternative of online TTS.

10.1.4.1 eSpeak

eSpeak was originally written as software for Acorn's RISC-OS (Reduced Instruction Set Computing – Operating System) computers in the late 1990s. Acorn's computers were advanced systems generally popular for home or education use, based around the ARM (Advanced RISC Machine) processor which was invented by Acorn. One reason why this history is interesting is that it is the ARM processor that has enabled the mobile computing revolution: almost all mobile telephones, modern tablets and portable media players are built around ARM processors today.

At present, eSpeak is available for almost all modern computing platforms,[2] and provides a standalone concatenative formant generator-based speech synthesiser that is capable of TTS, as well as having the ability to extend or modify the basic system for other applications. A number of voices are provided with the software, with more being available for download. The multilingual support is notable, with voices being supplied for over 80 languages, including several variants of English (including Northern, Received Pronunciation, Scottish, American and West Midlands) and Mandarin Chinese. Most languages provide multiple male and female voices.

eSpeak supports SSML (see Infobox 10.3) but can also read plain text documents and HTML (web pages), and is generally quite configurable. Having downloaded and installed `espeak` (or the standalone `speak` executable), the command line program operates according to the following examples for basic playback, and choice of either alternative voices (`-v`) such as West Midlands and Received Pronunciation:

```
speak "this is some text"
speak -ven-wm "I come from Birmingham"
speak -ven-rp "I don't come from Birmingham"
```

It is also possible to choose different male or female (+m or +f) voices, as well as separately adjust speaking rate (`-s`) and pitch (`-p`):

```
speak -ven-sc+f4 "Scottish female"
speak -ven-n "Newcastle or Liverpool?"
speak -s180 "fast speech"
speak -s80 -p 30 "slow and low speech"
```

Another fascinating feature of eSpeak is to output the decomposed phonemes that are being spoken. These (or in fact an output Wave file) can be captured and used for different applications:

[2] http://espeak.sourceforge.net

```
speak -x "show the phonemes"
speak --ipa "display phonemes in IPA format"
speak -w output.wav "write to a wave file"
```

As mentioned above, eSpeak is highly configurable, including in terms of creating or changing grammar rules, pronunciation rules and pronunciation exceptions. In addition the source code, written in C, can be modified if required.

10.1.4.2 The Festival speech synthesis system

The Festival speech synthesis system, created by the University of Edinburgh Centre for Speech Technology Research, is probably the most common synthesis system in use for research today. It is open source C++ software [146] which offers a highly configurable 'general framework' for speech synthesis research, as well as a fully working system that can synthesise speech in English, Welsh, American English and Spanish. Many other languages are available, although not currently Chinese.

Festival is capable of uttering whole sentences, which it can do through the use of a grammar or syntax structure. At a word level, a time-based sequence of words is defined inside Festival and associated with other attributes of each word. Words themselves are mapped to individual phones, which are the basic sound unit. Word functions are identified, and used to allow different rules to apply to those words. In this way Festival could, for example, utter the subject of a sentence at a slower speed, and a verb slightly louder than other words, since basic speaking information can be calculated, in real time, from syntactic information [147].

In common with many systems, including eSpeak, Festival uses a pronunciation lexicon to look up pronunciation of a word (or a word part, or find a near match). Unknown words are pronounced by a letter-to-sound rule which guesses at the correct pronunciation. A significant amount of work has also gone into the reading of various punctuation marks and abbreviations in Festival, which considerably improves the perceived naturalness when reading real-world text, such as emails. To further improve naturalness, Festival – like eSpeak – attempts to speak syllables rhythmically by adjusting speaking duration, modifying pitch to fit accents and adjusting any tonality which may be required. Users may also inform the synthesiser concerning the class of what is being spoken, such as words, phrases, raw text and so on, and provide relevant information to specify phrasing, timings and even tone changes [146].

With Festival installed, a basic command line interface is available:

```
echo "this is a test" | festival --tts
```

although it is better used either interactively (simply use `festival` with no arguments to launch the session) or by reading an entire file:

```
festival --tts textfile.txt
```

In practice, Festival (along with derivatives such as festvox and the efficient 'C' language alternative festival-life) is extremely configurable and provides excellent quality, at the expense of being rather more complex to operate.

10.1.4.3 Online TTS systems

Some of the highest-quality practical synthesis output is available by making use of very large scale and powerful cloud TTS systems that are accessible online from companies such as iFlytek, Google and Apple. This technology is almost the reverse of their ASR systems:

- **Online ASR** – a client application records a (usually short) audio segment locally, sending this to remote servers which are instructed to recognise the content. The servers then either send back the recognised text or act on the instruction, depending on the application. For example, when dictating a book, they would send back a sentence containing the recognised text, but if the request was 'what is the current time?' they would probably send back a synthesised time message (making use of online TTS; see below).
- **Online TTS** – a text string is sent to the remote servers to synthesise into a vocal representation, which is returned as an audio file (or perhaps as a phoneme string). The language, dialect and gender of the speaker can be specified along with various other attributes. The text string may be synthesised 'as is' or may contain a command, in which case the response to the command is synthesised.

At the time of writing, both of the above are confined to same-language processing (i.e. the same language is used for the input and the output of either service, although both are available in several languages). It does not take a giant leap of imagination to realise that this kind of service will eventually extend to translation (or interpretation) – which could allow either text or speech input in any language, and output to either text or speech in any other language.

Despite some nice-looking demonstrators from large companies in recent years (the developers of which tend to avoid mentioning the limitations), the best that current research systems can achieve is to translate simple and short fixed texts between a limited number of languages. However, this is an area of intense research in both academia and industry, so further advances are eagerly anticipated in the near future.

10.2 Stereo encoding

Stereo means something that is stored or reproduced in more than one channel. It can also be referred to as *stereophonic*. This was contrasted to the technical term 'binaural' in Section 4.2.13, which means something pertaining to both ears. In audio terms, stereo normally refers to two channels of audio, one for a right loudspeaker and one for a left loudspeaker, but many other multichannel arrangements are also possible.

The basic idea of stereo audio relates to the fact that most of us have two ears which we use to differentiate between sounds, particularly regarding source placement.

10.2 Stereo encoding

> **Box 10.3** Speech synthesis markup language
>
> Speech synthesis markup language (SSML) has been defined by the W3C (World Wide Web Consortium – the group that creates the standards that define how the Internet is used and grown). The same group defines and manages the hypertext markup language (HTML), which is the language that web browsers use to display information from websites. In the same way as HTML controls how information from a website will appear on the screen of a computer or mobile device (i.e. fonts, colours, emphasis and so on), SSML defines how textual information is spoken aloud. In effect, it controls or assists in the TTS of web information. In fact, SSML has many other uses apart from simply reading web pages, and has been used in applications as diverse as telephone response systems, basic dialogue systems and creating audio books.
>
> SSML is based on XML, and allows a content provider (such as a website) to specify the voice used (such as accent, age and gender), pronunciations, prosody, emphasis and so on. It defines which elements of a document are to be read, as well as how they are read. We will not consider SSML syntax here (instead, please refer to the standards document [148]); however, a few of the more useful SSML commands that can be used inside a `<speak>....</speak>` construct are shown below:
>
> | Command | Controlling attributes |
> |---|---|
> | `voice:` | |
> | `gender` | `male, female, neutral` |
> | `age` | *a numerical age of speaker to synthesise* |
> | `name` | *a short-cut named voice for convenience* |
> | `prosody:` | |
> | `pitch` | *specified base pitch in Hz or as a relative value* |
> | `range` | *overall pitch limits in Hz or as a relative value* |
> | `rate` | *adjust speaking rate from* `x-slow` *to* `x-fast` |
> | `volume` | *speech volume from* `silent` *to* `x-loud` |
> | `emphasis` | `strong, moderate, none, reduced` |
> | *Controls the stress applied to a word or phrase* | |
> | `break` | `x-strong, strong, medium, weak, x-weak, none` |
> | *Overrides default gaps in the spoken text* | |
> | `phoneme` | *Uses the* `ipa` *alphabet to specify a phonetic reading of text* |
> | `say-as` | *Uses the* `interpret-as` *and* `format` *attributes for enclosed text that is specialised in content, for example dates, URLs or when all items including punctuation should be read aloud.* |
> | `audio` | *allows a recording to be embedded in the speech* |
>
> It is likely, as speech interfaces become more common, that SSML will continue to evolve and improve, alongside alternatives like Sable and SGML (Standard Generalised Markup Language), and in-house speech markup languages from companies such as Apple, Google, Microsoft and IBM. Finally, it should be noted that extensions exist for other languages, most notably recent proposals for a `pinyin` alphabet as a complement to `ipa` to account for lexical tone information as well as regular pronunciation rules for languages such as Mandarin Chinese.

In nature, a sound to the left of our heads will be heard by the left ear slightly before it is heard by the right ear, and slightly louder. This tiny delay, or phase difference, between the two versions of a sound heard by the ears is sufficient to allow the brain to calculate where the sound originated from. We are not conscious of this calculation since it is performed automatically, but we can demonstrate the effect very easily using MATLAB.

First we will create a sound made from two related sinewaves. These need to be sampled at a higher frequency than some of our demonstrations because of the finer control over timing we can achieve. In this case we will use a 44.1 kHz sample rate to create a 440 Hz note in the usual way with our `tonegen()` function:

```
Fs=44100; %sample frequency
Ft=440;
note=tonegen(Ft, Fs, 0.5)
```

Next we will replay two versions of this using `soundsc`. In MATLAB, stereo sound is represented by a two-column matrix of samples. The first column holds samples for the left channel, while the second column holds samples for the right channel. We will create two two-column matrices of audio, one in which the left channel is delayed slightly and one in which the right channel is delayed slightly:

```
s1=[[zeros(1,20), note];[note, zeros(1,20)]];
s2=fliplr(s1);
```

In this case `fliplr()` is used to switch the left and right channels around. Next we will listen to those two sounds. For this you will definitely need a computer capable of replaying in stereo (fortunately, most computers are capable of doing so these days) and quite likely will require a pair of stereo headphones to really appreciate the stereo placement:

```
soundsc(s1, Fs);
soundsc(s2, Fs);
```

When listening, the first of these should sound like it is coming from a source located to the right, and the second should sound like it is from a source located to the left.

Also as mentioned in Section 4.2.13, if headphones are not used, then both ears will hear a mixture of the two sounds. This is a slightly more complicated proposition, but again, good quality loudspeakers should be able to accurately simulate the sound localisation. If in doubt, try it and see.

10.2.1 Stereo and noise

Interestingly, it is not necessary for the two sounds to be related to each other for the brain to hear them in stereo. In this example, we will create two vectors of random noise. First we will play the same vector to both ears:

```
r1=0.2*rand(1, Fs*0.5);
r2=0.2*rand(1, Fs*0.5);
%Note use sound not soundsc to save your ears
sound([r1;r1], Fs);
```

This should sound as if the noise is located centrally between our ears. Next, we will play different random noise signals to the two ears:

```
sound([r1;r2], Fs);
```

Hearing the uncorrelated signals causes the brain to assume that the signals are unrelated, and separated. The sound now appears to be spatial in nature – stereo noise in effect. As an aside, this is one reason why some audio purists still prefer to listen to music from a turntable: the noise from a stereo vinyl record is predominantly common to both channels, and thus is not stereo. The noise may appear to hover in between the loudspeakers, leaving the listener free to enjoy the more spatially separated sounds.

10.2.2 Stereo placement

An arbitrary spatial placement for stereo sounds can be calculated from a physical representation of the system we are attempting to model. Given a sound some distance d away from a listener, and at an angle θ away from the location directly ahead, sounds will need to travel further to reach one ear than to reach the other. This is shown in Figure 10.2, where the left ear will receive sound from the point identified slightly later than the right ear will.

If we denote the head radius as h (the distance between the two ears thus being $2h$), then we can use the cosine rule to calculate exactly the path difference travelled by audio reaching each ear:

$$l^2 = d^2 + h^2 - 2dh \times \cos(\pi/2 + \theta), \qquad (10.1)$$

$$r^2 = d^2 + h^2 - 2dh \times \cos(\pi/2 - \theta). \qquad (10.2)$$

Then, assuming location at sea level where the speed of sound is approximately 350 m/s, sounds will reach the left ear in $l/350$ s and the right ear in $r/350$ s. At a sample rate of Fs samples per second, the difference in number of samples between the two would be $Fs \times 350/(l-r)$. In MATLAB we could then replay a sound, in stereo, with one of the channels delayed by this many samples to cause the brain to interpret the sound as coming from angle θ.

This calculation has been used, in the MATLAB `stereo.m` demonstration code shown below, to simulate a sound moving around the head, additionally with a rudimentary scaling applied whereby the amplitude of signal heard in each ear is inversely

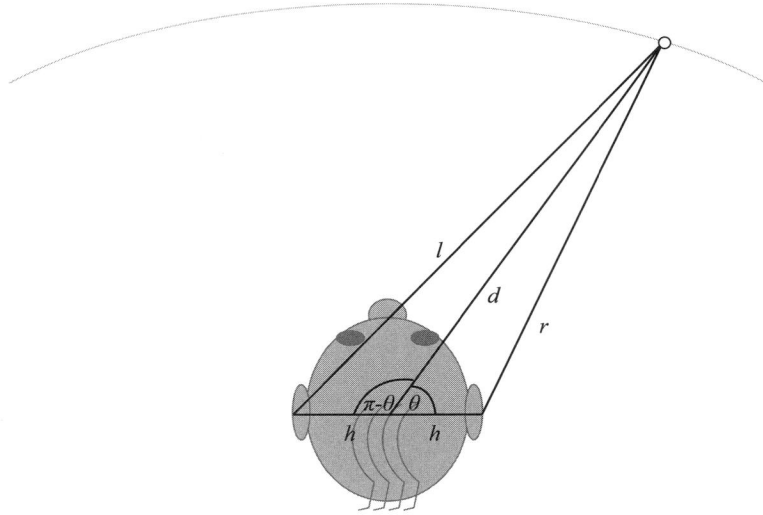

Figure 10.2 Diagram of angle-of-arrival of sound from a single source at distance d and angle θ causing a path distance difference for sound reaching the two ears.

proportional to the distance to the simulated sound source (so that the ear closer to the sound hears a slightly higher-amplitude signal than does the ear which is further away):

Listing 10.1 stereo.m

```
d=5;     %distance
h=0.1;   %head radius
Fs=44100;
Ft=600;
note=tonegen(Ft, Fs, 0.10);
note=note+tonegen(Ft*2, Fs, 0.1);
%Speed of sound
Vs=350;  %m/s
ln=length(note);
%Cosine rule constants
b2c2=d^2 + h^2;
b2c=2*d*h;
for theta=-pi:pi/20:pi
    %Calculate path differences
    lp= b2c2+b2c*cos((pi/2)+theta);
    rp= b2c2+b2c*cos((pi/2)-theta);
    %Calculate sound travel times
    lt= lp/Vs;
    rt= rp/Vs;
```

```
    %How many samples is this at sample rate Fs
    ls= round(Fs*lt);
    rs= round(Fs*rt);
    %Handle each side separately
    if(rs>ls)    %right is further
      df=rs-ls;
      left=[note, zeros(1,df)]/ls;
      right=[zeros(1,df),note]/rs;
    else         %left is further
      df=ls-rs;
      left=[zeros(1,df),note]/ls;
      right=[note, zeros(1,df)]/rs;
    end
    %Create the output matrix
    audio=[left;right];
    soundsc(audio, Fs);
    pause(0.1);
end
```

10.2.3 Stereo encoding

Stereo is normally stored in a file either as interleaved left and right samples or as a two-column matrix as in MATLAB, but in professional audio fields there are far more possibilities. Infobox 10.4 describes some of these.

Within audio systems, the traditional approach is to maintain left and right channels as separate streams of data. Gain, filtering and so on would be performed independently on each. However, systems which purport to enhance stereo separation, surround sound systems and spatial audio systems would each mix together the stereo channels with different, possibly dynamic, phase relationships. In compression systems, it is also common to encode stereo channels jointly. In many cases this is simply to encode middle and side channels, created from the difference between the left and right channels, separately (see Infobox 10.4: *Common stereo audio standards*, under Mid-side stereo, where the mid channel is $(R - L)/2$ and the side channel is $(L - R)/2$, performed to improve quantisation performance). On the other hand, some systems perform a type of joint stereo encoding in which the lower-frequency information is stripped from both the left and right channels (leaving them simpler, and thus able to be compressed using fewer bits), and then encoded separately.

Although the term 'joint stereo encoding' is frequently misused, especially in the speech compression community related to MP3 encoding, the idea of joint stereo encoding is that humans are not really able to localise low-frequency sounds, and hence there is no need to represent them in stereo.

> **Box 10.4** Common stereo audio standards
>
> There is a plethora of stereo audio standards, and many non-standards, in use around the world today. While the vast majority of stereo music is recorded in a straightforward R + L (right, left) arrangement, including in CDs, the digital audio coding used on DVD and more advanced media tends to be more complex. Professional audio systems, as used in well-equipped modern cinemas, add to the complexity.
>
> Classified under the general term 'surround sound', these advanced systems include the following notable arrangements (any trademarks and trade names mentioned are the property of their respective owners):
>
> | Name | Front | Middle | Rear | Notes |
> |---|---|---|---|---|
> | Mono | C | | | |
> | Stereo | L, R | | | |
> | Mid-side stereo | (R − L)/2, (L − R)/2 | | | |
> | Dolby surround | L, R | | S | |
> | Quadraphonic | L, R | | L, R | |
> | Dolby Pro Logic | L, R, C | | S | five speakers |
> | Dolby Pro Logic II | L, R, C | LFE | L, R | known as 5.1 |
> | Dolby Digital, DTS | L, R, C | LFE | L, R | known as 5.1 |
> | Dolby Pro Logic IIx | L, R, C | L, R | C | plus LFE |
> | Dolby Digital Plus | L, R, C | L, R, LFE | L, R | known as 7.1 |
> | 10.2 channel | 2 × L, C, 2 × R | 5 × S | 2 × LFE | 2 height channels |
> | 22.2 channel | arranged in 3 layers: 9 top, 10 middle, 3 lower, 2 woofers | | | |
>
> L = left channel, R = right channel, S = surround channel, C = centre channel
> LFE = low-frequency effects channel
>
> The urge to add more and more channels appears to be unstoppable, despite the fact that humans have only two ears with which to hear those sounds.

A very simple experiment to show this would be to repeat the MATLAB demonstration of the previous section with much-lower-frequency sounds, perhaps 40 Hz or so (if your computer can reproduce such a low frequency). Also, the prevalence of single subwoofer setups in modern audio systems is due to the fact that two would be redundant: a house-shaking bass sound permeates an entire room and the brain cannot easily localise the source.

10.2.4 More than two channels

We have already seen, in Infobox 10.4, how one channel became two channels, which in time became multiple channels, all in the race to better recorded sounds – especially in theatre environments.

In fact there are other reasons to have multiple audio channels, such as to better capture speech for ASR in situations such as meeting rooms, as discussed in Section 9.1.2.4.

10.2 Stereo encoding

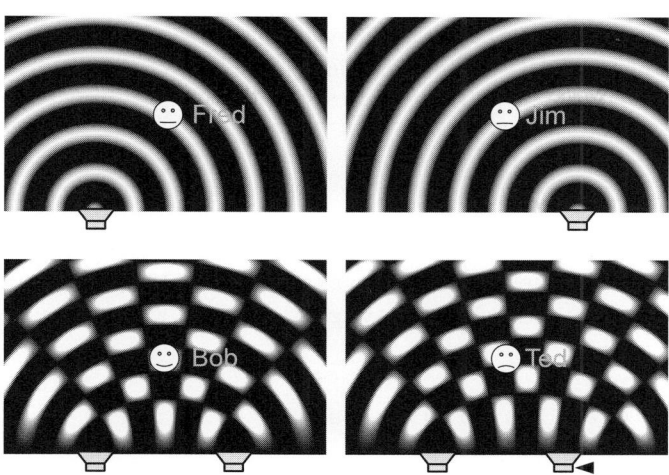

Figure 10.3 Fred and Jim have only a single loudspeaker so their ears hear the same sound with a small time delay. Bob has a stereo system and sits equidistant between the loudspeakers to hear sounds propagating to each ear, sounding as if they are coming from directly ahead. Meanwhile Ted accidentally shifted one of his loudspeakers, and now hears a louder sound in one ear than in the other.

To explore further, consider the diagrams in Figure 10.3, showing two monophonic and two stereophonic arrangements. In each case, only a fixed frequency sinewave is being output from the loudspeakers, and the signal amplitude is plotted in a rectangular space. These plots have been created using the simple function below:

```
function arr=pt_srce_sim(Lx,Ly,pix,Xp,Yp,W)
%Lx, Ly: horiz. and vert. dimensions, in m.
%pix: the size of one point to simulate, in m.
%Xp, Yp: coordinates of the sound source
%W: wavelength of the sound (where W=340/freq)
Nx=floor(Lx/pix);
Ny=floor(Ly/pix);
arr=zeros(Nx,Ny); %define the area
for x=1:Nx
  for y=1:Ny
    px=x*pix;
    py=y*pix;
    d=sqrt((px-Xp).^2+(py-Yp).^2);
    arr(x,y)=cos(2*pi*d/W);
  end
end
```

The usage of the code can be illustrated by the following, which was used to generate the simulated plot for Bob's stereo system in Figure 10.3:

```
Lx=1.6; Ly=1;
Xp=1; Yp=0.5;
W=0.2; % about 1.7 kHz
pix=1/1000; % 1 mm
arrR=point_source_sim(Lx,Ly,pix,0.45,0,W);
arrL=point_source_sim(Lx,Ly,pix,1.15,0,W);
%display image
imagesc(arrR+arrL)
colormap('gray') %using greyscale colours
%save to file
imwrite(arrR+arrL,'stereo_source.tiff');
```

The principle of superposition is applied to add the contributions in the simulated space from each loudspeaker – no matter how many are used. We can see that `arrR` comes from the right-hand loudspeaker (from Bob's perspective), located at the edge of the 'room', 45 cm from the wall. `arrL` is from the opposite side of the room (which has dimensions 1.6 m by 1.0 m). In each case, the walls are perfectly absorbing and do not reflect any sound.

Slightly more interesting are the cases in Figure 10.4. Five loudspeakers arranged in a non-uniform pattern creates a very complex and irregular sound field. Changing the frequency between speakers also creates a more interesting sound field.

Another adjustment we can make is illustrated in Figure 10.5, which shows the effect of a time delay between two loudspeakers. In this case the signal from the left loudspeaker is gradually delayed with respect to the signal from the right loudspeaker. We know that a delay could be introduced electronically, but could also have been achieved by moving the left loudspeaker backwards a few centimetres each time. The effect is to 'steer' the 'beam' of the sound so that it is heard differently by the listener.

There are many applications of these principles for loudspeakers. For example, a stereo system could 'steer' the sound so that it is centred on a listener who is not seated in the so-called sweet spot.[3] The sweet spot (depending upon who you ask) is equidistant between two loudspeakers that are positioned about 1.5 m apart, forward facing (in parallel, not angled) with respect to the listener who is seated at least 2 m further back, at a height which is vertically aligned with the high- or mid-frequency speakers.

In a particular room, if this seating arrangement is not possible, then an advanced music system could introduce a small delay so that the sound waves are steered enough

---

[3] The sweet spot is ideally where the superposition of sound waves from the loudspeakers always contributes to a maximum. The opposite is where sound waves from the loudspeakers coincide in a minimum, cancelling each other out. Bob was seated in the sweet spot in Figure 10.3, but Ted was not.

10.2 Stereo encoding

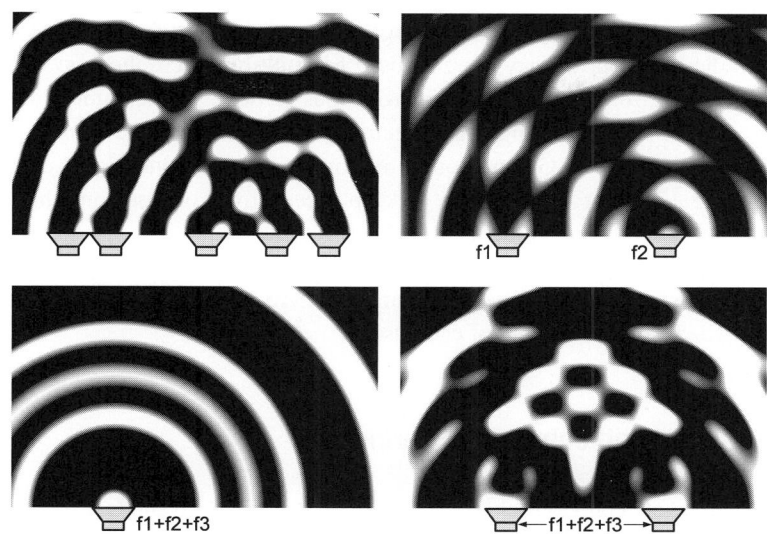

Figure 10.4 Top left is a system with five loudspeakers, delivering a complex sound field. Top right has two loudspeakers with a slight shift in frequency between them. Bottom left shows multiple frequency signals from a single loudspeaker. Bottom right is a stereo system outputting multiple frequencies from both channels.

Figure 10.5 Even with a simple stereo system at fixed frequency, we can change the time delay between the two loudspeakers to steer the sound field with respect to the fixed position of the listener, Alice.

to place a listener within the sweet spot. But what happens if there are two listeners, Bob and his wife, Alice? They can't both sit equidistant from the loudspeakers at the same time. This is where a complex sound field comes in, like that produced by five loudspeakers in Figure 10.4. With some mathematics and signal processing, it is possible to steer the waves so that there are two sweet spots, one for Alice and one for Bob. Furthermore, if Alice decides that she doesn't really want to listen to music, it is possible to steer a null spot towards her (i.e. a spot where the principle of superposition means that the waves cancel each other out), so that she hears a much quieter audio signal than Bob does.

At the time of writing, these techniques are not commonly found in listening systems, but exactly the same methods are commonly used in wireless data links with multiple antennas, in phased array radar systems, in synthetic aperture radar and in very large array radio-telescopes. They are also used for multiple microphone systems to record meetings. Electronically steering microphones is done in the same way as it is for loudspeakers: by introducing small electrical delays between the signals picked up by different microphones. Given enough microphones and an intelligent enough system, it is possible to steer the signal to pick up speech from one person while attenuating unwanted speech from another person.

It is highly likely that such techniques will be crucial to the success of future speech recognition systems in real-world situations where microphones are not guaranteed to be located close to the mouth of each user.

10.3 Formant strengthening and steering

As we know from previous chapters, LSPs are convertible to and from LPC coefficients, and, when in use in speech compression systems, LSPs will be quantised prior to transmission or storage. It is thus common to find that the decoded LPC coefficients differ from the original LPC coefficients in some ways. The use of an intermediate LSP representation for quantisation ensures that instabilities do not occur.

A comparison of the original and final spectra shows differences in the immediate frequency regions of the lines that were changed most. From these observations it has been found possible to alter the values of particular LSPs to change the underlying spectral information which they represent.

To illustrate this, Figure 10.6 plots the original LPC power spectrum from Figure 6.14, overlaid with a spectrum derived from an altered set of LSPs. The changes in the new spectrum came about through increasing the separation of LSP pair {1:2},

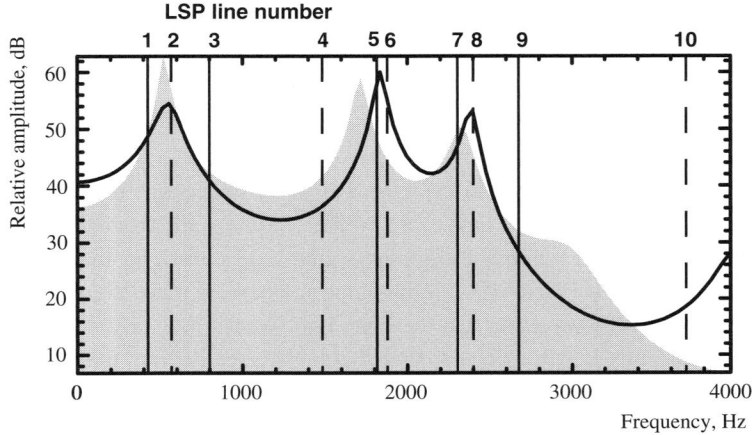

Figure 10.6 LPC power spectrum (in grey) overlaid with an altered set of LSPs (vertical lines) and the spectral envelope (solid curve) resulting from the alteration. The original spectrum was given in Figure 6.14 on page 166. The alterations include line widening and shifting.

decreasing the separation of line pair {5:6} and shifting line 10. The increased separation of pair {1:2} resulted in a wider, lower-amplitude spectral peak between them, whereas the decreased separation of pair {5:6}, plus a slight upward frequency translation, caused a sharper peak between them, now at a higher frequency. Finally, moving line 10 closer to the Nyquist frequency of 4 kHz caused a spectral peak to form at that frequency.

It was probably Paliwal [149] who first reported that the effects on the underlying spectrum of modifying a line are predominantly confined to the immediate frequency region of that line. However, this observation is correct only insofar as obvious large scale changes are concerned, since amplitude alterations in one spectral region will always cause compensatory power redistribution in other regions.

Despite this, as long as line alterations are minimal, the effects on other spectral regions will be negligible. This means that, for small movements, and small movements only, localised spectral adjustments can be made through careful LSP movement.

The example of Figure 10.6 showed a spectrum of voiced speech. The three spectral peaks thus represent formants, and as such we can see that the operations we performed affected formants directly. In fact, the LSP operations have demonstrably adjusted the formant bandwidths and positions.

The LSP operations can be formalised as follows. If ω_i are the LSP frequencies and ω'_i the altered frequencies, then narrowing line pair $\{i : i + 1\}$ by degree α would be achieved by:

$$\omega'_i = \omega_i + \alpha(\omega_{i+1} - \omega_i), \tag{10.3}$$

$$\omega'_{i+1} = \omega_{i+1} - \alpha(\omega_{i+1} - \omega_i), \tag{10.4}$$

and increasing the frequency of line k by degree γ may be achieved with:

$$\omega'_k = \omega_k + \omega_k(\gamma - 1)(\pi - \omega_k)/\pi. \tag{10.5}$$

When altering the line positions it is important to avoid forming unintentional resonances by narrowing the gaps between lines that were previously separated. This problem may be obviated either by moving the entire set of LSPs or by providing some checks to the adjustment process [150]. In the former case, movement of lines 1 and 10 closer to angular frequencies of 0 and π may also induce an unintentional resonance. Equation (10.5), designed for upward shifting, progressively limits the degree of formant shift as a frequency of π is neared. A similar method may be used for downward shifting.

In MATLAB, a simple program to detect the three narrowest pairs of lines in a set of LSPs (in vector lsp) and to narrow the pair spacing between each of the three pairs by degree sc is as given below:

Listing 10.2 lspnarrow.m

```
function nlsp=lspnarrow(lsp,sc)
p=length(lsp);
wid=diff(lsp);
```

```matlab
%which LSP pairs are narrowest?
n=[pi, pi, pi];
ni=[0, 0, 0];
for lp=1:p-1
    if(wid(lp) < n(3))
        if(wid(lp) < n(2))
            if(wid(lp) < n(1))
                n=[wid(lp), n(1:2)];
                ni=[lp, ni(1:2)];
            else
                n(2:3)=[wid(lp), n(2)];
                ni(2:3)=[lp, ni(2)];
            end
        else
            n(3)=wid(lp);
            ni(3)=lp;
        end
    end
end
%narrow the 3 narrowest pairs even more
nlsp=lsp;
for k=1:3
   nlsp(ni(k))   = lsp(ni(k))   + n(k)*sc;
   nlsp(ni(k)+1) = lsp(ni(k)+1) - n(k)*sc;
end
```

A scaling factor of $sc = 0.2$ or less is usually sufficient to noticeably accentuate spectral peaks, but the code can also be used in reverse, by scaling with $sc = 1.2$ to widen the spectral peaks. Thus, the MATLAB code shown was used to produce the LSPs and spectrum shown in Figure 10.7. Shifting, rather than narrowing, formant-related lines would involve simply replacing the subtraction in the final loop with an addition (although that would neither detect nor correct resonances caused by lines approaching either angular frequency extreme).

Definitely, it should be noted that adjusting lines in this way alters the frequency relationship between any underlying formants, and therefore will tend to degrade the quality of encoded speech. In fact the code shown, when applied to continuous speech recordings, will sometimes result in very unusual sounds once the degree of scaling becomes extreme (for narrowing with $sc \geq 0.5$ or widening with $sc \geq 1.5$), thus it is far better to make only small changes to a speech spectrum wherever possible.

10.3.1 Perceptual formant steering

In order to minimise perceived quality degradation, a perceptual basis can be used for line shifting. For example, frequencies can be altered by constant Bark, rather than by

10.3 Formant strengthening and steering

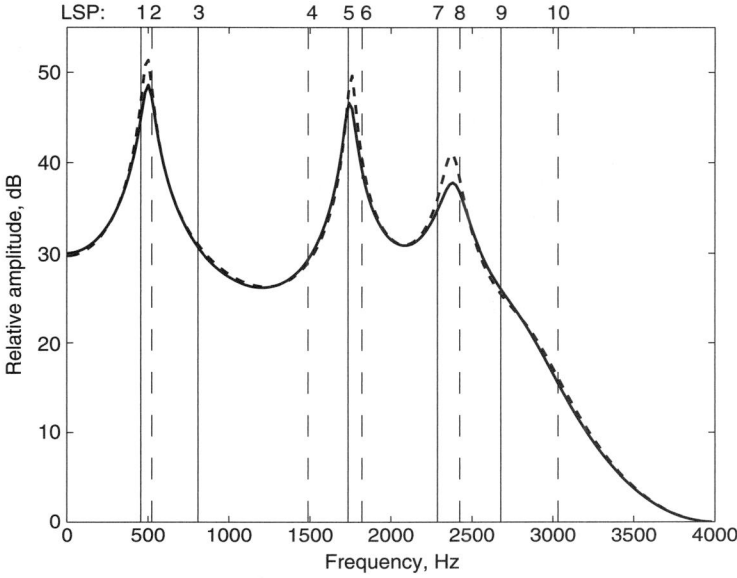

Figure 10.7 An original spectrum (dashed line) transformed to widen the bandwidth by spreading spectral peaks (solid line) through the use of LSP adjustment.

a fixed ratio. In this scheme, if B_k is the Bark corresponding to the frequency of a particular line, then that line shifted by degree δ is:

$$\omega'_k = 600 \sinh\{(B_k + \delta)/6\}. \tag{10.6}$$

To demonstrate this in MATLAB, we would very simply use the `bark2f()` and `f2bark()` functions developed in Section 4.3.2 to convert to and from Bark domain, and would apply these to each line in a set of LSPs. So, for an original LSP set `lsp`, we can shift by degree b Barks to derive a set of shifted lines `lsp2`. We can shift line n with:

```
lsp2(n)=bark2f(f2bark(lsp(n))+ b);
```

However, it would of course not be necessary to shift every line in the set of LSPs; only the lines directly relating to particular formants would need to be shifted. The hard limit that we applied above would still need to be used to prevent LSP values approaching the angular frequency extremes of 0 or π and checks made to prevent unintentional resonances caused by moving two lines too close together. A big disclaimer, mentioned previously in this book, is that applying anything other than small changes to a continuous speech spectrum is likely to lead to noticeably reduced quality.

10.3.2 Processing complexity

In terms of processing complexity, LSP narrowing requires three operations for every line (nine operations for a typical three-formant frame). Shifting using Equation (10.5) requires four operations per line, or 40 operations to shift all lines in a tenth-order analysis frame. LSP shifting using Equation (10.6) requires around six operations per line but, when implemented on a resource-constrained device, would usually necessitate a lookup table.

Similar formant processing effects can also be produced using adaptive filter techniques. Such a filter requires at least 2*NP* operations per *N*-sample *P*th-order analysis frame.

For tenth-order analysis operating on 240-sample frames, the LSP processes discussed here are between 40 and 400 times more efficient than an adaptive filter. Such figures are valid only where LSP data are available (which is the case for many CELP coders). If LSP data are not available, the overhead of transforming LPC coefficients to and from LSPs would be far greater than any possible efficiency gain.

The methods of LSP adjustment described here have been successfully applied in the intelligibility enhancement of speech [151]. In particular, the use of LSP shifting for altering the balance of spectral power between formant peaks and valleys shows promise. Figure 10.7 illustrates the effect of increasing the separation between the three most closely spaced line pairs by 20%, and was performed using the lspnarrow() MATLAB function on page 335. The figure plots both the resultant spectrum (drawn with a solid line) and the original spectrum (drawn with a dashed line). Traditional bandwidth-altering adaptive filters, such as that of Schaub and Straub [152], perform a similar task, but at a higher computational cost.

Further applications of LSP adjustment may include the quality enhancement of speech/audio and voice masking – described in Section 10.4.

10.4 Voice and pitch changer

If we record some speech at one sample rate and play it back at another, we may notice changes in the perceived frequency of the replayed speech. For example, recording a sentence at 8 kHz and replaying at 12 kHz in MATLAB would cause the output to be obviously different, speeded up in some way.

Let us try this – first record some speech (this records 2 seconds at 8 kHz sample rate):

```
ro=audiorecorder(8000,16,1);
record(ro); pause(2);
stop(ro);
speech=getaudiodata(ro, 'double');
```

We can try playing back at the normal sample rate, as well as both faster and slower:

```
soundsc(speech,8000);
soundsc(speech,16000);
soundsc(speech,4000);
```

Listening closely, it is noticeable that the output of the 16 kHz playback is both higher in frequency and spoken quicker (i.e. half the duration). Similarly the 4 kHz playback is both lower in frequency and spoken slower. The resulting speech in both faster and slower cases does not sound much like human speech – it has obviously been processed in some way. In fact, if you experiment with different sample rates, you might find that any change above about 10% to 15% results in unnaturally sounding speech.

The reason is related to the fact that speech is generated by the combination of several physiological processes as discussed in Chapter 3, not all of which are linearly scaleable. Put another way, although the voice of a child may have a pitch twice that of a man, the syllabic rate is unlikely to be twice as fast. Furthermore, the frequency locations of child formants are unlikely to be at frequencies as much as twice those of an adult (although they will most likely be increased in frequency to some degree).

So changing the frequency of a human voice is a non-trivial operation. We cannot simply double everything and expect the output to be convincingly human. Pitch must definitely be scaled to adjust voice 'frequency', but this needs to be adjusted differently from the other components of speech.

Practical methods of vocal frequency translation exist both in the time domain and in the linear-prediction domain. In either case, the important aspect of the technology is to *stretch* the voice of the speaker in some way. Having discussed the importance of pitch in Section 6.3, and the role of the particular shape of the pitch pulse, such stretching or contraction of pitch periods should be accomplished with as little damage to the pitch pulse shape as possible.

We will demonstrate two alternative ways of achieving this. The first, time-domain, method, is to detect the pitch periods and scale the waveform shape in a way which is sensitive to the importance of pitch. The second is to completely separate the pitch signal from the speech and then scale the pitch as required whilst either leaving the remaining speech components untouched or scaling them in a different way. Our first demonstration will be of a pitch-synchronous time-domain scaling, while the second demonstration is an LPC-based speech decomposition method. Both result in reasonable quality scaled speech that, unless the scaling ratios are very large, can be convincingly natural.

10.4.1 PSOLA

The primary traditional method for pitch scaling in audio is known as PSOLA (pitch synchronous overlap and add). This algorithm lives up to its pitch-synchronous name by first determining a fundamental pitch period. It then segments audio into frames of twice that size, windows them and reassembles the frames using an overlap–add method at a different rate (see Section 2.4 for a discussion on segmentation and overlap) [153].

The different rate of reassembly could either be faster or slower than the original, but, as with most such techniques, extreme adjustments can cause significant quality degradation. Figure 10.8 demonstrates the process of speeding up a recording of

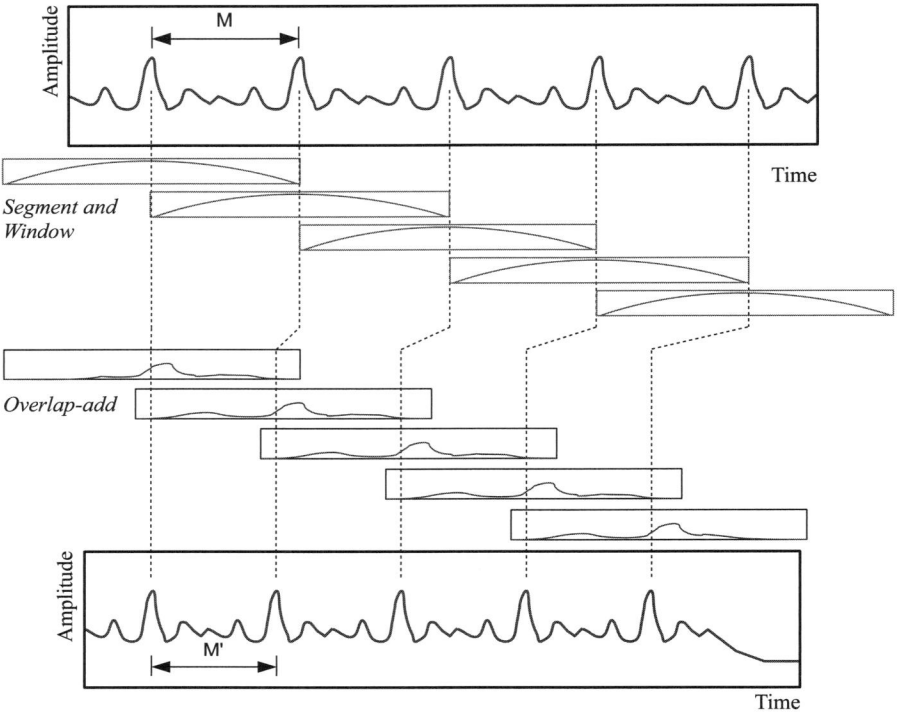

Figure 10.8 An illustration of the PSOLA algorithm, analysing the pitch period in the top waveform, segmenting it into double sized frames, one per pitch period, which are then windowed and reassembled by overlap-adding with reduced spacing to create a faster pitch signal result in the lower waveform.

speech. In this figure, a period of input speech (top waveform) is analysed to determine its fundamental pitch period, M. The speech is then segmented into frames of size $2M$ with a 50% overlap, ideally centred on the pitch pulse. Each frame is windowed (see Section 2.4.2), and the resultant audio is 'stitched together' at a different rate. In this case, the pitch pulses are more frequent, thus increasing the pitch rate of the resulting audio.

MATLAB code to demonstrate the effectiveness of the PSOLA algorithm is provided below. This relies upon the function `ltp()` for the pitch extraction method, which we developed in Section 6.3.2.1 to perform long-term prediction (LTP).

Within the code, a Hamming window is applied to N frames of size $2M$ extracted from input speech array `sp`. The array indexing, using variables `fr1`, `to1`, `fr2` and `to2` to denote the start and end indices of each array, is the heart of the method. This can be applied to a short recording of a couple of words of speech. In this case, a scaling of 0.7 will very clearly speed up the speech, whereas a scaling of perhaps 1.4 will slow down the speech. Note that the intelligibility of the speech remains high, although the characteristics of the speaker's voice will change.

Listing 10.3 psola.m

```
%Determine the pitch with a 1-tap LTP
[B, M] = ltp(sp);
%Scaling ratio
sc=0.35;
M2=round(M*sc);
out=zeros(N*M2+M,1);
win=hamming(1, 2*M);
%Segment the recording into N frames
N=floor(length(sp)/M);
%Window each and reconstruct
for n=1:N-1 %Indexing is important
    fr1=1+(n-1)*M;
    to1=n*M+M;
    seg=sp(fr1:to1).*win;
    fr2=1+(n-1)*M2-M;
    to2=(n-1)*M2+M;
    fr2b=max([1,fr2]);   %No negative indexing
    out(fr2b:to2)=out(fr2b:to2)+seg(1+fr2b-fr2:2*M);
end
```

Most probably, speech scaled by the PSOLA algorithm above will still sound reasonably true to human speech (in contrast to a straightforward adjustment of sample rate). PSOLA is also reported to work very well with music, which can be left as an exercise for the reader to try.

10.4.2 LSP-based method

For adjustment of speech, with potentially better performance than PSOLA, the speech signal can be decomposed, using a CELP-style analysis-by-synthesis system, into pitch and vocal tract components. The pitch can then be scaled linearly as required, and the vocal tract resonances can also be tugged upward or downward (by smaller amounts) to further scale the speech.

In the CELP vocoder, these alterations can be performed between the encode and the decode process, on the encoded speech parameters themselves. LSP parameter changes (on formant-describing line pairs) are used to tug formants either higher or lower in frequency. Scaling the pitch delay parameter in a one-tap LTP (see Section 6.3.2.1) similarly adjusts the pitch period. With these changes it is possible to shift vocal frequencies, either to change the pitch of a speaker's voice or to scramble their voice in some other way.

A block diagram of a CELP codec modified to perform pitch changing and voice scrambling is shown in Figure 10.9. It can be seen that the encoder and the decoder themselves are identical to the standard CELP coder of Chapter 6, although in practice

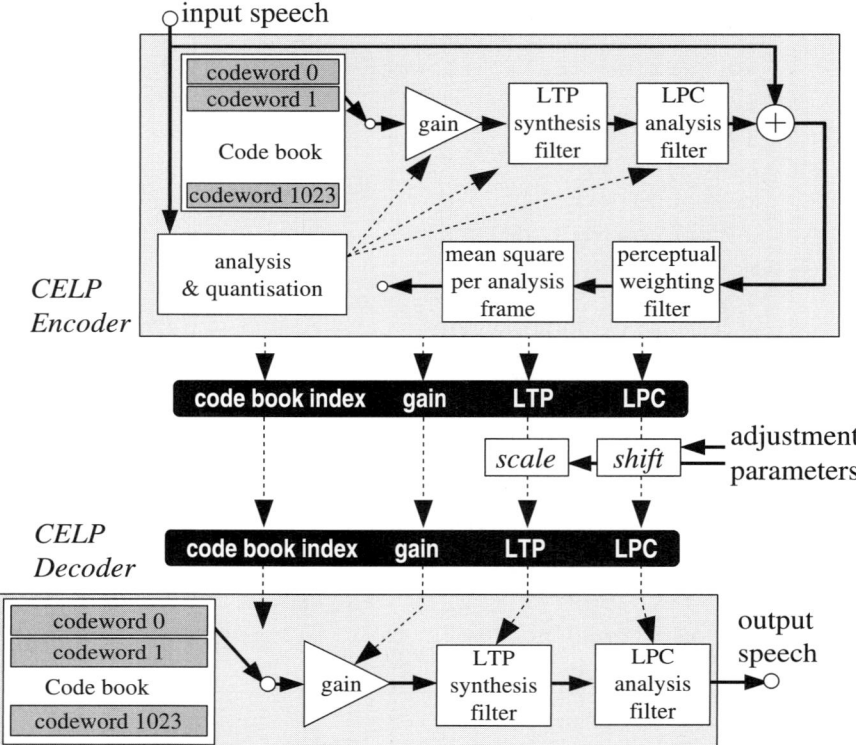

Figure 10.9 A voice changer based on adjusting intermediate speech representation parameters within a CELP codec, between the encoder and the decoder parts.

it would not be necessary to highly quantise the CELP analysis parameters as we would do in a speech compression system.

So, if these vocal parameters are not severely quantised, then the targeted adjustments made to the LSP and LTP parameters would cause changes to the processed speech, and these may well be the only perceptible changes made to the processed speech.

These changes are the scaling of the LTP delay parameter and the shifting of formant-describing LSP pairs. For both, a blanket shift throughout a speech recording would probably work, but a more intelligent system which shifted based upon an analysis of the underlying speech could provide better performance. This could, for example, scale the pitch and shift the formants of voiced speech, but leave the pitch (if any) and formants of unvoiced speech untouched.

Although this technique should properly be inserted into a continuous CELP-like analysis-by-synthesis structure, something of the potential can be demonstrated using the MATLAB code in Listing 10.4.

Listing 10.4 lsp_voicer.m

```
function Spout=lsp_voicer(speech,Fs,lsp_sc,ltp_sc)
%Good values for lsp_sc=0.8, ltp_sc=1.2
```

10.4 Voice and pitch changer

```matlab
Ws=floor(Fs*0.02);   %use 20 ms frames
Ol=Ws/2;   %with 50% overlap
L=length(speech);
Nf=fix(L/Ol)-1;
order=8;    %LPC order
Fn=Fs/2;    %Nyquist
speech=speech/max(speech); % normalise
SFS=4;   %pitch superframe size in frames
%-----------------------
%do pitch analysis for the 1st superframe
[B, M] = ltp(speech);
ltp_a=zeros(1,SFS*Ws);
ltp_a(1)=1;
ltp_a(M)=B;
%-----------------------
sfstart=1;
%Set up arrays and filter memory
SFo=zeros(1,order);
SFe=zeros(1,order);
SRo=zeros(1,order);
SRe=zeros(1,order);
PLhist=zeros(Ws,1);
Phist=zeros(Ws,1);
Spout=zeros(L,1);
Resid=zeros(L,1);
%-----------------------
%Loop through every frame
for ff=1:Nf
    n1 = (ff-1)*Ol+1;   % start of current speech frame
    n2 = (ff+1)*Ol;     % end of current speech frame
    wwf=speech(n1:n2).*hamming(Ws);
    %-----------------------
    %Get spectral response for each frame
    a=lpc(wwf,order);
   %Convert to LSPs
   lsp=lpc_lsp(a);
    %-----------------------
    %Remove the spectral response from the speech
%(filter memory for odd and even frames)
    if(rem(ff,2)==0)
        [Rwwf, SRo]=filter(a,1,wwf,SRo);
    else
        [Rwwf, SRe]=filter(a,1,wwf,SRe);
```

```
46      end
47         %construct an LPC residual array
48         Resid(n1:n2)=Resid(n1:n2)+Rwwf;
49      %-----------------------
50      %For every "superframe", do pitch analysis
51         if(rem(ff,SFS)==0)
52   %sfstart is where speech got to last time (i.e. non-
        overlapping pitch superframes)
53            [B, M] = ltp(Resid(sfstart:n2));
54            sfstart=n2; %store end filter location for next
                  superframe (save calculations)
55            % if the pitch > superframe, we've got problems,
                  so halve it
56            if(M>SFS*Ws/2) %prevent pitch doubling
57               M=round(M/2);
58            end
59         ltp_a=ltp_a-ltp_a;
60         ltp_a(1)=1;
61         ltp_a(M)=B;
62      end % end of superframe processing
63      %-----------------------
64      %Filter the residual to remove pitch
65         if(M > length(PLhist)+1)
66            [PLwwf, PLhist]=filter(ltp_a(1:M),1,Rwwf,
                  [PLhist;zeros(M-length(PLhist)-1,1)]);
                  %pad with zeros
67         else
68            [PLwwf, PLhist]=filter(ltp_a(1:M),1,Rwwf,PLhist
                  (1:M-1));      %cut short
69         end
70      %-----------------------
71      %Make the adjustment to LSPs
72      lsp0=lsp*lsp_sc;
73      Ra=lsp_lpc(lsp0);
74      %Make the adjustment to LTP tap
75      M2=round(M*ltp_sc);
76      %Create an LTP filter array
77      ltp_b=zeros(1,length(ltp_a));
78      ltp_b(1)=1;
79      ltp_b(M2)=B;%*0.95;
80      %-----------------------
81      %Now Reconstruct:
82         %1. Add pitch back to pitchless resid.
```

10.4 Voice and pitch changer

```
83      if(M2 > length(Phist)+1)
84          [Pwwf, Phist]=filter(1,ltp_b(1:M2),PLwwf,[Phist;
                zeros(M2-length(Phist)-1,1)]);
85          [Pwwf, Phist]=filter(1,ltp_b(1:M2),PLwwf,Phist
                (1:M2-1));
86      %-----------------------
87      %2. Add spectral response to resid. (filter memory
            for odd and even frames)
88      if(rem(ff,2)==0)
89          [Swwf, SFo]=filter(1,Ra,Pwwf,SFo);
90      else
91          [Swwf, SFe]=filter(1,Ra,Pwwf,SFe);
92      end
93      %-----------------------
94      %3. Rebuild speech output by overlap-add
95      Spout(n1:n2)=Spout(n1:n2)+Swwf;
96  end
```

This code is very easy to use – for example, if you have a recording called 'myspeech.wav', you can listen to adjusted speech (lowering the pitch) as follows:

```
[sp,fs]=audioread('myspeech.wav');
soundsc(sp,fs)
soundsc(lsp_voicer(sp,fs,0.8,1.2),fs);
```

If we wanted to increase the pitch instead, we would do something like this: lsp_voicer(sp,fs,1.12,0.8). Note that the code above has been shortened for simplicity – it does not take any account of problems caused by shifting the LSPs too far upwards or downwards. The only indication that such a thing has happened is when portions of speech turn 'squeaky' (or, in extreme cases, the playback sounds like a few isolated clicks instead of speech). It is a very rough and ready demonstration, although it is a good example of the use of 50% overlapping frames handled by alternating filters with individual filter memories.

Testing these kinds of methods, either alone or within a CELP coder, will quickly reveal that pitch scaling and LSP adjustments should remain relatively small. Within this range, they will result in fairly natural sounding speech. In very general terms, this method can change a male-sounding voice into a female-sounding one and vice versa. To apply larger changes, some more advanced LSP shifting code (such as lspnarrow() on page 335) and pitch adjustment code – perhaps using a PSOLA approach – would be necessary.

It should be noted at this point that the LSP-based voice changing method described above is really a modified CELP coder without the codebook. In fact the code was initially designed for use in a telecommunications application as an extension to an

346 Advanced topics

existing codec. Since the computational complexity of CELP is quite high, it would probably be an inefficient solution for applications where a CELP-style coder is not already present.

10.5 Statistical voice conversion

Section 10.4 introduced the use of PSOLA and LSP/pitch scaling methods as a means of adjusting some of the major parameters associated with a person's voice. Both of these were very much parametric methods of analysis and re-synthesis.

It should not be a surprise that the big data methods of Chapter 8 are also applicable to voice transformation and have been used to perform statistical voice conversion – including mapping one voice to another.

Although there are a number of possible approaches, all tend to work in similar ways to convert voice A to voice B, with reference to Figure 10.10:

- Start with a large 'training' database of voice A and voice B recordings with matching content. Normally this would mean having something like a set of sentences spoken by voice A, with the matching sentences being spoken by voice B. The important points are that (a) the same content is captured by both speakers and (b) there is enough training material to build an accurate statistical model (which also implies that everything we wish to convert – every type of speech – is well represented in the training data set).
- Use (typically) fixed sized frames, obtaining one feature vector for each frame. Common features are MFCCs, filterbank features and spectral envelopes.

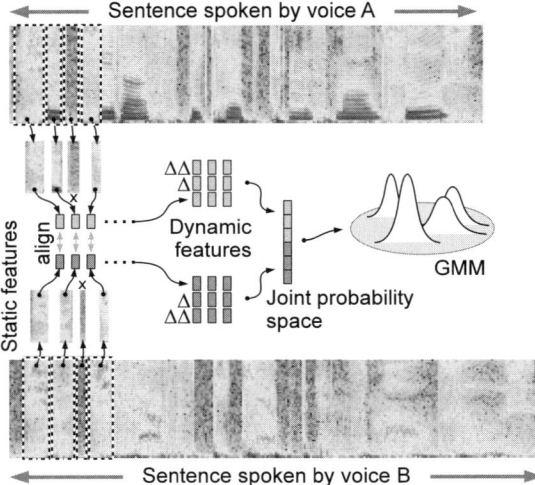

Figure 10.10 Block diagram of statistical voice conversion process, jointly modelling the dynamically aligned probability space of features from two speakers.

- Dynamic time alignment methods allow frames from one sentence to be matched to their corresponding frame in another sentence.
- A VAD or other technique is used to remove frames that do not contain recognisable speech content (there is no point in training a speech model on background noise).
- The dynamic Δ and $\Delta\Delta$ features are usually computed for each static feature vector (as described in Section 8.3.2), to double or triple the resulting feature vector length. This obviously increases complexity, but usually yields significant performance gains because it allows for a more sophisticated and accurate statistical mapping.
- A long, combined, feature vector is then constructed by stacking the static and dynamic features of corresponding elements from both speakers.
- A sequence of the combined feature vectors from all occupied frames of all test sentences is used to construct a Gaussian mixture model (GMM) of the combined probability space between the two speakers. The GMM is trained in this way using the procedures and processes outlined in Section 8.4.6.
- The trained GMM can then be used for reconstruction: given the static and dynamic feature vector from some previously unknown voice A speech, it will create the most likely corresponding (static) features of voice B.

In this way, it is possible to train a system that 'knows' how to map between voice A and voice B features. Once the system is trained, it can be presented with some speech from either voice and will convert it into the most likely corresponding features of the other voice.

Evidently, these features will now need to be reconstructed into output speech – which means that the choice of feature is important (and hence a high-dimensional spectral envelope tends to perform better than a system with 40 filterbank features, which in turn tends to outperform 13 MFCC features).

In practice, the features would usually be augmented with pitch information (for example one or more static pitch taps plus dynamic equivalents). There are also many improvements on the basic method described here, particularly because this is an active research area. Recent work has shown how such techniques can be used for voice transformation (such as converting male speech to female and vice versa), adjusting accents and conversion of singing voices. It has also been used for whisper-to-speech conversion (see Section 10.6.3), as well as being employed as one component in systems that can convert speech in one language to speech in another language.

10.6 Whisper-to-speech conversion

Whisper-to-speech conversion was first proposed as a solution to return the power of speech to post-laryngectomy patients (also known as laryngectomees). Since the larynx contains the vocal cords which are the source of pitch in normal speech, a laryngectomy leads to removal of the vocal cords, and thus removal of the pitch source.

Post-laryngectomy patients therefore – at best – produce speech that is devoid of pitch and is thus effectively the same as whispers.[4]

We discussed in Chapter 3 how normal speech is formed from a process of lung exhalation past taut vocal cords, which then resonate to create pitch. The periodic pitch signal then resonates through the vocal tract (VT) and out of the mouth and nose. For voiced speech, the resonances are called formants, but they are also present in unvoiced or whispered speech (although with significantly lower energy, because unvoiced sounds lack a pitch source). Different shapes of mouth and vocal tract lead to different resonances and thus the production of different phonemes.

For partial laryngectomees and during whispering, phonemes are formed in much the same way: they still exhale from the lungs, and the breath still flows through the VT and exits through the mouth and nose as normal after being shaped by the VT. The difference is the lack of pitch. So whispers and unvoiced phonemes are created by the same process as that which we discussed in Section 3.3.2. Figure 10.11 reproduces spectrograms of the same words being spoken and whispered. Two regions that emphasise the differences between the recordings in terms of low-frequency pitch energy and high-frequency fricative energy are identified on the figure.

Referring back to Chapter 6, we have seen how to model the speech process using linear prediction (and other techniques), including separately modelling both voiced and unvoiced speech in methods such as regular pulse excitation in Section 6.3.1. In general, we parameterise the content of speech (see Section 6.2) using models for the lung exhalation, the pitch excitation and the VT filtering effect. In whispers, the main difference to the model is the absence of pitch. These models are shown in Figure 10.12, which additionally proposes a reconstruction method (at the bottom of the figure) that takes whispers and adds the pitch at the end rather than at the beginning of the process. Since the models for pitch, VT and gain are all linear time invariant (LTI), it does not matter in theory in which sequence they are applied. Reconstruction could thus simply take the approach of adding pitch back into whispers. While this sounds like common sense, and is possible to some extent, the low power and noise-like nature of recorded whispers does present a number of problems with the application of this technique in practice, particularly by making it susceptible to interfering noise. We will briefly survey three methods of reconstructing whispers below. The first two (MELP/CELP-based conversion and parametric conversion) effectively take the approach of adding pitch to the whisper signal. However, the third, statistical, method of conversion is based on the statistical voice conversion methods of Section 10.5, which generate speech without reusing the components of this whisper (although it does so through analysis of the whisper).

---

[4] Only after a partial laryngectomy: a total laryngectomy, by contrast, includes re-routing the airway to a stoma (a hole in the throat area), to bypass the vocal tract when breathing. This complicates the situation further for voice reconstruction, and may require the use of an electrolarynx to allow speaking – a hand-held vibration device that is held against the throat to create an artificial pitch signal.

10.6 Whisper-to-speech conversion

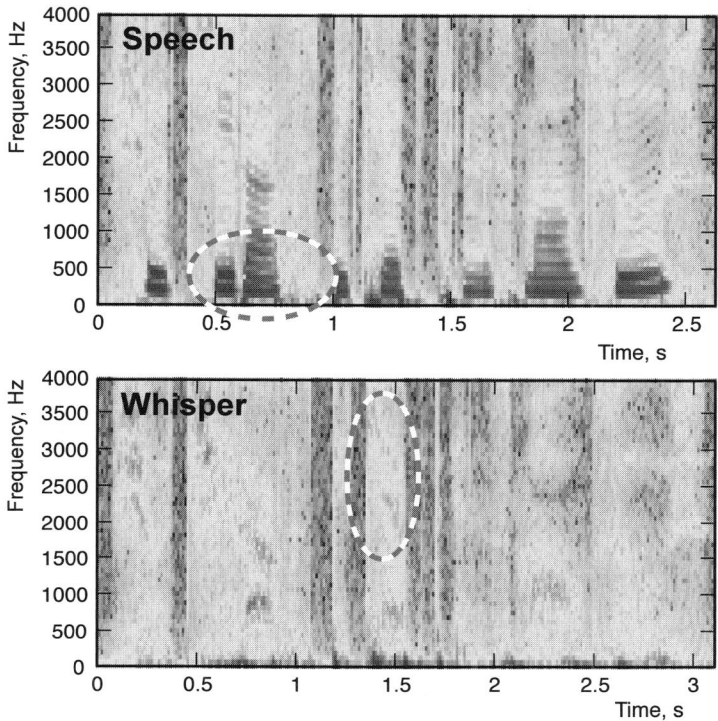

Figure 10.11 Spectrograms of the same words being spoken (top) and whispered (bottom), clearly revealing the presence of low-frequency pitch regions in the spoken signal (circled) that are absent in the whispered signal, while some high-frequency detail in the whispered signal (circled) is less prevalent in the speech signal.

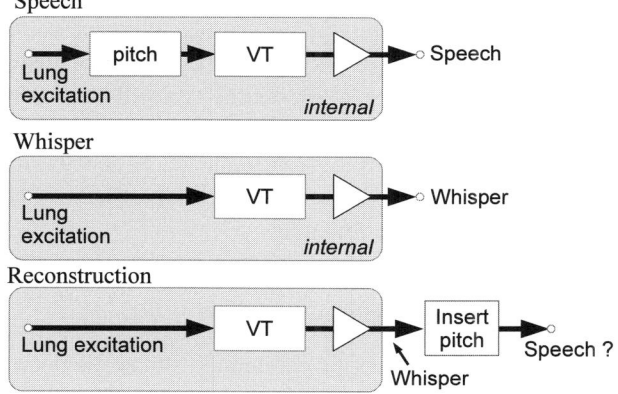

Figure 10.12 Models of the normal speaking process (top), whispering process (middle) and one possible reconstruction method (bottom), given that the components such as the pitch source and vocal tract filter are LTI.

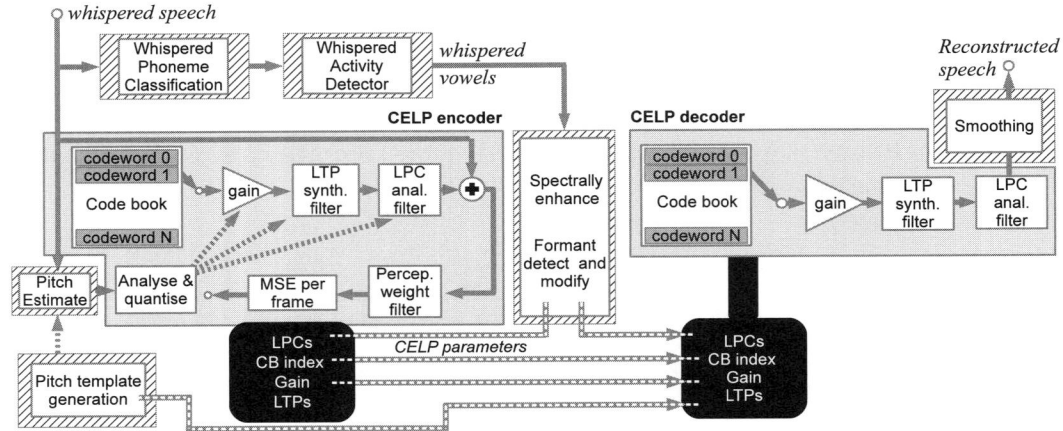

Figure 10.13 Block diagram of a CELP-based whisper-to-speech conversion process using analysis-by-synthesis.

10.6.1 MELP- and CELP-based conversion

MELP (mixed excitation linear prediction) and CELP (codebook excited linear prediction) were speech compression methods introduced in Chapter 6. The important aspect of both is that they decompose the speech signal into different components that are relevant to the human speech production system, namely lung, pitch and vocal tract components.

Both have been used as a basis for reconstruction of voiced speech from whispers [154, 155]. Taking the example of the CELP method, the encoder (Figure 10.13, left) is used to decompose input whispers into the usual components: gain, codebook (CB) index, linear prediction coefficients (LPCs) and pitch long-term prediction (LTP) values. Because we know that the pitch is absent, the extracted LTP information is discarded. Instead, a pitch generation template is used to create a new pitch signal which combines with the other parameters in the decoder to reconstruct voiced speech. Meanwhile, a spectral enhancement block is used to modify the LPC information (perhaps in the form of LSPs) to strengthen formants and shift them in frequency to account for the known difference between voiced and whispered formant frequencies. The method works reasonably well, but suffers from two main difficulties:

- Handling phonemes that are *supposed* to be unvoiced.
- Deciding what level of pitch to apply to the speech.

In fact we will find that both difficulties are common to almost all methods of reconstructing speech from whispers. The handheld electrolarynx prosthesis commonly used by laryngectomees applies a fixed pitch constantly while it is operating. In other words it overcomes the difficulties by assuming that *all* phonemes should be voiced (in addition to gaps between words and other white space) and applying the same level of pitch to everything. Any other reconstruction solution needs to overcome these difficulties. Both the MELP and CELP methods apply curvilinear pitch contours, which means slowly varying pitch levels (that are designed to change in the same way as pitch alters across the average sentence), and use a voiced/unvoiced (V/UV) detection process that attempts to identify when a phoneme occurs that should be voiced, using some kind of frequency or energy metric. Due to the low signal-to-noise ratio of whispers, both methods end up making some quite erroneous decisions, which become the limiting factor in the quality of both approaches.

10.6.2 Parametric conversion

Parametric conversion is a relatively new method [156] that aims to address the two main difficulties above by firstly avoiding any need to switch from a voiced to unvoiced mode and secondly applying a *plausible* pitch contour rather than a fixed, curvilinear contour or one of some other fixed shape.

The idea is that, although it may not be possible to determine what the correct pitch signal would be if the input were spoken, we know that the pitch changes at a phonetic rate in time with formant changes. In some way, pitch strength and frequency map to formant strength and frequency in speech: it is obvious that strong pitch forms strong formant resonances, and that weak pitch forms weak formant resonances. It is also obvious that formants – as resonances of the pitch signal – are related in some way to pitch. Unfortunately, testing has revealed that the actual relationship between pitch and formants is not one-to-one (i.e. there are many possible formant combinations that derive from a single pitch frequency). Fortunately, it does not rule out a one-to-one mapping. Thus, we can define a plausible one-to-one mapping between formants and pitch, in effect deriving a pitch signal from an analysis of the formants, and deriving the pitch strength from an analysis of the formant strength.

This is precisely the method used for plausible pitch generation in the parametric systems shown diagramatically in Figure 10.14. The system works on a frame-by-frame basis, beginning with an Lth-order LPC representation of a windowed analysis frame:

$$F(z) = \sum_{k=1}^{L} a_k z^{-k}, \tag{10.7}$$

given L LPC coefficients, a_k. The roots of the LPC equation are then found for each frame, and the highest peak positions are tracked over time to yield N formant tracks (which are smoothed versions of the frame-to-frame peaks).

For each frame, formant frequencies F and magnitudes M are determined within predefined ranges bounded by λ_{FXlow} and λ_{FXhigh} such that $F_X \in [\lambda_{FXlow}, \lambda_{FXhigh}]$ (for formants $X = 1, 2, 3, \ldots, N$). Obviously, not every analysis frame from recorded whispers will contain correct formants, especially non-speech periods between words. For this reason, a judgement must be made as to whether a particular formant candidate is genuine, by comparing the instantaneous average magnitude with the long-term average magnitude $\bar{M}_X$ for each formant candidate X:

$$F'_X(n) = \begin{cases} F_X(n) & \text{when} \quad M_X(n) \geq \eta_X \bar{M}_X, \\ 0 & \text{when} \quad M_X(n) < \eta_X \bar{M}_X. \end{cases} \tag{10.8}$$

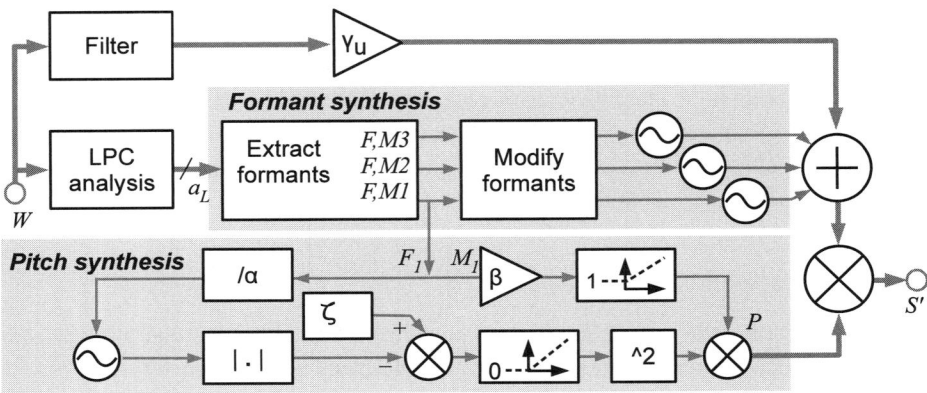

Figure 10.14 Signal diagram of a parametric whisper-to-speech conversion process.

This removes formants with very low power, and in practice $\eta_X = 2^{(X-5)}$ works reasonably well – it quadratically decreases the ability of formants to survive based upon their frequency, taking into account the much reduced energy of higher formants (and hence their lower SNR when corrupted by flat noise). An absent formant is represented by a zero.

Formant shifting is then performed to account for the difference in average formant frequency between speech and whispers. Thus $F''_X(n) = F'_X(n) - \epsilon(n)$, where ϵ is empirically determined from the mean vowel formant shift (e.g. using values such as those in [11], which are approximately {200, 150, 0, 0} Hz for the first four formants).

The formant frequencies derived from analysis of the whispers, pruned and shifted as mentioned above, are now used to create a kind of sinewave speech S', speech composed of the formant sine waves only:

$$S' = \left\{ \sum_{X=1}^{N_s} M_X \cos(F''_X) + \gamma_U W \right\} .P, \quad (10.9)$$

where P represents a glottal pitch modulation, which will be defined below. γ_U is a scalar constant multiplier of the whisper signal (allowing the inclusion of wide band excitation present in the original whispers, W, to be carried forward to the reconstructed speech, which is useful for sibilant and fricative sounds). Glottal modulation, P, is synthesised from a cosine waveform:

$$P = \max\{M_1 \beta, 1\} . \max\{\zeta - |\cos(f_0)|, 0\}^2, \quad (10.10)$$

where β relates the depth of the pitch modulation frequency f_0 to the formant energy. In this way a less obvious formant presence, i.e. reduced voicing, results naturally in reduced modulation depth – which is normally the case in speech. The degree of clipping, ζ, affects the overall pitch energy contribution to the equation.

The main contribution of this parametric reconstruction method is to derive a plausible pitch signal, f_0, rather than attempt to derive an accurate pitch (or constant, or curvilinear pitch). This was based upon an observation that, in voiced speech, changes in F_1 and f_0 trajectory tend to occur simultaneously at phoneme transitions, with the hypothesis being that maintaining this synchronicity is more important than ensuring the f_0 signal is perfectly accurate in mean level. Hence f_0 is synthesised as an integer sub-multiple of the smoothed formant frequenies:

$$f_0 = \xi|F_3 - F_2| + \alpha|F_2 - F_1|, \qquad (10.11)$$

where α and ξ are constants which are empirically determined to yield a mean pitch frequency within a suitable range (a value of around 20 for both seems to work well).

The motivation for this approach is to derive a plausibly varying pitch frequency. In fact, the resulting pitch contour changes slowly when formants are smooth, but exhibits much more rapid changes at phoneme boundaries. It means that the pitch varies in a way that is related to the underlying speech content (unlike the EL, which has fixed pitch, or the MELP/CELP methods). The situation in real speech is, of course, different since formants are not simple harmonics of the pitch frequency and pitch is not a pure tone. However, both pitch and formants do vary in time with speech content – and it is this variation which we are attempting to replicate.

In summary, reconstruction begins with a summation of pure cosine formants with frequencies specified by shifted smoothed extracted whisper resonances. The cosines have amplitude determined by the detected formant energy levels. They are augmented by the addition of the scaled whisper signal to impart high-frequency wideband energy that is difficult to model with fundamental cosines. It is important to note that no decision process is implemented between V/UV frames: γ_U does not vary because hard decisions derived from whispers do not tend to work well in practice – they are often incorrect due to the presence of corrupting acoustic noise. The combined signal is then modulated by a clipped, raised cosine glottal 'excitation' which is harmonically related to the formant frequencies, with depth of modulation reduced during low energy frames.

10.6.3 Statistical conversion

We discussed the method of statistical voice conversion (SVC) in Section 10.5, including a basic description of the steps involved in training such a model. Unsurprisingly, this method has been applied to whisper and post-laryngectomy speech. One of the earlier proponents was Professor Tomoki Toda of Nara Institute of Science and Technology in Japan, who used SVC to reconstruct non-audible murmur (NAM) microphone signals into speech, and later to convert both electrolarynx speech and whispers into more natural sounding speech.

NAM devices are typically attached to, or placed in proximity to, the bone behind the ear (they can be incorporated in glasses, headsets or hearing aids). During speech, sound is conducted from the VT, through the jaw bone, back to the microphone, where it can be recorded as a very low-amplitude signal. Apart from being low energy, the sound is an extremely muffled and distorted version of the speech being produced, although it does include all of the speech components to some extent. In an SVC approach, a GMM models the joint probability between normal speech and the NAM features, i.e. in Figure 10.10, voice A would be recorded normally at the lips while voice B would be the NAM signal – since these can be obtained and recorded simultaneously for a single speaker, the dynamic time alignment step is unnecessary (and hence the data collection and training process is much simplified). Results from this technique are very good, with near-natural speech reconstruction being possible.

Extending the method to whisper SVC, the data collection and training steps are now more complex: a single test subject must record large amounts of scripted speech, plus record the same material using whispers (this is called 'parallel' training material). During training, dynamic time alignment is important to match the equivalent parts of speech from both speaking modes. The process is further complicated because the voice activity detection step, although likely to work well for speech, is often problematic for the much-lower-energy and noise-like whispers. However, fairly good performance has been demonstrated from the same basic technique by Dr Toda and others, including this author.

Despite encouraging performance, the disadvantages of SVC-based whisper conversion are probably quite predictable:

1. A new model must be constructed for each person using the system, which is time-consuming and expensive. It also requires extensive training data.
2. The model requires training with parallel sentences of speech and whisper. Unfortunately most of the post-laryngectomy or voice-loss patients who most need this technology will have lost the ability to speak before they reach this point, and thus are unable to record the speech sentences.
3. The entire method is computationally complex and potentially slow – making it difficult to implement in real time (which many applications would require).

As with the alternative whisper reconstruction methods, much research is continuing in this area, particularly in the combination of statistical and parametric techniques.

10.7 Whisperisation

Although we have spent a large portion of this chapter trying hard to convert whispers into speech, we are occasionally faced with the converse problem: converting speech into whispers. We call this whisperisation. The task may be viewed as simply another type of voice conversion, perhaps used for transforming continuous speech or changing the way a voice sounds to mask a person's voice. However, it could also be useful in creating parallel whisper and speech data (which was identified in the section, where the requirement to obtain such data for training and testing was described as being a particular disadvantage of statistical methods).

We have seen, for whisper-to-speech conversion, that statistical methods performed best – and this would probably be true of speech-to-whisper conversion too, except that no researchers appear to have evaluated this due to lack of sufficient motivation, or perhaps lack of parallel data. However, it is not difficult to build a parametric method based on the techniques we have examined up to now.

10.7 Whisperisation

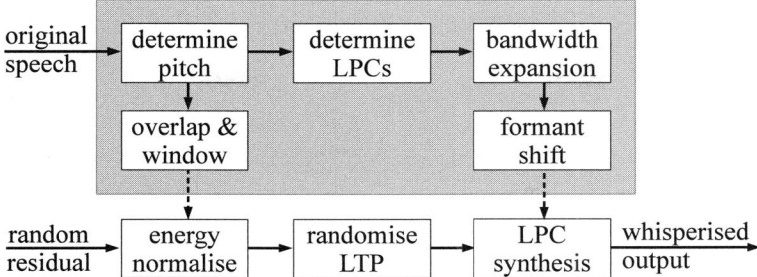

Figure 10.15 Signal block diagram of an LPC-based whisperer.

Knowing that whispers are effectively (a) pitchless speech, (b) lower power than speech, (c) with shifted and widened formants, then we can model each of those steps parametrically, as shown in Figure 10.15. A method of achieving most of these objectives is given in Listing 10.5. What this does *not* do at present is shift the formant locations; this task is left as an exercise for the reader (a quick and easy way to achieve the shift would be to use the LSP-based formant shifting method of Section 10.4.2, which also forms the basis for this whisperisation code).

Listing 10.5 whisperise.m

```
function wout=whisperise(speech,fs)
% divide into  50% overlapping frames
Ws=floor(fs*0.02); %20 ms frames
Ol=Ws/2;           %50% overlap
L=length(speech);
Nf=fix(L/Ol)-1;    %No. of frames
order=10;          %LPC order
Fn=fs/2;           %Nyquist
speech=speech/max(speech); %Normalise speech
SFS=4;             %superframe size (frames)
gamma=1.03;        %bandwidth expansion

%Get pitch for 1st superframe
[B, M] = ltp(speech);
ltp_a=zeros(1,SFS*Ws);
ltp_a(1)=1; ltp_a(M)=B;
%--------------------
sfstart=1;
SFo=zeros(1,order);
SFe=zeros(1,order);
SRo=zeros(1,order);
SRe=zeros(1,order);
wout=zeros(L,1);
```

```matlab
PLhist=zeros(Ws,1);
Phist=zeros(Ws,1);
%--------------------
for ff=1:Nf
   n1 = (ff-1)*Ol+1;    % start of current frame
   n2 = (ff+1)*Ol;      % end of current frame
   wwf=speech(n1:n2).*hamming(Ws);

   %For every superframe, get LTP
   if(rem(ff,SFS)==0)
     %sfstart = where we finished last time
     [B, M] = ltp(speech(sfstart:n2));
     ltp_a=ltp_a-ltp_a;
     ltp_a(1)=1; ltp_a(M)=B;
     sfstart=n2; %save calculations
     % if the pitch > superframe, halve it
     if(M>SFS*Ws/2)
       M=round(M/2);
     end
     ltp_a(1)=1; ltp_a(M)=0;
     ltp_a(10+fix(rand(1)*200))=rand(1);
   end

   %Get spectral response for each frame
   a=lpc(wwf,order);

   %Remove the spectral response
   if(rem(ff,2)==0)
     [Rwwf, SRo]=filter(a,1,wwf,SRo);
   else
     [Rwwf, SRe]=filter(a,1,wwf,SRe);
   end

   %Filter residual to remove pitch
   if(M > length(PLhist)+1)   %pad with zeros
     [PLwwf, PLhist]=filter(ltp_a(1:M),1,Rwwf,[PLhist;
         zeros(M-length(PLhist)-1,1)]);
   else  %cut short
     [PLwwf, PLhist]=filter(ltp_a(1:M),1,Rwwf,PLhist(1:M
         -1));
   end
   %--------------------
   PLL=length(PLwwf);
```

```matlab
65    PLenerg=sum(PLwwf.^2);
66    PLwwf=0.5 - rand(PLL,1);
67    NPLenerg=sum(PLwwf.^2);
68    PLwwf=PLwwf*(PLenerg/NPLenerg);
69    %--------------------
70    %Add pitch back to pitchless residual
71    if(M > length(Phist)+1)    %pad with zeros
72      [Pwwf, Phist]=filter(1,ltp_a(1:M),PLwwf,[Phist;
           zeros(M-length(Phist)-1,1)]);
73    else   %cut short
74      [Pwwf, Phist]=filter(1,ltp_a(1:M),PLwwf,
75         Phist(1:M-1));
76    end
77    %--------------------
78    Ra=Pwwf;     %reconstruct LPC residual
79    a=a*gamma;   %bandwidth expansion
80    %--------------------
81
82    %Add in the spectral response
83    if(rem(ff,2)==0)
84      [Swwf, SFo]=filter(1,a,Ra,SFo);
85    else
86      [Swwf, SFe]=filter(1,a,Ra,SFe);
87    end
88
89    %Rebuild speech output array by overlap-add
90    wout(n1:n2)=wout(n1:n2)+Swwf;
91  end
```

Please note that the output whispers are not really whispers even though they have perceptually similar characteristics (i.e. they sound like whispers). For example, the formants are not shifted sufficiently, the energy distribution is different and they have the same timing as the input speech – whereas, in reality, whispered words are almost always produced at a slower rate than the equivalent speech.

10.8 Super-audible speech

The super-audible frequency (SAF) range is a recently named region lying just above the upper frequency limit of normal hearing. It is thus at the extreme low-frequency (LF) end of the ultrasonic range (which spans a vast frequency range from about 20 kHz up to 1 GHz or more). A few applications make use of such frequencies, including some

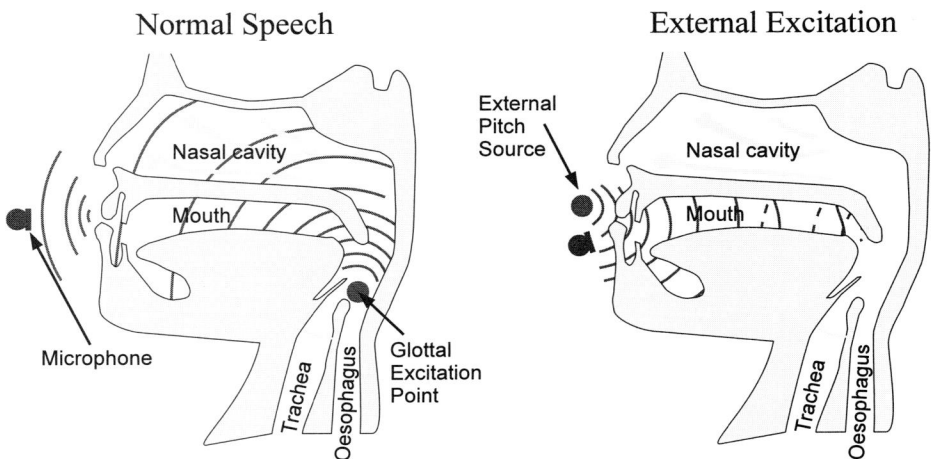

Figure 10.16 VT pitch excitation from the vocal cords in normal speech (left) and from an external low-frequency ultrasonic source (right).

industrial ultrasound devices, but in general the region is relatively 'unoccupied' – meaning that it is mostly quiet in many environments. Part of the SAF range is quite attractive for exploitation in future applications because it lies within the range of modern digital audio hardware, including most smartphones and mobile devices which can sample at up to 48 kHz. If we assume that the upper range of human hearing is 20 kHz, there is a 4 kHz band up to the Nyquist frequency of 24 kHz, which is inaudible but can be output and input by such devices. Note that, although the devices themselves can operate at such frequencies, it is likely that there may be some degree of signal attenuation due to the physical characteristics of microphones and loudspeakers that are in use (particularly in the miniature loudspeakers currently used in some smartphones). Two recent occupiers of the SAF range have been a system that allows mobile devices to communicate using short bursts of inaudible sound and a system that uses reflection to determine the 'mouth state' of a human during a mobile telephone call, effectively an inaudible voice activity detector (VAD – see Section 9.2). It is the latter that we will explore here.

SaVAD, the super-audible VAD, uses low-frequency ultrasound to detect whether the mouth is open or closed during an analysis frame, and can do so even in the presence of extremely loud background noise levels. It builds upon earlier research to examine vocal tract resonances using frequency sweeps generated from near the lips [157] (at both audible and LF ultrasonic frequencies), which led to the observation that ultrasound reflections from the face are very different when the mouth is open from when the mouth is closed [158, 159]. The excitation arrangement is shown, compared with that for normal speech, in Figure 10.16.

In operation, a SAF excitation signal is generated and then output from a loudspeaker pointing at the mouth region. The excitation is a linear cosine chirp of duration τ which spans a frequency range 20 kHz to 24 kHz:

$$x(t) = \cos\{2\pi(20000t + 4000t^2/2\tau)\}. \tag{10.12}$$

10.8 Super-audible speech

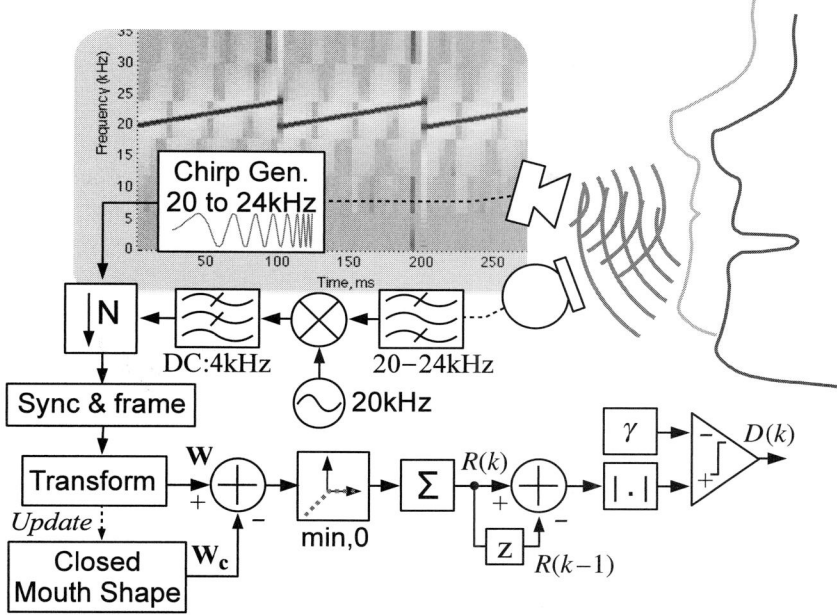

Figure 10.17 Arrangement of microphone and loudspeaker for the low-frequency ultrasonic excitation and measurement.

This signal is repeated continually, typically at a repetition rate of $\tau = 0.1$ s as shown above. A spectrogram of the chirp is illustrated at the top left of the block diagram in Figure 10.17, which shows the repetition and frequency span.

In MATLAB, at sampling frequency Fs, this is done as follows:

```
ChirpLen=0.1; %seconds
Fs=96000;
Fc=20000;
Fe=24000;
tt=0:1/Fs:ChirpLen;
ch=chirp(tt,Fc,ChirpLen,Fe);
ch=ch(1:length(ch)-1);
%Repeat forever (or as long as test lasts)
out=repmat(ch, 1, ceil(TotalRecordingLen/ChirpLen));
%Or we could do it in stereo:
%out=repmat(ch, 2, ceil(TotalRecordingLen/ChirpLen));
```

To test the method, experiments were conducted with subjects seated so that their lips were a few centimetres in front of the SAF signal source (typically between 1 and 6 cm away). A microphone was located slightly to the side of their face, at a similar distance (typically between 2 and 8 cm). In fact, earlier research [160] showed that exact positioning of the microphone is not critical, and a wide range of microphone distances and angles with respect to the mouth are also possible [161]. Subjects were requested to remain as stationary as possible during the experiments.

As shown in Figure 10.16, the SAF excitation was able to enter the mouth of a subject when their lips are open and travel through the VT – but it would be reflected straight back from the face when their lips are closed. The difference in reflection means that signal characteristics measured at the microphone differ very markedly between when the lips are open and when they are closed. It is therefore possible to derive an automated measure which is capable of detecting those two states and to use this as the basis for a VAD, assuming that speaking is defined as a pattern of mouth opening and closing.

The signal from the microphone, $m(t)$, sampled at $F_s = 96\,\text{kHz}$ (although 48 kHz could have been used) is first bandpass filtered using a 33-order infinite impulse response (IIR) filter with a passband of 20 to 24 kHz before being demodulated to the baseband, $m'(t) = m(t).\sin(2\pi \times 20000t)$, and down sampled (to rate $F'_s = 8\,\text{kHz}$ in these experiments.

A frame of received reflected signal samples of length $L_W > \tau + \delta$ is first cross-correlated against the transmitted chirp (to account for timing delays between output and input of up to δ samples), to align the input signal with the exact chirp timing. Vector **w**, of length τ, is the resulting time-aligned output from this process. Because a linear chirp was used as the excitation, the amplitude envelope of this vector (of length 800 for a $\tau = 0.1$ s chirp) corresponds to its frequency response. The resonances (peaks and valleys) in this vector reveal much about the lip state information, but the size of the vector needs to be reduced before an automated measure can be derived from it. In practice, a very simple technique in MATLAB is to use a Savitzky–Golay smoothing filter [162] with a third-order polynomial over a 41-sample window. Given a $\tau = 0.1$ s analysis window w, it is easily obtained for frame n:

```
Wbig(n)=sgolayfilt(abs(w),3,41);
```

Note that this type of filter is a surprisingly useful practical method of obtaining a smoothed envelope for speech-related signals. The command above effectively computes an amplitude envelope estimate of the reflected chirp which should then be averaged and down sampled to yield a shorter vector **W** (of size $L_{PS} = 133$). To determine mouth state, vector **W** is compared with the expectation of the closed-lip response vector, $\mathbf{W_c}$, for each analysis frame ($\mathbf{W_c}$ would be periodically adapted in practice to account for natural changes in positioning of the sensors and the face, but is essentially the average **W** signal during periods when the mouth is closed). A scalar signature, R, is found by comparing the current and expected response vectors:

$$R = \sum_{i=0}^{L_{PS}} \min\{\mathbf{W}(i) - \mathbf{W_c}(i), 0\}. \qquad (10.13)$$

In MATLAB this would be:

```
Wd=W-Wc;
R=sum(Wd.*(Wd<0));
```

Notice that R is computed from the negative going portions of the spectral envelope difference. The explanation for this is that mouth opening couples the vocal tract (VT) in parallel with a static reflection from the face. When the mouth opens, some of the energy that was previously reflected in the closed state now enters the vocal tract, which shows up as zeros in the system transfer function [163]. So we are effectively trying to detect the presence of these zeros which are visible as sharp valleys in the frequency response. These are indicative of the lips being opened. Peaks in the spectral difference, by contrast, are primarily resonances with the surface of the face, and can be ignored because they are always present.

Finally, we are looking for the state of opening or closing the mouth, which means a change in the signal. Such changes are usually determined by computing a difference between the current and previous frame, and comparing this with a threshold γ. Thus the detection **D**, for each frame k, can be computed as:

$$D(k) = \{|R(k) - R(k-1)| > \gamma\}. \tag{10.14}$$

A simple way to determine γ is to set it to some convenient fixed value, but it is more reliable to relate it to the actual data in some way. Thus, twice the mean of $\{R(k) - R(k-1)\}$ seems to work reasonably well in practice. In MATLAB, a similar process to accomplish this would be:

```
D=abs(diff(R)) > 2*average(diff(R));
```

In academic papers which evaluate this technique using long-duration recordings, the threshold γ is set to a moving average instead of a fixed mean (so that it adapts slowly to changes in $R(k)$ over time), and we average D over three neighbouring frames to make the scoring more accurate [159].

10.8.1 Signal analysis

Figure 10.18 plots a contour spectrogram of the Wd signal in the SAF region underneath a waveform plot recorded at the microphone. The speaker is reciting digits from 'five' to 'fourteen' in this test. All of the odd numbered digits are spoken normally, whereas the first three even numbered digits are mimed (i.e. the mouth moves exactly as if speaking but no sound is produced) and the last two even digits are whispered. The digits are identified on the waveform plot, and changes that correspond to those digits being produced should be evident in the spectrogram. Note how the spectrogram changes in time with the digits, but also shows resonances that gradually shift over time due to

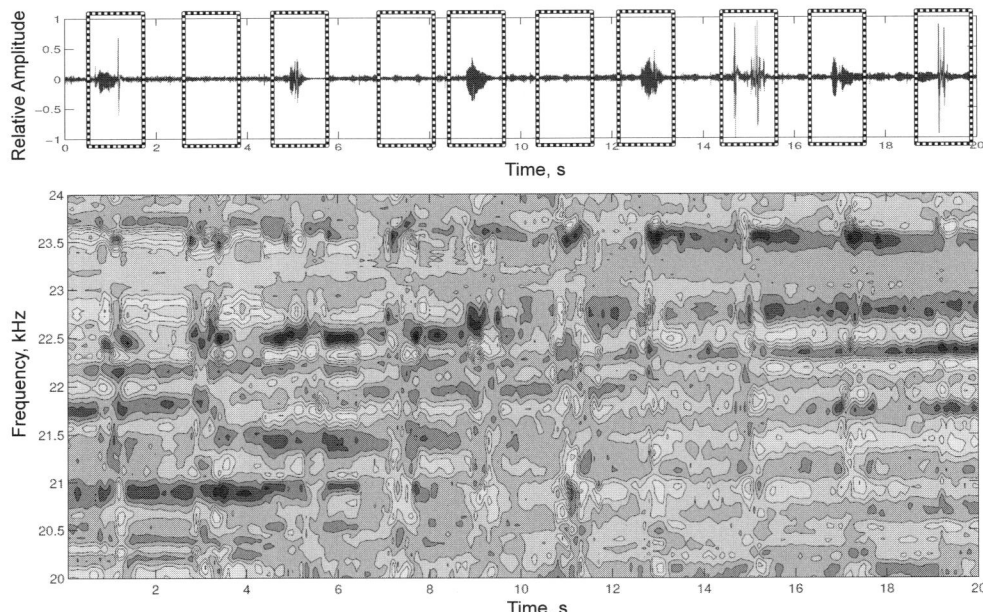

Figure 10.18 Waveform (top) and contour spectrogram (bottom) from a recitation of the ten digits from 'five' to 'fourteen' inclusive. All odd numbers are spoken normally, but 'six', 'eight' and 'ten' are mimed (no sound) whereas 'twelve' and 'fourteen' are whispered.

the changing physical arrangement between sounder, sensor and face. The important feature here is that the changes due to mouth opening happen irrespective of the *sound* being produced – i.e. spoken, whispered and mimed digits each cause similar visible changes to the spectrogram.

The metrics developed in the previous subsection were used as a VAD in [159] (and from the lower part of Figure 10.18 we can start to visualise how the VAD, or 'mouth movement detector', can work, even if the amount of sound being produced is small or non-existent). However, further questions arise from a plot like this one: does the pattern of the spectrogram say anything about *what* is being spoken? Might this be usable for recognising the content of speech (not just the presence of speech)? Perhaps even more interesting is the idea that the patterns might reveal *who* is speaking.

10.8.2 Applications

We have already mentioned the SaVAD application, although we did not describe how or why such an application might be useful.

Firstly, improved voice activity detection is advantageous for speech applications that are required to operate in acoustically noisy environments (for example speech communications devices or ASR systems). At present, virtually all speech communications systems (and human–computer speech interfaces) include a VAD which operates on the basis of audible acoustic input signals. Since speaking is related to, and synchronous with, patterns of lip opening and closing, knowledge of mouth movement can potentially lead to a more reliable VAD than ones that rely only on audible sound (which become confused in a noisy environment – especially in an environment with a lot of babble noise from multiple speakers). Note that the converse is also true: when lips remain closed, speech is unlikely. Knowing this reliably in a mobile device may be advantageous for power saving if, for example, complex and power-hungry speech analysis/transmission algorithms employed during a voice call can be paused while a speaker's mouth remains closed. The system will thus be able to ignore loud background noises which would otherwise trigger operation of a traditional VAD.[5]

Secondly, there are potential uses for voice regeneration in speech impaired patients. This is something we have considered elsewhere in this chapter, where we investigated the reconstruction of voiced speech from whispers. Note the problem: whispers are lower in energy and more noise-like in their characteristics than speech. It is therefore very difficult to create a working VAD for whispers, but the SaVAD technique works just as well for whispers as it does for speech. All of this solves one of the major problems with speech impaired patients: detecting automatically when they are speaking.

We could extend the idea further by using SaVAD as a hands-free lip-synchronous control for an electrolarynx, as part of a speech therapy aid or for speech prosthesis design.

Finally, the technique also has potential for speaker validation, as we hinted at the end of the previous subsection. A biometric authentication system might add non-audible mouth movement data to the recorded speech of a subject speaking into a system. This would authenticate a speaker based not only on their audible speech, but also upon the patterns of mouth opening and closing that they produce. It would, for example, require that the subject is physically present during authentication, rather than just a recording of the subject's speech.

10.9 Summary

This chapter has presented a collection of techniques, ideas and applications that have built upon the speech and audio analysis and processing foundations laid in previous chapters. We began with speech synthesis and text-to-speech (TTS), which are areas of growing importance in the speech business world (and of ongoing importance to researchers). Playback and transcription approaches were considered, and some very practical tools discussed, including a mention of the increasingly important cloud-based speech service insfrastructure. We discussed stereo speech, and gave an example of controlling stereo positioning (incidentally, this method and code was used by the

---

[5] Note that ultrasonic VAD devices do exist at present, making use of specialised sensors at frequencies around 40 kHz which sense Doppler frequency changes due to the velocity of lip movement. These are useful in situations such as driving a vehicle, which are not constrained by power or sensor size, and operate at distances of more than about 10 cm.

author to investigate speech intelligibility in noisy smart homes [47]). Next was the idea of formant strengthening and steering using LSPs, and perceptually based steering, with the basic LSP technique then being extended for a voice changing application, which is compared with the common pitch-synchronous overlap and add (PSOLA) method. Voice changing was also achieved through the means of statistical voice conversion (SVC) – which we extended into whisper-to-speech conversion, and compared against several alternative methods such as a parametric reconstruction framework. Naturally, we would then wish to consider the opposite transformation of speech-to-whisper, although we presented only a parametric approach, not a statistical method. Finally, we described super-audible speech, a very new method of assisting speech processing by using information located just beyond the upper limit of human hearing. This was developed for a voice activity detector, but could be explored for other applications such as improving voiceprint biometric authentication through using mouth movement and in silent or whispered speech recognition. The latter is part of the exciting research area known as silent speech interface (SSI) technology, which is expected to be an enabler for future human–computer interaction with wearable technology, much of which is likely to rely upon speech interfacing. SSIs are particularly useful for times when speech cannot be used (such as during a lecture or in a library), or when the information being communicated is sensitive (such as personal information, or when speaking an identification code). They are also an important accessibility tool for those who cannot speak audibly, including post-laryngectomy patients.

Bibliography

- *Speech Enhancement*
  Ed. J. S. Lim (Prentice-Hall, 1983)

- *Speech Communications, Human and Machine*
  D. O'Shaughnessy (Addison-Wesley, 1987)
  A rather expensive book, but one with over 500 pages describing the speech communications field, from basic topics extending to more state-of-the-art coverage of speech enhancement, speech recognition and even a final chapter dedicated to speaker recognition.

- *Survey of the State of the Art in Human Language Technology*
  Eds. R. Cole, J. Mariani, H. Uszkoreit, G. Batista Varile, A. Zaenen, A. Zampolli and V. Zue (Cambridge University Press and Giardini, 1997)
  Also available online from www.dfki.de/~hansu/HLT-Survey.pdf. This book is the result of a joint project between the European Commission and the National Science Foundation of the USA. As the name implies, it describes the state of the art in the human language fields, including speech recognition, speaker recognition and so on.

Questions

Q10.1 Describe the major features of a concatenative text-to-speech (TTS) system.

Q10.2 When a TTS system is presented with an unfamiliar word (perhaps a made-up word), briefly discuss the main options it has regarding how to pronounce that word.

Q10.3 Apart from basic lexical information (the dictionary meaning of the words), list other types or classes of information than can be conveyed by spoken language.

Q10.4 Identify two advantages and two disadvantages with online ASR and TTS services as opposed to standalone (i.e. local) systems

Q10.5 Repeat the stereo noise example given in Section 10.2.1 which plays the same noise to both ears: try altering the noise in one ear in terms of amplitude, changing the sign (i.e. use `[-r1;r1]` in place of `[r1;r1]`), and in time-domain sequencing (e.g. reversing the array (i.e. `r1(length(r1):-1:r1)`). How do these transformations affect the perceived noise compared with the situation where a different random signal is used in both ears?

Q10.6 Referring to Section 10.3.1, develop perceptual LSP-based formant shifting code using the `bark2f()`, `f2bark()` which can make small perceptual shifts in formant frequencies using a test array of LPC values from Chapter 6 (this array is available on the website along with all of the required MATLAB functions). Plot the resulting spectra and lines using the `lspsp()` function. Test a few shift amounts and from your results identify a range of realistic degrees of shift.

Q10.7 Create a few seconds of linear super-audible chirp signal at 44.1 kHz sample rate and listen to the output using `soundsc()` – can you hear the chirp? Replay the signal using a lower sample rate such as 32 kHz, 24 kHz or 16 kHz and briefly describe what you hear (if anything) at the transition point when one chirp transitions to the next one.

Q10.8 When converting whispers to speech, any reconstruction method must overcome the issue of missing information. Identify what is the major missing information in whispered speech compared with fully voiced speech.

Q10.9 In the whisperisation code, `whisperise()`, what is the purpose of the array called `PLhist`? What would happen if this was not used?

Q10.10 Experiment with the PSOLA code from Section 10.4.1 and identify the approximate extent of scaling factors that produce reasonable quality speech (i.e. slowing down beyond which quality is severely degraded, and likewise for speeding up).

11 Conclusion

You approach the doorway ahead of you. There is no visible lock or keyhole, no keypad, not even a fingerprint or retina scanner but as you come closer you hear a voice saying 'Welcome back sir'. You respond with 'Hello computer', which prompts the door to unlock and swing open to admit you. You enter through the doorway and the door closes behind you as lights turn on in the hallway you have just entered.

As you remove and hang up your coat and hat, the same voice asks you 'Can I prepare a cup of tea or coffee for you, sir?', to which you reply 'Yes, Earl Grey in the living room please'. A minute or two later, sitting in a comfortable chair with your feet up and sipping a perfect cup of tea, the voice continues, 'While you were out, you received several messages, do you wish to review them now?'

Reluctantly you reply, 'Just summarise the most important messages for me, computer'.

'There was a message from your brother asking if you would like to play squash tomorrow night, as well as a reminder from your secretary about the meeting early tomorrow morning. In addition, you still haven't replied to Great Aunt Sophie regarding...'

'Skip that message,' you interject.

The voice continues, 'Okay, then the final message is a reminder that your car is due for servicing next week. Would you like me to schedule an appointment for that?'

'Yes computer, make it Monday morning, around 10am. Oh, and confirm the squash for tomorrow evening at 5.30pm. If his computer hasn't booked a court yet, make sure you book one right away.'

Later you are watching a re-run of Star Trek: The Next Generation *on 3D television and have a sudden idea that might help you in your job as a speech technology researcher. Before you forget it, you speak out, 'Computer, take a brief note'. The television programme pauses before a voice announces 'ready', and you then speak a few sentences terminating in 'end note'. At a command of 'resume', you are again confronted with Captain Jean-Luc Picard and his crew.*

Does this passage sound like science fiction? It certainly does at the time of writing – but is definitely not as great a leap of imagination as it was before the technology of Siri (real) and Iron Man's J.A.R.V.I.S. (imaginary) as we had discussed back in Chapter 1. Consider this: the nature of a printed book is that it essentially lasts forever. So, while some will read this chapter in 2016, others may read it in 2026 or even in 2036. Maybe by 2026 most readers will have it read out to them by virtual personal assistants that communicate through natural sounding dialogue and speech? Perhaps 2036's teenagers will be laughing over how primitive the technology described above sounds?

We do not know for certain what the future will bring, but we can make three statements with a fair degree of certainty: (i) the current state of speech technology is not far from being able to achieve much of the scenario above (we will consider how in a little more detail below); (ii) the trend of human–computer interaction is towards greater naturalness, and what could be more natural than communicating using speech?; and (iii) if you are serious about speech, audio and hearing research (and I hope you are serious about it, having read all the way through this book up to this point), then it will be you – my readers – who will make such a future possible.

Technology overview

This book began by considering some everyday speech systems, looking at current trends and hinting at the future, although not quite as described above. Various computational tools were introduced, such as Sox, Audacity and of course MATLAB, which we used extensively for processing, analysis and visualisation of speech and audio.

It is important never to forget the human scale of what we are working with: audio, speech and hearing, and thus three chapters were devoted to audio processing, speech and hearing, before we delved into the mysteries of psychoacoustics. Speech communications and audio analysis followed before we tackled the combined topics of big data and machine learning, which unleashed the power of statistical approaches which we could then apply to automatic speech recognition (ASR). ASR itself implies quite a bit of technology, ranging from voice activity detection through to recognition of differences between voices, natural language processing and dialogue systems, not forgetting speaker identification, language identification and topic or keyword spotting. A chapter on some diverse and interesting advanced topics brought many of the earlier threads together in different applications, which included whisper-to-speech conversion (and its converse) plus the important topic of text-to-speech (TTS).

Now consider how these topics could grow into the pseudo-science-fiction scenario that opened this chapter:

- **Greeting** – TTS technology is often used for greetings, even today, such as in automated telephone systems and hands-free telephone systems.
- **Authentication** – we have discussed the topics of speaker detection, validation and verification in previous chapters. Voiceprint technology has been a staple of science fiction for decades, but is really possible, even today. For example, it has been widely reported that some countries (perhaps many), have microphones installed at immigration checkpoints which are connected to systems able to obtain and match voiceprints. This is possibly one reason why immigration officials will often greet new arrivals with a few meaningless questions. Continuing the theme of public security, vast computing resources such as the officially

unacknowledged ECHELON (said to be part of the UKUSA Security Agreement, implemented by GCHQ in the UK, by GCSB in New Zealand, by the NSA in the USA as well as counterparts in Canada, Australia and probably elsewhere). The popular press is probably correct in assuming that this extensive system is able to identify and track particular targets through their mobile and landline voice communications.

- **Dialogue** – discussed briefly in Chapter 9, dialogue systems incorporate elements of ASR and TTS in a real-time response system, often aided by a little natural language processing (see below). We can see the beginnings of viable dialogue systems in the most advanced automated telephone call systems where one is prompted to 'state what you are calling about' after the tone.[1] Unfortunately, most of us are destined to interact much more with such systems in the future, simply because they are less expensive than employing humans to deal with customers. Meanwhile systems such as Apple's Siri and Google's Now are pushing the boundaries of dialogue using cloud technology.

- **Natural language processing** – NLP is the understanding of semantic content and context, and may even extend to being able to prioritise messages. Examples are the ability to know that a phrase such as 'call him back' refers to a previously mentioned subject, as well as decode a sentence such as 'Fred told Jim to book it, can you check with him?'. To whom does the 'him' refer, Jim or Fred? Apple's Siri was reputed to have originally derived from Wolfram Alpha, a so-called computational knowledge engine that attempts to interpret and decode the meaning of input text in order to act upon near-natural language queries. In fact NLP is not confined to speech, it is primarily related to understanding the written word.

- **Noise** – when watching a noisy television programme, presumably a device is listening for your voice. As soon as your voice is detected, directed at the computer, the television programme is paused and dialogue begins. This requires a system able to determine when you are speaking, and which can operate in noise. Most likely this would involve some form of VAD (which we have also discussed). Knowing who is speaking and to whom the speech is directed (i.e. to the computer or to another person or even to oneself) may require a combination of speaker identification (SID) and keyword spotting technologies – both of which are popular research topics that have demonstrated promising progress over recent years.

Where to from here

Thank you for reading this book: I hope that you have enjoyed it as much as I enjoyed writing it, and hope it took you a lot less time to read than it took me to write. From the first, I was attracted to speech and audio technology because it is just so accessible – in

---

[1] The author has recently spent many unhappy hours 'talking' to systems like this which are used by Her Majesty's Revenue and Customs office (which handles UK income tax) as a gatekeeper between the public and the few remaining human tax officers.

a few minutes with MATLAB I learned to create almost any type of sound, experiment with it, change it, transform it, test it out using my own (and other victims') ears, and visualise it in new ways. Almost all of us enjoy the privilege of being able to hear, and I personally enjoy the privilege of being able to hear my own creations.

At its heart, digitised sound is simply a vector or sequence of numbers that has been converted to a series of analogue voltages, amplified and then output from a loudspeaker. In MATLAB we can create or read in new data and experiment with it very quickly and conveniently. My preferred way of using MATLAB is at the command prompt, because it is fast, gives immediate feedback, and provides a tangible connection to the underlying data. Using MATLAB in this way allows a hands-on, experiential learning mechanism that is both fun and informative. Just connect a pair of headphones or loudspeakers to your computer and you can listen to your creations. Add a microphone, then you can capture and play with your own speech. In education circles having fun with experimentation and hands-on activity has long been recognised as the best way to learn. This is characterised by the Montessori Method, which is the modern name given to the learning approaches of Maria Montessori.

So go ahead and play with sounds and discover how easy and fun it is to manipulate audio in MATLAB. Of course it is also easy to mangle it too, and create something that sounds appalling (and if we are honest we can admit that we have all done that from time to time). However, the important point is that when you do fail, just try again, and again, and never stop experimenting. In so doing, you will learn and maybe discover something new.

I hope that you can start to use your new-found knowledge to innovate and create something better. World peace and global income equality are definitely on my wishlist, but in truth I will be happy if I have inspired a few readers to invent new speech and audio techniques and create applications that I can probably only dream of at present.

References

[1] M. Frigo and S. G. Johnson. The design and implementation of FFTW3. *Proc. IEEE*, 93(2):216–231, February 2005. doi: 10.1109/JPROC.2004.840301.

[2] S. W Smith. *Digital Signal Processing: A Practical Guide for Engineers and Scientists*. Newnes, 2000. www.dspguide.com.

[3] J. W. Gibbs. Fourier series. *Nature*, 59:606, 1899.

[4] R. W. Schaefer and L. R. Rabiner. Digital representation of speech signals. *Proc. IEEE*, 63(4):662–677, 1975.

[5] B. P. Bogert, M. J. R. Healy, and J. W Tukey. The quefrency analysis of time series for echoes: cepstrum, pseudo-autocovarience, cross-cepstrum and saphe cracking. In M. Rosenblatt, editor, *Proc. Symposium on Time-Series Analysis*, pages 209–243. Wiley, 1963.

[6] D. G. Childers, D. P. Skinner, and R. C. Kemerait. The cepstrum: a guide to processing. *Proc. IEEE*, 65(10):1428–1443, October 1977.

[7] F. Zheng, G. Zhang, and Z. Song. Comparison of different implementations of MFCC. *J. Computer Science and Technology*, 16(6):582–589, September 2001.

[8] J. Barkera and M. Cooke. Is the sine-wave speech cocktail party worth attending? *Speech Communication*, 27(3–4):159–174, April 1999.

[9] M. R. Schroeder, B. S. Atal, and J. L. Hall. Optimizing digital speech coders by exploiting masking properties of the human ear. *J. Acoustical Society of America*, 66(6):1647–1652, 1979.

[10] I. Witten. *Principles of Computer Speech*. Academic Press, 1982.

[11] H. R. Sharifzadeh, I. V. McLoughlin, and M. J. Russell. A comprehensive vowel space for whispered speech. *Journal of Voice*, 26(2):e49–e56, 2012.

[12] B. C. J. Moore. *An Introduction to the Psychology of Hearing*. Academic Press, 1992.

[13] I. B. Thomas. The influence of first and second formants on the intelligibility of clipped speech. *J. Acoustical Society of America*, 16(2):182–185, 1968.

[14] J. Pickett. *The Sounds of Speech Communication*. Allyn and Bacon, 1980.

[15] Z. Li, E. C. Tan, I. McLoughlin, and T. T. Teo. Proposal of standards for intelligibility tests of Chinese speech. *IEE Proc. Vision Image and Signal Processing*, 147(3):254–260, June 2000.

[16] F. L. Chong, I. McLoughlin, and K. Pawlikowski. A methodology for improving PESQ accuracy for Chinese speech. In *Proc. IEEE TENCON*, Melbourne, November 2005.

[17] K. Kryter. *The Handbook of Hearing and the Effects of Noise*. Academic Press, 1994.

[18] L. L. Beranek. The design of speech communications systems. *Proc. IRE*, 35(9):880–890, September 1947.

[19] W. Tempest, editor. *The Noise Handbook*. Academic Press, 1985.

[20] M. Mourjopoulos, J. Tsoukalas, and D. Paraskevas. Speech enhancement using psychoacoustic criteria. In *Proc. Int. Conf. on Acoustics, Speech and Signal Processing*, pages 359–362, 1991.

[21] F. White. *Our Acoustic Environment*. John Wiley & Sons, 1976.

[22] P. J. Blamey, R. C. Dowell, and G. M. Clark. Acoustic parameters measured by a formant estimating speech processor for a multiple-channel cochlear implant. *J. Acoustical Society of America*, 82(1):38–47, 1987.

[23] I. V. McLoughlin, Y. Xu, and Y. Song. Tone confusion in spoken and whispered Mandarin Chinese. In *Chinese Spoken Language Processing (ISCSLP), 2014 9th Int. Symp. on*, pages 313–316. IEEE, 2014.

[24] I. B. Thomas. Perceived pitch of whispered vowels. *J. Acoustical Society of America*, 46: 468–470, 1969.

[25] Y. Swerdlin, J. Smith, and J. Wolfe. The effect of whisper and creak vocal mechanisms on vocal tract resonances. *J. Acoustical Society of America*, 127(4):2590–2598, 2010.

[26] P. C. Loizou. *Speech Enhancement: Theory and Practice*. CRC Press, 2013.

[27] N. Kitawaki, H. Nagabuchi, and K. Itoh. Objective quality evaluation for low-bit-rate speech coding systems. *IEEE J. Selected Areas in Communications*, 6(2): 242–248, 1988.

[28] Y. Hu and P. C Loizou. Evaluation of objective quality measures for speech enhancement. *IEEE Trans. Audio, Speech, and Language Processing*, 16(1):229–238, 2008.

[29] A. D. Sharpley. Dynastat webpages, 1996 to 2006. www.dynastat.com/SpeechIntelligibility.htm.

[30] S. F. Boll. Suppression of acoustic noise in speech using spectral subtraction. *IEEE Trans. Acoustics, Speech and Signal Processing*, 27(2):113–120, 1979.

[31] R. E. P. Dowling and L. F. Turner. Modelling the detectability of changes in auditory signals. *Proc. Int. Conf. on Acoustics, Speech, and Signal Processing*, vol. 1, pages 133–136, 1993.

[32] J. I. Alcantera, G. J. Dooley, P. J. Blamey, and P. M. Seligman. Preliminary evaluation of a formant enhancement algorithm on the perception of speech in noise for normally hearing listeners. *J. Audiology*, 33(1):15–24, 1994.

[33] J. G. van Velden and G. F. Smoorenburg. Vowel recognition in noise for male, female and child voices. *Proc. Int. Conf. on Acoustics, Speech, and Signal Processing*, pages 796–799, 1995.

[34] G. A. Miller, G. A. Heise, and W. Lichten. The intelligibility of speech as a function of the context of the test materials. *Experimental Psychology*, 41:329–335, 1951.

[35] W. G. Sears. *Anatomy and Physiology for Nurses and Students of Human Biology*. Arnold, 4th edition, 1967.

[36] J. Simner, C. Cuskley, and S. Kirby. What sound does that taste? Cross-modal mappings across gustation and audition. *Perception*, 39(4):553, 2010.

[37] R. Duncan-Luce. *Sound and Hearing: A Conceptual Introduction*. Lawrence Erlbaum and Associates, 1993.

[38] W. F. Ganong. *Review of Medical Physiology*. Lange Medical Publications, 9th edition, 1979.

[39] H. Fletcher and W. A. Munson. Loudness, its definition, measurement and calculation. *Bell System Technical Journal*, 12(4):377–430, 1933.

[40] K. Kryter. *The Effects of Noise on Man*. Academic Press, 2nd edition, 1985.

[41] R. Plomp. Detectability threshold for combination tones. *J. Acoustical Society of America*, 37(6):1110–1123, 1965.

[42] K. Ashihara. Combination tone: absent but audible component. *Acoustical Science and Technology*, 27(6):332, 2006.

[43] J. C. R. Licklider. *Auditory Feature Analysis*. Academic Press, 1956.

[44] Y. M. Cheng and D. O'Shaughnessy. Speech enhancement based conceptually on auditory evidence. *Proc. Int. Conf. on Acoustics, Speech, and Signal Processing*, pages 961–963, 1991.

[45] N. Virag. Speech enhancement based on masking properties of the auditory system. *Proc. Int. Conf. on Acoustics, Speech, and Signal Processing*, pages 796–799, 1995.

[46] Y. Gao, T. Huang, and J. P. Haton. Central auditory model for spectral processing. *Proc. Int. Conf. on Acoustics, Speech and Signal Processing*, pages 704–707, 1993.

[47] I. V. McLoughlin and Z.-P. Xie. Speech playback geometry for smart homes. In *Consumer Electronics (ISCE 2014), 18th IEEE Int. Symp. on*, pages 1–2. IEEE, 2014.

[48] C. R. Darwin and R. B. Gardner. Mistuning a harmonic of a vowel: grouping and phase effects on vowel quality. *J. Acoustical Society of America*, 79:838–845, 1986.

[49] D. Sen, D. H. Irving, and W. H. Holmes. Use of an auditory model to improve speech coders. *Proc. Int. Conf. on Acoustics, Speech, and Signal Processing*, vol. 2, pages 411–415, 1993.

[50] H. Hermansky. Perceptual linear predictive (PLP) analysis of speech. *J. Acoustical Society of America*, 87(4):1738–1752, April 1990.

[51] S. S. Stevens, J. Volkmann, and E. B. Newman. A scale for the measurement of the psychological magnitude pitch. *J. Acoustical Society of America*, 8(3):185–190, 1937. doi: http://dx.doi.org/10.1121/1.1915893. http://scitation.aip.org/content/asa/journal/jasa/8/3/10.1121/1.1915893.

[52] D. O'Shaughnessy. *Speech Communication: Human and Machine*. Addison-Wesley, 1987.

[53] G. Fant. Analysis and synthesis of speech processes. In B. Malmberg, editor, *Manual of Phonetics*, pages 173–177. North-Holland, 1968.

[54] ISO/MPEG–Audio Standard layers. Editorial pages. *Sound Studio Magazine*, pages 40–41, July 1992.

[55] A. Azirani, R. Jeannes, and G. Faucon. Optimizing speech enhancement by exploiting masking properties of the human ear. *Proc. Int. Conf. on Acoustics, Speech, and Signal Processing*, pages 800–803, 1995.

[56] A. S. Bregman. *Auditory Scene Analysis*. MIT Press, 1990.

[57] H. Purwins, B. Blankertz, and K. Obermayer. Computing auditory perception. *Organised Sound*, 5(3):159–171, 2000.

[58] C. M. M. Tio, I. V. McLoughlin, and R. W. Adi. Perceptual audio data concealment and watermarking scheme using direct frequency domain substitution. *IEE Proc. Vision, Image & Signal Processing*, 149(6):335–340, 2002.

[59] I. V. McLoughlin and R. J. Chance. Method and apparatus for speech enhancement in a speech communications system. *PCT international patent* (PCT/GB98/01936), July 1998.

[60] Y. M. Cheng and D. O'Shaughnessy. Speech enhancement based conceptually on auditory evidence. *IEEE Trans. Signal Processing*, 39(9):1943–1954, 1991.

[61] N. Jayant, J. Johnston, and R. Safranek. Signal compression based on models of human perception. *Proc. IEEE*, 81(10):1383–1421, 1993.

[62] D. Sen and W. H. Holmes. Perceptual enhancement of CELP speech coders. In *Proc. Int. Conf. on Acoustics, Speech and Signal Processing*, pages 105–108, 1993.

[63] J. Markel and A. Grey. *Linear Prediction of Speech*. Springer-Verlag, 1976.

[64] J. Makhoul. Linear prediction: a tutorial review. *Proc. IEEE*, 63(4):561–580, April 1975.

[65] S. Saito and K. Nakata. *Fundamentals of Speech Signal Processing*. Academic Press, 1985.

[66] J. L. Kelly and C. C. Lochbaum. Speech synthesis. *Proc. Fourth Int. Congress on Acoustics*, pages 1–4, September 1962.

[67] B. H. Story, I. R. Titze, and E. A. Hoffman. Vocal tract area functions from magnetic resonance imaging. *J. Acoustical Society of America*, 100(1):537–554, 1996.

[68] N. Sugamura and N. Favardin. Quantizer design in LSP speech analysis–synthesis. *IEEE J. Selected Areas in Communications*, 6(2):432–440, February 1988.

[69] S. Saoudi, J. Boucher, and A. Guyader. A new efficient algorithm to compute the LSP parameters for speech coding. *Signal Processing*, 28(2):201–212, 1995.

[70] T. I. and M. I. T. *TIMIT database. A CD-ROM database of phonetically classified recordings of sentences spoken by a number of different male and female speakers*, disc 1-1.1, 1990.

[71] N. Sugamura and F. Itakura. Speech analysis and synthesis methods developed at ECL in NTT – from LPC to LSP. *Speech Communications*, pages 213–229, 1986.

[72] J. S. Collura and T. E. Tremain. Vector quantizer design for the coding of LSF parameters. In *Proc. Int. Conf. on Acoustics, Speech and Signal Processing*, pages 29–32, 1993.

[73] I. V. McLoughlin. LSP parameter interpretation for speech classification. In *Proc. 2nd IEEE Int. Conf. on Information, Communications and Signal Processing*, December 1999.

[74] I. V. McLoughlin and F. Hui. Adaptive bit allocation for LSP parameter quantization. In *Proc. IEEE Asia–Pacific Conf. on Circuits and Systems*, paper number 231, December 2000.

[75] Q. Zhao and J. Suzuki. Efficient quantization of LSF by utilising dynamic interpolation. In *IEEE Int. Symp. on Circuits and Systems*, pages 2629–2632, June 1997.

[76] European Telecommunications Standards Institute. Trans-European trunked radio system (TETRA) standard. 1994.

[77] K. K. Paliwal and B. S. Atal. Efficient vector quantization of LPC parameters at 24 bits per frame. In *Proc. Int. Conf. on Acoustics, Speech and Signal Processing*, pages 661–664, 1991.

[78] D.-I. Chang, S. Ann, and C. W. Lee. A classified split vector quantization of LSF parameters. *Signal Processing*, 59(3):267–273, June 1997.

[79] R. Laroia, N. Phamdo, and N. Farvardin. Robust and efficient quantization of speech LSP parameters using structured vector quantizers. In *Proc. Int. Conf. on Acoustics, Speech and Signal Processing*, pages 641–644, 1991.

[80] H. Zarrinkoub and P. Mermelstein. Switched prediction and quantization of LSP frequencies. In *Proc. Int. Conf. on Acoustics, Speech and Signal Processing*, pages 757–760, 1996.

[81] C. S. Xydeas and K. K. M. So. A long history quantization approach to scalar and vector quantization of LSP coefficients. In *Proc. Int. Conf. on Acoustics, Speech and Signal Processing*, pages 1–4, 1993.

[82] J.-H. Chen, R. V. Cox, Y.-C. Lin, N. Jayant, and M. J. Melchner. A low-delay CELP coder for the CCITT 16 kb/s speech coding standard. *IEEE J. Selected Areas in Communications*, 10(5):830–849, June 1992.

[83] B. S. Atal. Predictive coding of speech at low bitrates. *IEEE Trans. Communications*, 30(4):600–614, 1982.

[84] M. R. Schroeder and B. S. Atal. Code-excited linear prediction CELP: high-quality speech at very low bit rates. In *Proc. Int. Conf. on Acoustics, Speech and Signal Processing*, pages 937–940, 1985.

[85] L. M. Supplee, R. P. Cohn, J. S. Collura, and A. V. McCree. MELP: the new Federal standard at 2400 bps. In *Proc. IEEE Int. Conf. Acoustics Speech and Signal Processing*, vol. 2, pages 1591–1594, April 1997.

[86] I. A. Gerson and M. A. Jasiuk. Vector sum excited linear prediction (VSELP) speech coding at 8 kbps. In *Proc. IEEE Int. Conf. Acoustics Speech and Signal Processing*, vol. 1, pages 461–464, April 1990.

[87] L. R. Rabiner and R. W. Schaefer. *Digital Processing of Speech Signals*. Prentice-Hall, 1978.

[88] I. V. McLoughlin. LSP parameter interpretation for speech classification. In *Proc. 6th IEEE Int. Conf. on Electronics, Circuits and Systems*, paper number 113, September 1999.

[89] K. K. Paliwal. A study of LSF representation for speaker-dependent and speaker-independent HMM-based speech recognition systems. In *Proc. Int. Conf. on Acoustics, Speech and Signal Processing*, vol. 2, pages 801–804, 1990.

[90] J. Parry, I. Burnett, and J. Chicharo. Linguistic mapping in LSF space for low-bit rate coding. *Proc. Int. Conf. on Acoustics, Speech, and Signal Processing*, vol. 2, pages 653–656, March 1999.

[91] L. R. Rabiner, M. Cheng, A. Rosenberg, and C. McGonegal. a comparative performance study of several pitch detection algorithms. *IEEE Trans. Acoustics, Speech and Signal Processing*, 24(5):399–418, October 1976.

[92] L. Cohen. *Time–Frequency Analysis*. Prentice-Hall, 1995.

[93] Z. Q. Ding, I. V. McLoughlin, and E. C. Tan. How to track pitch pulse in LP residual – joint time-frequency distribution approach. In *Proc. IEEE Pacific Rim Conf. on Communications, Computers and Signal Processing*, August 2001.

[94] A. G. Krishna and T. V. Sreenivas. Musical instrument recognition: from isolated notes to solo phrases. In *Proc. IEEE Int. Conf. on Acoustics Speech and Signal Processing*, vol. 4, pages 265–268, 2004.

[95] The British Broadcasting Corporation (BBC). BBC Radio 4: Brett Westwood's guide to garden birdsong, May 2007. www.bbc.co.uk/radio4/science/birdsong.shtml.

[96] A. Harma and P. Somervuo. Classification of the harmonic structure in bird vocalization. In *Proc. IEEE Int. Conf. on Acoustics, Speech, and Signal Processing*, vol. 5, pages 701–704, 2004.

[97] I. McLoughlin, M.-M. Zhang, Z.-P. Xie, Y. Song, and W. Xiao. Robust sound event classification using deep neural networks. *IEEE Trans. Audio, Speech, and Language Processing*, PP(99), 2015. doi: dx.doi.org/10.1109/TASLP.2015.2389618.

[98] R. F. Lyon. Machine hearing: an emerging field. *IEEE Signal Processing Magazine*, 42: 1414–1416, 2010.

[99] T. C. Walters. Auditory-based processing of communication sounds. PhD thesis, University of Cambridge, 2011.

[100] A. Kanagasundaram, R. Vogt, D. B. Dean, S. Sridharan, and M. W. Mason. i-vector based speaker recognition on short utterances. In *Interspeech 2011*, pages 2341–2344, Firenze Fiera, Florence, August 2011. International Speech Communication Association (ISCA). http://eprints.qut.edu.au/46313/.

[101] C. Cortes and V. Vapnik. Support-vector networks. In *Machine Learning*, 20(3):273–297, 1995.

[102] C.-C. Chang and C.-J. Lin. LIBSVM: a library for support vector machines. *ACM Transactions on Intelligent Systems and Technology*, 2(3), article 27, software available at www.csie.ntu.edu.tw/~cjlin/libsvm.

[103] S. Balakrishnama and A. Ganapathiraju. Linear discriminant analysis – a brief tutorial. Institute for Signal and information Processing, 1998. www.isip.piconepress.com/publications/reports/1998/isip/lda

[104] A. Hyvärinen and E. Oja. Independent component analysis: algorithms and applications. *Neural Networks*, 13(4):411–430, 2000.

[105] C. M. Bishop. *Pattern Recognition and Machine Learning*. Springer, 2006.

[106] G. E. Hinton, S. Osindero, and Y.-W. Teh. A fast learning algorithm for deep belief nets. *Neural Computation*, 18(7):1527–1554, 2006.

[107] R. B. Palm. Prediction as a candidate for learning deep hierarchical models of data. Master's thesis, Technical University of Denmark, 2012.

[108] Y. LeCun and Y. Bengio. Convolutional networks for images, speech, and time series. In *The Handbook of Brain Theory and Neural Networks*, page 3361, MIT Press, 1995.

[109] J. Bouvrie. Notes on convolutional neural networks. 2006. http://cogprints.org/5869/.

[110] Y. LeCun, L. Bottou, Y. Bengio, and P. Haffner. Gradient-based learning applied to document recognition. *Proc. IEEE*, 86(11):2278–2324, 1998.

[111] O. Abdel-Hamid, A.-R. Mohamed, H. Jiang, and G. Penn. Applying convolutional neural networks concepts to hybrid NN-HMM model for speech recognition. In *Acoustics, Speech and Signal Processing (ICASSP), 2012 IEEE Int. Conf. on*, pages 4277–4280. IEEE, 2012.

[112] T. N. Sainath, A.-R. Mohamed, B. Kingsbury, and B. Ramabhadran. Deep convolutional neural networks for LVCSR. In *Acoustics, Speech and Signal Processing (ICASSP), 2013 IEEE Int. Conf. on*, pages 8614–8618. IEEE, 2013.

[113] H.-M. Zhang, I. McLoughlin, and Y. Song. Robust sound event recognition using convolutional neural networks. In *Proc. ICASSP*, paper number 2635. IEEE, 2015.

[114] R. Cole, J. Mariani, H. Uszkoreit, G. B. Varile, A. Zaenen, A. Zampolli, and V. Zue, editors. *Survey of the State of the Art in Human Language Technology*. Cambridge University Press, 2007.

[115] C. A. Kamm, K. M. Yang, C. R. Shamieh, and S. Singhal. Speech recognition issues for directory assistance applications. In *Proc. 2nd IEEE Workshop on Interactive Voice Technology for Telecommunications Applications IVTTA94*, pages 15–19, Kyoto, September 1994.

[116] J. G. Fiscus, J. Ajot, and J. S. Garofolo. The rich transcription. 2007 meeting recognition evaluation. In *Multimodal Technologies for Perception of Humans*, pages 373–389. Springer, 2008.

[117] H. B. Yu and M.-W. Mak. Comparison of voice activity detectors for interview speech in NIST speaker recognition evaluation. In *Interspeech 2011*, pages 2353–2356, August 2011.

[118] F. Beritelli, S. Casale, and A. Cavallaro. A robust voice activity detector for wireless communications using soft computing. *IEEE J. Selected Areas in Communications,*, 16(9):1818–1829, 1998.

[119] M.-Y. Hwang and X. Huang. Subphonetic modeling with Markov states-Senone. In *Acoustics, Speech, and Signal Processing, 1992. ICASSP-92, 1992 IEEE Int. Conf. on*, vol. 1, pages 33–36. IEEE, 1992.

[120] Y. Song, B. Jiang, Y. Bao, S. Wei, and L.-R. Dai. i-vector representation based on bottleneck features for language identification. *Electronics Letters*, 49(24):1569–1570, November 2013. doi: 10.1049/el.2013.1721.

[121] B. Jiang, Y. Song, S. Wei, J.-H. Liu, I. V. McLoughlin, and L.-R. Dai. Deep bottleneck features for spoken language identification. *PLoS ONE*, 9(7):e100795, July 2014. doi: 10.1371/journal.pone.0100795. http://dx.doi.org/10.1371%2Fjournal.pone.0100795.

[122] B. Jiang, Y. Song, S. Wei, M.-G. Wang, I. McLoughlin, and L.-R. Dai. Performance evaluation of deep bottleneck features for spoken language identification. In *Chinese Spoken Language Processing (ISCSLP), 2014 9th Int. Symp. on*, pages 143–147, September 2014. doi: 10.1109/ISCSLP.2014.6936580.

[123] S. Xue, O. Abdel-Hamid, H. Jiang, and L. Dai. Direct adaptation of hybrid DNN/HMM model for fast speaker adaptation in LVCSR based on speaker code. In *Proc. ICASSP*, pages 6339–6343, 2014.

[124] C. Kong, S. Xue, J. Gao, W. Guo, L. Dai, and H. Jiang. Speaker adaptive bottleneck features extraction for LVCSR based on discriminative learning of speaker codes. In *Chinese Spoken Language Processing (ISCSLP), 2014 9th Int. Symp. on*, pages 83–87. IEEE, 2014.

[125] S. Young, G. Evermann, M. Gales, T. Hain, D. Kershaw, X. Liu, G. Moore, J. Odell, D. Ollason, D. Povey et al. *The HTK book*, vol. 2. Entropic Cambridge Research Laboratory, Cambridge, 1997.

[126] L. R. Rabiner. A tutorial on hidden Markov models and selected applications in speech recognition. *Proc. IEEE*, 77(2):257–286, 1989.

[127] M. Gales and S. Young. The application of hidden Markov models in speech recognition. *Foundations and Trends in Signal Processing*, 1(3):195–304, 2008.

[128] L. F. Uebel and P. C. Woodland. An investigation into vocal tract length normalisation. In *Eurospeech*, 1999.

[129] D. Povey, A. Ghoshal, G. Boulianne, L. Burget, O. Glembek, N. Goel, M. Hannemann, P. Motlicek, Y. Qian, P. Schwarz et al. The Kaldi speech recognition toolkit. In *IEEE 2011 Workshop on Automatic Speech Recognition and Understanding*. IEEE Signal Processing Society, December 2011. IEEE Catalog No.: CFP11SRW-USB.

[130] W. Walker, P. Lamere, P. Kwok, B. Raj, R. Singh, E. Gouvea, P. Wolf, and J. Woelfel. Sphinx-4: a flexible open source framework for speech recognition, 2004. cmusphinx.sourceforge.net/sphinx4/doc/Sphinx4Whitepaper.pdf.

[131] A. P. A. Broeders. Forensic speech and audio analysis, forensic linguistics 1998 to 2001 – a review. In *Proc. 13th INTERPOL Forensic Science Symposium*, pages 51–84, Lyon, October 2001.

[132] A. P. A. Broeders. Forensic speech and audio analysis, forensic linguistics 2001 to 2004 – a review. In *Proc. 14th INTERPOL Forensic Science Symposium*, pages 171–188, Lyon, 2004.

[133] R. Togneri and D. Pullella. An overview of speaker identification: accuracy and robustness issues. *IEEE Circuits and Systems Magazine*, 11(2):23–61, 2011. doi: 10.1109/MCAS.2011.941079.

[134] S. Furui. Recent advances in speaker recognition. *Pattern Recognition Letters*, 18:859–872, 1997.

[135] S. Furui. Speaker-dependent-feature extraction, recognition and processing techniques. *Speech Communication*, 10:505–520, 1991.

[136] G. Doddington, W. Liggett, A. Martin, M. Przybocki, and D. A. Reynolds. Sheep, goats, lambs and wolves: a statistical analysis of speaker performance in the NIST 1998 speaker recognition evaluation. In *Proc. 5th Int. Conf. on Spoken Language Processing*, vol. 0608, November 1998.

[137] M. A. Zissman and K. M. Berkling. Automatic language identification. *Speech Communication*, 35:115–124, 2001.

[138] K. Wu, Y. Song, W. Guo, and L. Dai. Intra-conversation intra-speaker variability compensation for speaker clustering. In *Chinese Spoken Language Processing (ISCSLP), 2012 8th Int. Symp. on*, pages 330–334. IEEE, 2012.

[139] S. Meignier and T. Merlin. LIUM SpkDiarization: an open source toolkit for diarization. In *CMU SPUD Workshop*, 2010.

[140] D.-C. Lyu, T. P. Tan, E. Chang, and H. Li. SEAME: a Mandarin-English code-switching speech corpus in South-East Asia. In *INTERSPEECH*, volume 10, pages 1986–1989, 2010.

[141] M. Edgington. Investigating the limitations of concatenative synthesis. In *EUROSPEECH-1997*, pages 593–596, Rhodes, September 1997.

[142] C. K. Ogden. *Basic English: A General Introduction with Rules and Grammar*. Number 29. K. Paul, Trench, Trubner, 1944.

[143] T. Dutoit. High quality text-to-speech synthesis: an overview. *J. Electrical & Electronics Engineering, Australia: Special Issue on Speech Recognition and Synthesis*, 17(1):25–36, March 1997.

[144] T. B. Amin, P. Marziliano, and J. S. German. Glottal and vocal tract characteristics of voice impersonators. *IEEE Trans. on Multimedia*, 16(3):668–678, 2014.

[145] I. V. McLoughlin. The art of public speaking for engineers. *IEEE Potentials*, 25(3):18–21, 2006.

[146] The University of Edinburgh, The Centre for Speech Technology Research. The festival speech synthesis system, 2004. www.cstr.ed.ac.uk/projects/festival/.

[147] P. Taylor, A. Black, and R. Caley. The architecture of the Festival speech synthesis system. In *Third International Workshop on Speech Synthesis*, Sydney, November 1998.

[148] Voice Browser Working Group. Speech synthesis markup language (SSML) version 1.0. *W3C Recommendation*, September 2004.

[149] K. K. Paliwal. On the use of line spectral frequency parameters for speech recognition. *Digital Signal Processing*, 2:80–87, 1992.

[150] I. V. McLoughlin and R. J. Chance. LSP-based speech modification for intelligibility enhancement. In *13th Int. Conf. on DSP*, Santorini, July 1997.

[151] I. V. McLoughlin and R. J. Chance. LSP analysis and processing for speech coders. *IEE Electronics Letters*, 33(99):743–744, 1997.

[152] A. Schaub and P. Straub. Spectral sharpening for speech enhancement/noise reduction. In *Proc. Int. Conf. on Acoustics, Speech and Signal Processing*, pages 993–996, 1991.

[153] H. Valbret, E. Moulines, and J. P. Tubach. Voice transformation using PSOLA technique. In *IEEE Int. Conf. Acoustics Speech and Signal Proc.*, pages 145–148, San Francisco, CA, March 1992.

[154] R. W Morris and M. A. Clements. Reconstruction of speech from whispers. *Medical Engineering & Physics*, 24(7):515–520, 2002.

[155] H. R. Sharifzadeh, I. V. McLoughlin, and F. Ahmadi. Reconstruction of normal sounding speech for laryngectomy patients through a modified CELP codec. *IEEE Trans. Biomedical Engineering*, 57:2448–2458, October 2010.

[156] J. Li, I. V. McLoughlin, and Y. Song. Reconstruction of pitch for whisper-to-speech conversion of Chinese. In *Chinese Spoken Language Processing (ISCSLP), 2014 9th Int. Symp. on*, pages 206–210. IEEE, 2014.

References

[157] F. Ahmadi and I. V. McLoughlin. Measuring resonances of the vocal tract using frequency sweeps at the lips. In *2012 5th Int. Symp. on Communications Control and Signal Processing (ISCCSP)*, 2012.

[158] F. Ahmadi and I. McLoughlin, The use of low-frequency ultrasonics in speech processing. In *Signal Proceesing*, S. Miron (ed.). InTech, 2010, pp. 503–528.

[159] I. V. McLoughlin. Super-audible voice activity detection. *IEEE/ACM Trans. on Audio, Speech, and Language Processing*, 22(9):1424–1433, 2014.

[160] F. Ahmadi, I. V. McLoughlin, and H. R. Sharifzadeh. Autoregressive modelling for linear prediction of ultrasonic speech. In *INTERSPEECH*, pages 1616–1619, 2010.

[161] I. V. McLoughlin and Y. Song. Mouth state detection from low-frequency ultrasonic reflection. *Circuits, Systems, and Signal Processing*, 34(4):1279–1304, 2015.

[162] R. W. Schafer. What is a Savitzky–Golay filter? [lecture notes]. *IEEE Signal Processing Magazine*, 28(4):111–117, 2011. doi: 10.1109/MSP.2011.941097.

[163] F. Ahmadi, M. Ahmadi, and I. V. McLoughlin. Human mouth state detection using low frequency ultrasound. In *INTERSPEECH*, pages 1806–1810, 2013.

Index

μ-law, 6, 14, 148
.wav file format, 10, 13, 14

A-law, 6, 14, 19, 109, 148
A-weighting, 89, 131, 191
absolute
    pitch, 97
    scaling, 20
accelerometer, to measure pitch, 208
accent, 303
accuracy, 234
ACELP, *see* codebook excited linear prediction, algebraic
Adams, Douglas, 307
adaptive differential PCM, 144, 148, 177
    sub-band ADPCM, 147
ADC, *see* analogue-to-digital converter
ADPCM, *see* adaptive differential PCM
affricative, 64
ageing, effect of, 303
allophone, 63
allotone, 63
AMDF, *see* average magnitude difference function
analogue-to-digital converter, 4, 5, 13, 141
analysis, 149, 158, 180, 196
analysis algorithim, 33
analysis-by-synthesis, 182
angular frequency, 40
animal noises, analysis of, 219
aphonia, 311
Apple
    iPod, 1, 109
    Siri, 2, 366
articulation index, 82, 270
ASA, *see* auditory scene analysis
ASR, *see* automatic speech recognition
audio
    waveform, 17
audio fidelity, 6, 122, 148, 149
audioread(), 13, 229
audiorecorder(), 10
audiowrite(), 16
auditory adaptation, 94, 103

auditory cortex, 93
auditory fatigue, 93
auditory image, 231
auditory scene analysis, 112
authentication, 363
autocorrelation, 41, 160
automatic speech recognition, 224, 267–269, 286, 298, 301, 310
average magnitude difference function, 199, 200, 209, 211

Babbage, Charles, 225
back propagation, 239
backwards speech, 102
bandwidth, 142
Bark, 104, 124, 129, 131, 137
bark2f(), 105, 106, 337
basilar membrane, 94
Bayes' theorem, 293
Bayes, Thomas, 293
Bayesian inference criteria, 309
Bell, Alexander Graham, 1
Bernoulli–Bernoulli RBM, 254
big data
    rationale, 225
    sources of, 226
    storage, 227
big endian, 15, 17, 19
binaural, 97, 98, 324
biometric, 363
birdsong, 219
blind source separation, 242
Blu-ray, 1
bronchial tract, 55
buffer(), 27

cadence, of speech, 216
calibrated reference speaker, 88
candidate vectors, 184, 186
cceps(), 204
cceps(),icceps(), 42
CD, 6, 121, 330
CELP, *see* codebook excited linear prediction
cepstrum, 40, 42, 202, 204, 209, 211, 299

Chinese, 60, 61, 63, 70, 216, 272, 288, 299, 306, 308, 310, 317
chirp
  exponential, 46
  generation of, 46
  linear, 46
chirp(), 34, 46, 90
chord, musical, 50, 100, 118
classification, 230, 232, 235, 239, 255, 292
classification error, 215
click, 27
closure, of sounds, 114
clustering, 247, 250
CNN, *see* convolutional neural network
co-modulation masking release, 95
code excited linear prediction, 183
code switching, 310
codebook, 286, 350
codebook excited linear prediction, 140, 153, 165, 172, 174, 183–186, 188, 191, 207, 338, 341, 348
  in speech synthesis, 316
  algebraic, 186
  computational complexity of, 186
  forward–backward, 189
  latency, 189
  split codebooks, 187
  standards, 190
coffee, 304
combination tones, 90
comfort noise, 116
common fate, of sounds, 116
compression, 228
concatenative synthesis
  of speech, 316
concert pitch, 45
consonant, 63, 64
contextual information, 80
continuous processing, 28
continuous speech recognition, 268
contour plot, 213
contour spectral plot, 214
convergence, 230
converison between reflection coefficients and LPCs, 161
conversion of LPC to LSP, 168
convolutional neural network, 259
correlogram, 40, 41
covariance, 160
critical band, 94, 95, 110, 123, 124, 127–129, 134, 137, 148
  filters, 104
CVSDM, *see* continuously variable delta modulation

DAB, *see* digital audio broadcast
DAC, *see* digital-to-analogue converter

data
  storage cost, 223
  storage, 227
  value of, 223
database
  public, 228
dBA, 62
DBF, 309
DBN, *see* deep bottleneck network
DCT, *see* discrete cosine transform
decoding, 293
deep belief network, 257
deep bottleneck features, 305
deep bottleneck network, 286
deep neural network, 253, 256, 257, 305
DeepLearn Toolbox, 258, 261
delta modulation, 143
  adaptive, 144
  continuously variable slope, 144
  slew rate, 143, 144
  slope overload, 143
development platform, 3
DFT
  discrete Fourier transform, 36
diagnostic rhyme test, 79, 210
dialogue, 368
diarization, 231, 302, 308, 309
dictionary, 299
differential PCM, 145
digital audio broadcast, 1
digital compact cassette, 109
digital-to-analogue converter, 4, 5, 13
dimensionality, 229
diphthong, 64
dir(), 229
directory name, 229
discontinuities, 27
discrete cosine transform, 33
discrete digital signal processing, 142
discrete Fourier transform, 36
DNN, *see* deep neural network
dog bark, 219, 220
Dolby, 330
DRT, *see* diagnostic rhyme test
DTMF, *see* dual tone multiple frequency
DTW, *see* dynamic time warping
dual tone multiple frequency, 182
Durbin–Levinson–Itakura method, 161
DVD, 1
dynamic range, 6, 142
dynamic time warping, 303
dysarthria, 311
dysphonia, 311

ear
  basilar membrane, 85
  cochlea, 85

drum, 17, 85
  human, 85
  organs of Corti, 85
  protection, 86
  Reissner's membrane, 85
earache, 87
ECHELON, 368
echoes, 101
EGG, *see* electroglottograph
electroencephalogram, 268
electroglottograph, 208
electrolarynx, 348, 350, 353, 363
emotion, 310
endian, 15, 19
enhancement
  of sounds, 110
  of speech, 111
equal loudness, 89, 123, 131, 133
equal-loudness contours, 87, 103, 109

f2bark(), 105, 106, 337
f2mel(), 106
fast Fourier transform, 21, 23, 32–36, 39, 40, 75, 128, 195, 199, 201, 202
fclose(), 16
feature transformation, 227
features, 227, 229, 230, 232, 278, 284, 346
Festival speech synthesis system, 323
FFT, *see* fast Fourier transform
fft(), 21, 34, 130, 203
FFTW, 22
file path, 229
filter, 21
  analysis, 151, 152, 184
  continuity of, 28
  FIR, 21, 30, 158
  history, 31
  IIR, 21, 151, 178
  internal state of, 31
  LPC, 165, 167
  pitch, 179
  pole-zero, 21
  stability, 157
  synthesis, 151, 152, 191
filter(), 21, 29, 31, 158
filterbank, 231
finite impulse response, 21
Fletcher–Munson curves, 87
fopen(), 15
formant strengthening, 334
formants, 58, 152, 279, 351
Fourier transform, 21, 36, 37, 42, 128, 211
frame, 231
frame power, 198, 211
fread(), 15, 17
freqgen(), 47, 48, 115, 119

frequency discrimination, 96, 104, 123
frequency resolution, 22
frequency selectivity, 86
freqz(), 152, 170
fricative, 64
front-end clipping, 281
FS1015, 190
FS1016, 190
fwrite(), 16

G.711, 148
G.721, 148
G.722, 147, 148
G.723, 148, 190
G.726, 148
G.728, 182, 190
G.729, 190
Gaussian
  process, 242
Gaussian–Bernoulli RBM, 254
Gaussian mixture model, 250, 283, 347
gestures, 274
getaudio(), 12
getaudiodata(), 11
Gibbs phenomena, 27
glide, 64
glissando, 119
global system for mobile communications, *see* GSM
glottis, 56, 69, 150, 162, 165, 166, 176, 208, 303, 352
GMM, *see* Gaussian mixture model
golden ears, 6, 7
good continuation, of sounds, 119
GPU, *see* graphics processing unit
graphics processing unit, 301
ground truth, 281
Global System for Moblie Communications, *see* GSM
GSM, 2, 6, 116, 177, 190
Gulliver's Travels, 19

Haas effect, 100
handphone, 2
hang time, 280
harmonics, 92, 100, 118, 217
HAS, *see* human auditory system
hearing, 86, 87
hearing loss, 93
Hermansky, 124, 127, 129, 132
hidden Markov model, 207, 288, 299, 301, 303
hidden Markov model toolkit, *see* HTK
HMM, *see* hidden Markov model
HTK, 105, 250, 301, 311
human auditory system, 18, 112, 116, 143
human ethics, 226, 227

i-vector, 231, 285
ICA, see independent component analysis
icceps(), 204
ifft(), 203
iFlytek, 225, 226, 263
IIR, see infinite impulse response
imagesc(), 212
impersonation, 306
independent component analysis, 242
induction, auditory, 114
infinite impulse response, 21
infinite impulse response filter, 151
international phonetic alphabet, 63, 320
International Telecommunications Union, 72, 148
IPA, see international phonetic alphabet
IS, see Itakuro–Saito distance
Itakura–Saito distance, 77, 78
ITU, see International Telecommunications Union

J.A.R.V.I.S, 366
JND, see just-noticeable difference
just-noticeable difference, 105

K–L tube model, see Kelly–Lochbaum model
k-means, 247, 249
Kaldi, 301, 311
Kelly–Lochbaum model, 162
key-phones, 307
keyword spotting, 268
Kirk, Captain, 319

language
   n-gram model, 300
   classification of, 305
   grammar, 303
   identification, 231, 305
   learning, 310
   model, 299
   morphology, 306
   phonology, 306
   syntax, 306
LAR, see log area ratios
large vocabulary continuous speech recognition, 300
laryngectomy, 347, 348, 350, 353
   partial, 348
larynx, 303
lateral inhibition function, 124
LDA, see linear discriminant analysis
Le Roux method, 161
learning, 235, 239
line spectral frequencies, see line spectral pairs
line spectral pairs, 159, 165, 166, 168, 170–174, 204, 208, 209, 211, 216, 217, 219, 299, 334, 338, 341
linear discriminant analysis, 242

linear prediction, 149, 299
linear predictive coding, 43, 76, 150–152, 154, 157–162, 165, 168, 171–174, 177, 178, 183–191, 195, 209, 219, 350
linear time invariant, 348
lips, 303
lisp, 311
little endian, 15, 17, 19
LLR, see log-likelihood ratio
load, 16
log area ratios, 157, 177
log-likelihood ratio, 76, 77
Loizou, Philip, 77
long-play audio, 6
long-term prediction, 177, 178, 183–190, 208, 340, 350
long-term predictor, 60
loudness, 103, 104
LPC, see linear predictive coding, 231
LPC cepstral distance, 76, 77
lpc(), 151, 217
lpc_code(), 154
lpcsp(), 165
lpsp(), 171
LSF, see line spectral pairs
LSF interpolation, 172
LSP, see line spectral pairs
   analysis, 204, 207
   instantaneous analysis, 204
   time-evolved analysis, 207
lsp_bias(), 205
lsp_dev(), 205
lsp_voicer(), 345
lspnarrow(), 336, 338
LTI, see linear time invariant
LTP, see long-term predictor
ltp(), 340
lung, 55, 56
   excitation, 176, 184
LVCSR, see large vocabulary continuous speech recognition

machine learning, 224, 227, 230, 232, 272
magnetic resonance imaging, 86, 268
Mandarin, see Chinese, 300
masking, 95, 103, 109, 123, 137
   binaural, 97, 111
   non-simultaneous, 96, 121
   post-stimulatory, 96, 121
   pre-stimulatory, 96
   simultaneous, 93, 94, 148
   temporal masking release, 111
   two-tone suppression, 111
max pooling, 259
max(), 153

McGurk effect, 86, 112
mean opinion score, 72, 142
mean-squared error, 73, 160, 179
mel frequency, 43, 105, 126, 130, 134, 207
mel frequency cepstral coefficients, 43, 123
mel2f(), 106
MELP, *see* mixed excitation linear prediction
MFCC, *see* mel frequency cepstral coefficients, 231, 286, 299, 301, 304, 309, 347
microphone
    directional, 273
mid-speech clipping, 281
mid-side stereo, 330
mind control, 268
MiniDisc, 109
mixed excitation linear prediction, 188, 190, 348
MNB, 73
mobile telephone, 2
modified rhyme test, 79
monaural, 97
mono-phonemes, 307
mood, 310
MOS, *see* mean opinion score
MP3, 1, 14, 15, 109, 121, 122, 137, 216
MP4, 16
MPEG, 15
MRI, *see* magnetic resonance imaging
MRT, *see* modified rhyme test
MSE, *see* mean-squared error
multi-layer perceptron, 239, 253, 260
multilingual, 310
music
    analysis of, 216
musical notes, 45

nasal, 64, 99
NATO phonetic alphabet, 80
natural frequency, 40
natural language, 267
natural language processing, 269, 320, 368
Newton–Raphson approximation, 167
NIST, 271, 305, 307
NLP, *see* natural language processing
noise
    detected as speech, 281
    effect of correlation, 326
    background, 57, 368
    cancellation, 122
    characteristics, 61
    generation of, 46
    masking, 124
    perception, 89, 93, 94, 111
    reduction through correlation, 111
    response of speaker to, 61
non-audible murmur, 353
normal distribution, 242

normalisation, 18
Nyquist, 5, 22, 35, 46, 127, 152

objective function, 247
occupancy rate, of channel, 215
Octave, 229
Ogg Vorbis, 14–16, 109
orthotelephonic gain, 89
overlap, 25, 27, 231
overlap–window–add, 27

parallel processing, 229
parameterisation, 157
parameterisation of speech, 148, 150, 154
PARCOR, *see* partial correlation
partial correlation, 159, 160
path, 229
pause(), 11
PCA, *see* principal component analysis
PCM, *see* pulse coded modulation
perceptron, 234, 239
perceptual error, 184
perceptual error weighting filter, 191
perceptual weighting, 183, 191
perfect pitch, 97
PESQ, 73
PEWF, *see* perceptual error weighting filter
phase locking, 92
phone, 63, 307, 323
phoneme, 57, 63, 66, 67, 79, 208, 299, 302, 317
phonemes, 69, 162, 350
phonetic spelling, 317
phonograph, 1
phonology, 306
pitch, 33, 231, 279, 348, 350–352
pitch lag, fractional, 178
pitch perception, 90, 91, 96, 100, 110
pitch synchronous overlap and add, 339, 341
plain old telephone service, 2
play(), 11
plosive, 57, 64
PLP, 309
Pocketsphinx, 312
poly-phones, 307
POTS, *see* plain old telephone service
pre-emphasis of the speech signal, 158
precedence effect, 100
precision, 228
principal component analysis, 242, 285
private mobile radio, 6
pronunciation, 299
    in speech synthesis, 317, 320
    in speaker classification, 303
    in speech synthesis, 303
    of phonemes, 63
prosody, 306

proximity, 113
pseudo-stationarity, 32, 67, 150, 159
pseudo-Wigner–Ville distribution, 210
PSOLA, *see* pitch synchronous overlap and add
PSQM, 73
psychoacoustics, 87, 94, 109, 121–124, 129–131, 191
puberty, 303
pulse coded modulation, 7, 13, 14, 85, 141, 144, 145
    standards, 148
PWVD, *see* pseudo-Wigner–Ville distribution

quantisation, 6, 140, 145, 146, 148, 154, 159, 171, 182, 185, 228
    of stereo audio, 329
    of LSPs, 173
    of audio samples, 11, 17
    of LPC parameters, 157, 171
    of LSPs, 171, 172, 174
    of speech, 150
    split vector, 176
    vector, 175
quantisation error, 142–144, 146

RBM, *see* restricted Boltzmann machine
reassigned smoothed pseudo-Wigner–Ville distribution, 210
recall, 232
record(), 10
redundancy, 80
reflection coefficients, 159, 161
regular pulse excitation, 177, 183, 184, 190, 348
    standards, 190
reinforcement learning, 232
relative scaling, 20
reshape(), 101
residual, 153
resonance conditions, 165, 166
restricted Boltzmann machine, 253
resume(), 11
RPE, *see* regular pulse excitation
RSPWVD, *see* reassigned smoothed pseudo-Wigner–Ville distribution

sample rate, 6, 18, 142
sampling, 5
save, 16
scaling of amplitude, 20
SD, *see* spectral distortion
segmental signal-to-noise ratio, 74, 171
segmentation, 21, 24, 25, 27, 28, 31, 32, 57, 177, 274
    in CELP coder, 183
SEGSNR, *see* segmental signal-to-noise ratio
semi-supervised learning, 232
semiology(), 39

sensing, 223
Shatner, William, 319
short-time Fourier transform, 38, 209
SIFT, *see* simplified inverse filtering technique
signal processing, 93
signal-to-noise ratio, 74
silent speech interface, 310, 364
simplified inverse filtering technique, 209
size constraints, 140
slope overload, 143
smartphone, 2
SNR, *see* signal-to-noise ratio
softmax, 255
sone, 103
sound detection, 231
sound perception, 92, 93
sound pressure level, 61
sound strengthening, 110
sound(), 11, 18, 49
soundsc(), 12, 18, 29, 49, 90, 92, 95, 100, 101, 114, 326
spatial placement, 327
speaker
    authentication, 363
    identification, 231, 302
    validation, 363
    verification, 302
speaker identification, 43
specgram(), 38
spectogram(), 38
spectral distortion, 75, 172, 183
spectrogram, 30, 38, 211, 214, 231
spectrogram time–frequency distribution, 209
spectrogram(), 34, 38, 212
speech
    activity detection, 275
    concatenative synthesis, 315
    pitch changer, 338
    pitch period, 339
    transformation of, 346
    amplitude, 61, 62, 304
    analysis of, 224
    apraxia, 311
    articulation, 64, 66
    atypical, 70
    backwards, 102
    cadence, 216
    characteristics, 57
    characteristics of, 149
    classification, 58, 208
    codec, 142, 311
    coding algorithms, 149
    colloquial, 310
    compression of, 122, 142, 150, 183, 190, 191, 311
    disorder, 311
    energy, 58, 65, 231
    formants, 58, 59, 65, 66, 110, 138, 191, 200, 337

frequency distribution, 65
impediment, 311
intelligibility, 58, 65, 71, 81, 82, 97, 102, 122, 138, 178
intelligibility testing, 79
intelligibility vs. quality, 71
lexical, 62
perception, 101
pitch contour, 60
pitch doubling, 181, 182
pitch extraction, 178, 179, 182, 208
pitch halving, 182
pitch models, 176
pitch period, 182, 202
pitch synthesis, 150
power, 65
production, 55
quality, 71, 102, 148
quality testing, 72
recognition, 43, 110, 267, 303
repetition, 81
reversed, 102
shouting, 68
sinewave, 54, 196
spectrum, 199, 201
statistics, 303
super-audible, 362
synthesis, 314
to whisper, 354
unvoiced, 64
voiced, 64
voicing, 215
waveform, 199, 214
speech recognition, 231
Sphinx, 301, 312
SPL, *see* sound pressure level
split vector quantisation, 176
spread_hz(), 125
spreading function, 124
SSI, *see* silent speech interface
stammer, 311
Star Trek, 307, 319, 366
stationarity, 32, 67
statistical voice conversion, 346, 353
statistics, of speech, 213, 303
steganography, 122
stepsize doubling and halving, 144
stereo, 4, 98, 324
stereo encoding
    joint, 329
Stewart, Patrick, 319
STFD, *see* spectrogram time–frequency distribution
STFT, *see* short-time Fourier transform
stop(), 11
stress, on words, 275
stutter, 311
subsampling, 259

super-vector, 285
superposition, principle of, 332
supervised learning, 232
support vector machine, 239, 255
surround sound, 330
SVM, *see* support vector machine
swapbytes(), 17, 18
sweet spot, 332
syllabic rate, 67, 215, 303
syllable, 63, 66, 67, 79, 215
synthesis, 149
synthesiser
    of speech, 314

TCR, *see* threshold crossing rate
telephone, 1
    mobile, 2
temporal integration in hearing, 93
temporary threshold shift, 93, 111
TETRA, *see* trans-European trunked radio
text-to-speech, 367
TFD, *see* time–frequency distribution
three-dimensional mesh spectral plot, 214
threshold crossing rate, 198, 211
timbre, 92
time-domain waveform, 37, 200
time–frequency distribution, 209–211
TIMIT, 173
toll-quality, 6
tone
    generation of, 44, 47
    induction, 110
tonegen(), 44, 48, 88, 90, 91, 95, 97, 100, 113, 326
tongue, placement of, 303
topic, 272, 310
training, 230, 232
trans-European trunked radio, 190
transcription, 274
transcription systems, 269
transducer, 273
transfer learning, 286
tri-phone states, 299
TTS, *see* temporary threshold shift *or* text-to-speech
tube model, 162–166
Turk, the, 314

UBM, *see* universal background model
universal background model, 283
unsupervised learning, 232

VAD, *see* voice activity detection
variable tone generation, 47
vector quantisation, 175, 286, 303
vector sum excited linear prediction, 183
velum, 55, 56

Index

violin, 217, 218
  analysis of, 216
visualisation, 34, 37
Viterbi algorithm, 293, 309
Viterbi, Andrew, 293
vocabulary, 272
vocal chord, *see* glottis
vocal tract, 58, 63, 64, 176, 177, 191, 303
  filter, 150
  parameters, 184
  resonances, 183, 184
voice
  cloud, 226
  scrambler, 338
voice activity detection, 231, 274, 275, 309, 362, 368
voice operated switch, 275
Voicebox, 164, 250
VOS, *see* voice operated switch
vowel, 57, 58, 63, 64, 303
VSELP, *see* vector sum excited linear prediction

waterfall(), 34
wave file format, 10, 13, 14, 212
waveread(), 13
wavrecord(), 12
wavwrite(), 16
weighting, 89
whisperise(), 357
whisperiser, 354
whispers, 228, 348, 349
white noise, 46, 55, 111, 123, 184
wideband coding, 148
Wigner–Ville distribution, 209, 210
windowing, 27, 28, 161, 231
  in CELP coder, 183
  window functions, 27–29
  window size, 32, 35
WVD, *see* Wigner–Ville distribution

xcorr(), 40

ZCR, *see* zero-crossing rate
zero-crossing rate, 196, 207, 209, 211